Mastering
Crystal Reports 2008

Mastering
Crystal Reports 2008

David McAmis

Kuiper Publishing Pty Ltd
Melbourne, Australia

Mastering Crystal Reports 2008

Kuiper Publishing

454 Collins Street
Melbourne VIC 3000
Australia

Find us on the web at
http://www.kuiperpublishing.com

To report errors, please visit our web site and use the errata form — we appreciate your feedback.

To arrange bulk purchase discounts for sales promotions, premiums or user groups, please contact us.

IBSN-13: 978-0-9807458-0-1
ISBN: 0-9807458-0-2

To Madeleine, age 5, who always notes
"You wouldn't make much of an opera singer."
(Guess I better stick with writing computer books.)

About the Author

David McAmis is an IT consultant, software developer, journalist and writer who lives and works in Melbourne, Australia and tries to live by the adage of "The way to love life is to know many things" (Vincent Van Gogh). As a writer, David's work spans multiple genres with each one representing new challenges and opportunities for growth. In the past decade, while holding down a full-time job, he has written over a dozen books translated into 8 different languages and has written over 600 magazine articles appearing in a wide range of publications.

Acknowledgements

First and foremost, a big "Thank You" to Karina H McInnes, who managed the production of this book. It has been an amazing effort and I am truly thankful for her patience, persistence and keen eye to detail.

To the entire team at Avantis, thank you for the support and encouragement, as well as your feedback and all the amazing work you do with our clients. You continually impress, amaze and astound me with the work you do.

A special thanks to Alice, Colin and Dave for their support over the years, including this latest endeavour. As Avantis celebrates its 10th year, I couldn't think of a better group of people to be working with.

And to the worldwide network of family and friends who have to put up with me while I am writing a book, I appreciate your love and support.

A percentage of the proceeds from this book are donated to the
Blue Ridge Community College Foundation
to fund a scholarship for students.

You can find out more about Blue Ridge Community College from
http://www.blueridge.edu

Is this book right for you?

The first time I used Crystal Reports, I was fresh out of college and had started my first "real" job. I was working as an IT consultant and there was a customer who had some data they wanted to present in a formatted report. I installed Crystal Reports 4.0 from a series of floppy disks and through trial-and-error, worked my way through creating my first report.

Since that day 14 years ago, I have worked with Crystal Reports almost every day of my working life. Over the years, as a trainer and consultant, I have taught over 3,000 students and delivered more reports than I care to count to customers all around the world, including the US, Australia, Hong Kong, Thailand, Korea and more.

What keeps me interested in working with Crystal Reports is that each data set I encounter is different, each set of requirements unique and Crystal Reports has the power and flexibility to deliver almost any report you can imagine. And that is what this book is about — giving you the skills you need to master Crystal Reports and create reports from your own data source.

With that said, Crystal Reports is always a difficult topic to write about and in my e-mail every week there is an e-mail from someone (near or far) who has read one of my books and complains "Why didn't you cover using Crystal Reports with XXX?" (where "XXX" is one of the tens of thousands of applications or data sources that Crystal Reports can be used with). Or another standard question is "Can you tell me how to create a YYY report?" (again, where YYY is one of the millions of reports that could be created with Crystal Reports).

So I am here to tell you now — no computer book is going to cover every single usage and application of piece of software. If it did, the book would have so many pages and be so heavy, you wouldn't be able to carry it around!

However, what a good book can do is give you the skills that you can apply to your own unique situation and requirements. And in the case of Crystal Reports, regardless of the data source you are using or the type of report you want to create, the same report development skills apply. So while the samples in the book are from an ODBC data source, you may be reporting directly from a database using a native driver, or connecting to an Excel spreadsheet, or even connecting directly to SAP R/3 or BW, and all of the same skills and techniques apply.

To add to your mastery of Crystal Reports, I have also included a chapter on "Advanced Report Techniques" where different features and functions are combined within a single report to produce some of the most common types of reports (Summary Reports, Month-To-Date, Year-to-Date, etc).

Another feature is that this book includes over 100 sample reports and step-by-step instructions that you can follow along with. From training Crystal Reports users for so long, I know that the best way to learn an application is through hands-on experience.

So now that you know a little bit about what this book does cover, we'll also look at what it doesn't. Specifically, we aren't covering the development of custom reporting applications using the .NET or Java SDK. If you do need to integrate Crystal Reports into your application, I recommend looking at the server-based technology delivered in Crystal Reports, BusinessObjects Edge or BusinessObjects Enterprise.

All three of these platforms are built on the same framework and feature an "out of the box" web portal and direct links to reports and other content. We will be looking at Crystal Reports Server towards the end of the book, as well as how it can be used to deliver your reports to a wider audience.

For developers who would like to integrate Crystal Reports into your applications, either through the Crystal Reports .NET or Java SDK's or through any of the BusinessObjects server technologies, there is a section in Appendix A that lists the developer resources and links to sample code and more.

And finally, I hope this book will be a good reference for your own report development and that you will be as excited about working with Crystal Reports as I have been.

David McAmis

Setting up the Sample Files

This book comes with a number of sample files, to be used in conjunction with the book text. These sample files include database, reports, code listings, etc. and can be used to follow along with the step-by-step instructions found in each chapter.

You can download a copy of these sample files by visiting **http://www.kuiperpublishing.com.** Navigate to the web page for this book "Mastering Crystal Reports 2008" and look for the "Download Files" link.

The download files are provided in a WinZip© file format. You will need to unzip these files to a folder on your hard drive. When you have finished unzipping the files, the directory structure should look something like the one shown below.

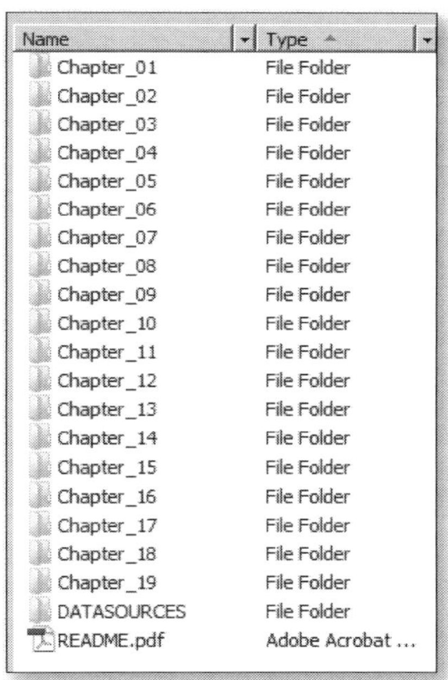

Name
Type
Chapter_01 File Folder
Chapter_02 File Folder
Chapter_03 File Folder
Chapter_04 File Folder
Chapter_05 File Folder
Chapter_06 File Folder
Chapter_07 File Folder
Chapter_08 File Folder
Chapter_09 File Folder
Chapter_10 File Folder
Chapter_11 File Folder
Chapter_12 File Folder
Chapter_13 File Folder
Chapter_14 File Folder
Chapter_15 File Folder
Chapter_16 File Folder
Chapter_17 File Folder
Chapter_18 File Folder
Chapter_19 File Folder
DATASOURCES File Folder
README.pdf Adobe Acrobat ...

There is a folder for each chapter of the book which contains the sample report files, code examples, etc. for that chapter.

In addition, in root folder there is a file called README.PDF which lists all of the filenames, as well as a brief description of each file. Also included in the folder structure is a folder named "DATASOURCES" that contains the data sources used for the sample reports.

To use some of the Crystal Reports demonstrated in this book, you will need to create an ODBC System Data Source Name (DSN) named "Galaxy" to point to this sample database.

We have included instructions on the following page on how to set up an ODBC driver on your PC.

These instructions refer to setting up an ODBC system DSN on Windows Vista™. The instructions for setting up an ODBC data source on your computer may be slightly different, depending on your operating system.

To set up the ODBC system DSN on your computer, follow these steps:

1. Download and unzip the book download files from the Kuiper Publishing web site. Make a note of where you have unzipped the files, as you will need this location a little later when you specify the location of the sample database.
2. From the Windows Start Menu, select Settings > Control Panel > Administration Tools > Data Sources (ODBC). This will open the ODBC Data Source Administrator.
3. Click on the "System DSN" tab and click the "Add" button and then select "Microsoft Access Driver (*.MDB)" to open the dialog shown below:

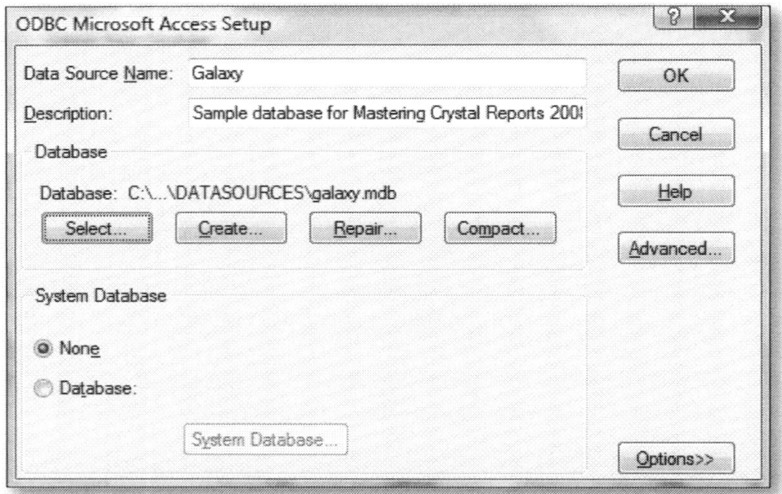

4. For the Data Source Name, enter "Galaxy" and then use the Select button to browse to your book download files, to the DATASOURCES folder and select "galaxy.mdb".
5. When you have finished selecting the database, click OK to save your DSN setup.

You should now be able to use the Sample reports in this book and create your own reports for the step-by-step activities. If you have problems using the sample database, double-check your ODBC setup and remember, it may be slightly different depending on your version of Windows.

Contents at a Glance

Contents

Chapter 1
What is Crystal Reports?

In This Chapter

- Why Crystal Reports?
- What's New in This Release?
- Which BusinessObjects Reporting Tool Should I Use?

INTRODUCTION

Crystal Reports, simply put, is a powerful Windows-based report design tool that has been designed to take your raw data and create a formatted report. Using the tools within Crystal Reports, you can not only display that data, but also group and analyze it, add standard calculations, simple to complex formulas, charts, geographic mapping and more.

With Crystal Reports, you can connect to virtually any data source and create presentation-quality reports which leverage the power and sophistication that the platform has to offer. As a Crystal Reports designer, you have the ability to create any report you can imagine (providing you have the data available).

Over the years, Crystal Reports has been used to create everything from simple column-based reports, to complex analysis reports, to print invoices, purchase orders and other corporate documents, to summarizing large volumes of data in a concise format, like the reports shown in Figure 1.1 on the next page.

Figure 1.1 Typical Crystal Reports

What makes Crystal Reports unique is that it has the ability to create almost any report format you require, through the combination of a powerful reporting engine and flexible formatting options.

When it comes time to distribute those reports, you have the ability to distribute them with saved data, to be viewed through the free Crystal Reports 2008 Report Viewer, or alternately you can export your report to one of 18 different export formats, including Microsoft Excel, CSV, PDF, etc.

For application developers, you can integrate Crystal Reports into your .NET or Java applications, as well as publish these reports to any of the SAP BusinessObjects server platforms. This includes SAP's software-as-a-service (SaaS) offerings, CrystalReports.com and BI On-Demand, as well as the server-based technologies found in Crystal Reports Server, BusinessObjects Edge and BusinessObjects Enterprise.

Using Crystal Reports Server for example, your reports can be securely shared through a web browser, refreshed on demand, scheduled and distributed through a secure web portal like the one shown in Figure 1.2, via e-mail, FTP, or sent directly to a printer, etc.

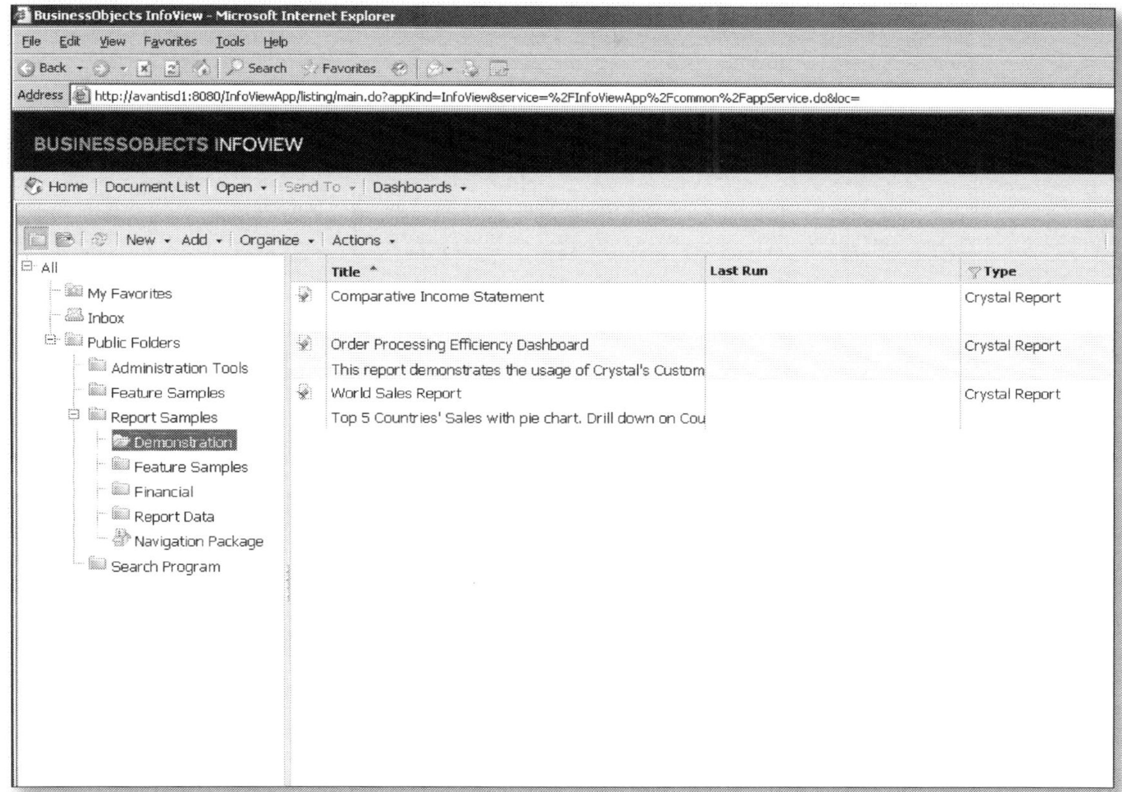

Figure 1.2 Crystal Reports Server "InfoView" portal

If you need to scale up beyond just Crystal Reports to leverage more of the BusinessObjects tools (including Universes, Web Intelligence, etc.), you can publish your reports to BusinessObjects Edge or BusinessObjects Enterprise. This provides a robust-scalable framework for delivering formatted Crystal Reports, as well as all of the other features and functionality the wider BusinessObjects platform provides.

WHY CRYSTAL REPORTS?

When considering Crystal Reports as a reporting tool, you may wonder what all the fuss is about. In the section below, we're going to look at some of the reasons why Crystal Reports remains the market leader for formatted reporting.

HISTORY

Crystal Reports was the first Windows report writer, and since its development there have been over a dozen releases of Crystal Reports. It is one of the most stable software products you will find on the market and is continually enhanced and updated to take advantage of the latest technologies, including XML, Web Services, and more. As data sources progressed and the way we accessed data changed, Crystal Reports has kept up with the times.

4.5 → 5.0 → 6.0 → 7.0 → 8.0 → 8.5 → 9.0 →10.0 → XI R1 → XI R2 → 2008

ANY DATA SOURCE

Crystal Reports features the widest data source support of any tool on the market. The product ships with a number of native and ODBC drivers, as well as application-specific drivers and integration kits that allow it to natively access application data, as well as data held in spreadsheets, local databases and more.

REPORT DESIGNER

The report designer interface has been updated over the years with an emphasis on developer productivity, with most features and functionality available in several places, including right-click menus, icons and file menus. With a number of report-design wizards to help jump-start report development, a new report designer can get up to speed quickly and start creating their own reports within a few minutes.

REPORT FORMATTING

They say that the only limit to the reports you can create with Crystal Reports is your imagination, and this is true (you just need to have the data for your report). With advanced section and formatting options, as well as the ability to use conditional formatting, subreports and complex formatting formulas, you can control the look and feel of your report, as well as its behavior when users are viewing the report, drilling down, changing interactive parameters, etc.

SUMMARIES & CALCULATIONS

It's no good just presenting raw data to users — you want to be able to add some value to the data you are presenting. The in-built summaries within Crystal Reports are a great way to do that. Encapsulating the most popular summarizations, these summary fields can be quickly added to your report without having to write any formula code.

But for more complex calculations (or where you just want to get your hands dirty) Crystal Reports features a powerful formula language with not one, but two syntaxes (Crystal and Basic) which you can use to create complex formula and calculations that you can display on your report, or just use behind the scenes.

These formulas can leverage the full set of functions that ship with Crystal Reports, and you can extend these with your own custom functions which you can re-use time and time again. So there is no need to re-invent the wheel each time.

In addition, Crystal Reports can leverage SQL expressions to create number-crunching expressions that are evaluated with your SQL on your database server when the data is retrieved. By leveraging the power of the server, you can perform complex SQL calculations in record time.

ANALYSIS

Since most people actually want to analyze the data that they are reporting, Crystal Reports features powerful analysis features that will allow you to uncover trends and deep-dive into your data. Using drill-down, report users are able to drill down to the details of the data, all the while highlighting exceptions and key figures with conditional formatting.

CHARTING & MAPPING

If a picture is worth a thousand words, then a well-designed chart should probably be worth TEN thousand. Crystal Reports 2008 features a powerful charting engine that can be used to visualize report data, using chart formats that are familiar to business users and easy to read at-a-glance. In addition, Crystal Reports ships with technology from MapInfo that allows you to integrate geographic maps into your reports.

STUNNING VISUALIZATIONS

With the integration of Adobe flash content, you can now embed stunning visualizations and dashboards that were created with Xcelsius. These visualizations can consume Crystal Reports data and retain their interactivity when they are viewed. This integration adds to the already formidable charting and mapping technology in Crystal Reports to allow you to visualize your report data using any of the components that Xcelsius provides. An example of this integration is shown in Figure 1.3, where an Xcelsius dashboard has been inserted into a Crystal Report.

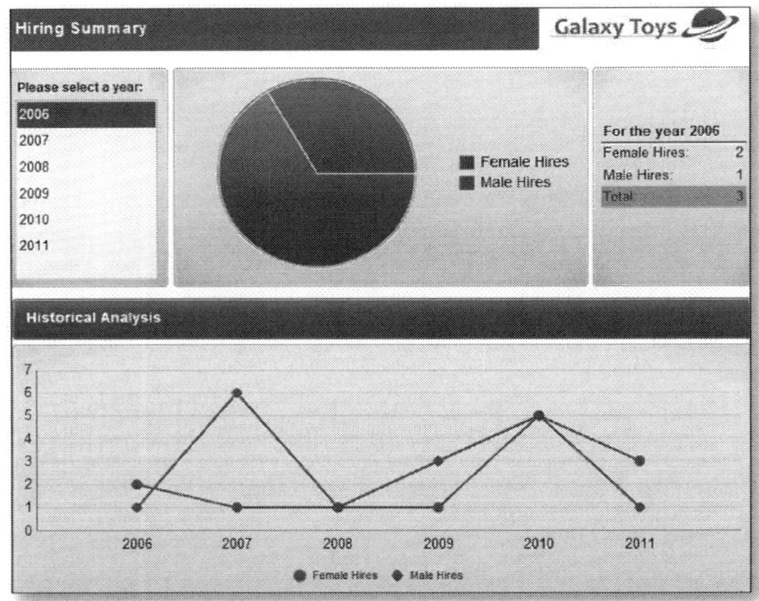

Figure 1.3 Xcelsius visualization integrated in a Crystal Report

REPORT DISTRIBUTION

When it comes time to distribute your report, there are so many options it is hard to choose between them! To start, you can save your report with data, and distribute it to users who can then view the report interactively using the free Crystal Reports 2008 viewer, which is available for both Windows and Mac, and shown in Figure 1.4.

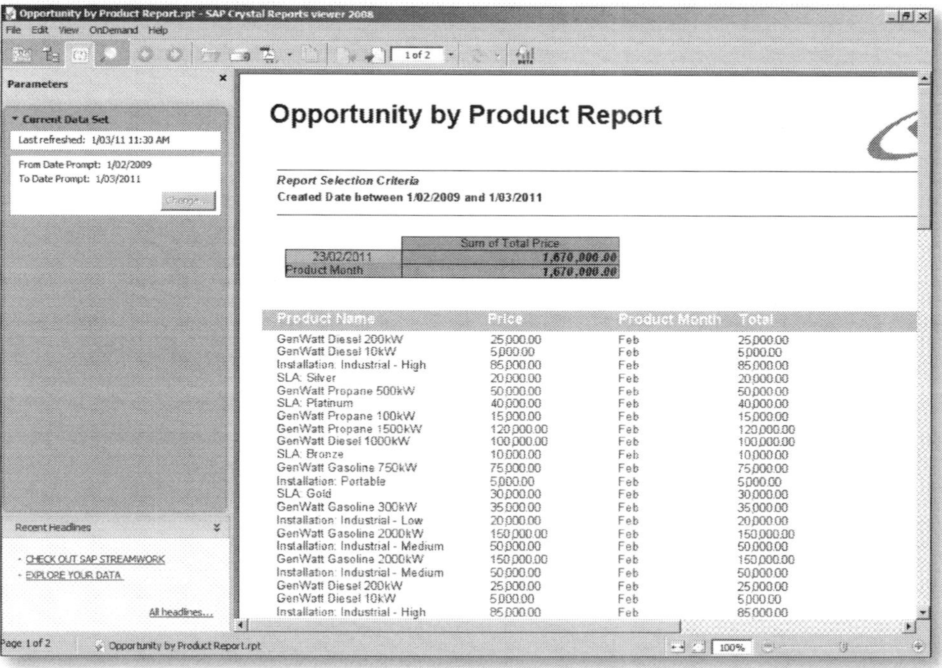

Figure 1.4 Crystal Reports 2008 Viewer

In the case of users who may not have the viewer installed, the report can also be exported to one of the many export formats that Crystal Reports supports. These formats range from Adobe PDF for presentation-quality reports, to a variety of Excel and text file formats, where you may want to use the data in another application. Regardless of what you need to do with your report or its data, there is an export format to suit.

For ease of distribution, you can't beat the server technologies provided by BusinessObjects, which include both SaaS and on-premise options. We'll be looking at Crystal Reports Server a little later in the book, but there are still a number of options available for you to deploy your reports to a wider audience, through a secure, web-based environment.

WHAT'S NEW IN THIS RELEASE?

With Crystal Reports, there are "integration" releases and "innovation" releases of the product — and with Crystal Reports 2008, we definitely have the latter. This new version features a number of features and functionality designed to make report design easier, in addition to introducing some new innovative features to take your reports to the next level. In the following sections, we are going to be taking a quick look at some of these new features (and don't worry — we'll be covering them in detail later in the book).

PRODUCT VERSIONS/PACKAGING

The first thing that you will notice is that there is now only a single edition of Crystal Reports 2008. In the past, there were Standard, Developer and Professional editions, but in this release there is only one edition. In addition, the product setup files have a much smaller footprint, so there is no waiting around for your software to download. Another key point is that Crystal Reports no longer ships with the sample data files and sample reports, but these are just a click away from the Crystal Reports start page. There is a direct link to download these files, as well as some additional DataDirect drivers that are included with the product.

INTERACTIVE REPORTING

With Crystal Reports 2008, you have even more options for interactive reporting — you can add sort controls to your report to allow end-users to sort by report columns when they are viewing the report, in addition to a new parameter panel, shown in Figure 1.5 that can be used to further filter the data and control report features, like turning formatting or highlighting on or off.

Figure 1.5 Interactive parameters at work

Parameter fields can also now be optional, which gives even more flexibility when creating reports for a wider audience. When you combine this with the functionality found in the parameter panel, you can create a single report that can be filtered based on the user's unique requirements.

You can also add sort controls to your report, which allow the user to sort the report contents at run-time using a single click. This feature alleviates the need for users to export the report data to another format just to change the sorting of the columns.

FLEXIBLE PAGE FORMATTING

You can now also mix portrait and landscape pages within a single report, offering the ultimate flexibility in how you present your report. Using this feature, you can quickly create reporting "packs" where you can combine information from multiple sources, in multiple page layouts in a single report. We'll be looking at this a little later in Chapter 6, when we have a look at creating sections in Crystal reports.

BUILT-IN BARCODE SUPPORT

Previously if you wanted to create a barcode in Crystal Reports, you would have needed to create a formula and then download a barcode font separately to format this formula field as a barcode. With Crystal Reports 2008, this feature is now included with the product and available from a right-click menu, as shown in Figure 1.6.

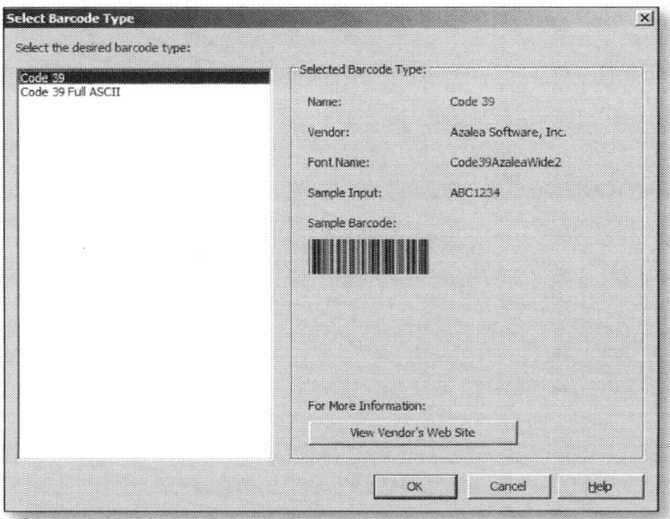

Figure 1.6 Barcode option

NEW EXPORT OPTIONS

With Crystal Reports 2008, there is a new XML export option that allows more control over the export, including the ability to associate a style sheet with the exported XML file. There are also a number of sample style sheets and examples for download. This new export format allows report developers to more easily share Crystal Reports data with other applications. In addition, exporting has been made easier with better exporting options for existing formats.

CROSS-TAB ENHANCEMENTS

Cross-tabs are an important element of reporting, allowing developers to quickly create a table that can be used to summarize data by rows and columns (similar to a Microsoft Excel pivot table). With this release, cross-tabs have been enhanced to include the ability to insert custom calculations, including summary, counts, variances, etc. which can then be used in corresponding cross-tab charts.

HYPERLINKING

The BusinessObjects platform provides a powerful API called "OpenDocument" for linking to content published to Crystal Reports Server, BusinessObjects Edge or BusinessObjects Enterprise. This API provides the ability for report developers to link to different reports (even if they are off a different data source or a different reporting tool) as well as provide further detail from summary reports.

The OpenDocument API has always relied on developers to understand how to build up a URL and query string to access reports, dashboards, etc. With Crystal Reports 2008, you can now browse your server environment and select the document you want to link to. This allows developers to quickly link their reports to other server documents, without having to create the OpenDocument URL by hand.

WHICH BUSINESSOBJECTS REPORTING TOOL SHOULD I USE?

One of the most common questions that report developers will ask is "Which BusinessObjects tool should I use for reporting?" There are two primary reporting tools in the BusinessObjects family — Crystal Reports and Web Intelligence. It can be difficult trying to decide which is best for your needs, so I have put together the handy guide below. Keep in mind that this guide was put together based on my own experience, and you may find different uses for the different products.

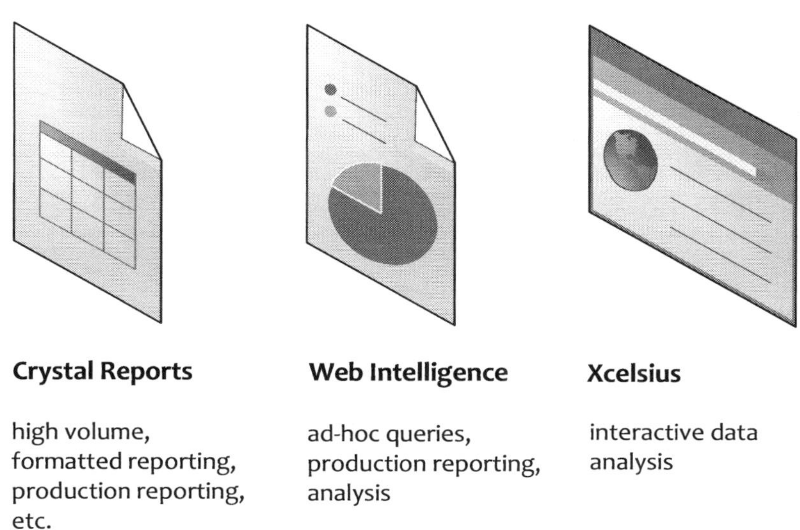

Figure 1.7 BusinessObjects reporting options

CRYSTAL REPORTS

Crystal Reports has its origins in print-based reporting and can be used to create pixel-perfect reports. Crystal Reports is a great choice where you have a specific format in mind, or where the formatting is important. For example, Crystal Reports is often used to produce Invoices, Statements, Bills, etc. where the formatting may include logos, barcodes and a specific format. Crystal Reports also features the ability to conditionally format most elements of the report, including sections, objects, etc.

Crystal Reports is also useful where large volumes of data need to be processed, as the Crystal Reports print engine can handle the volume, especially when used in conjunction with any of the server technologies (Crystal Reports Server, BusinessObjects Edge, BusinessObjects Enterprise, etc).

Crystal Reports can be developed directly from a data source, or alternately from a Universe query. A lot of existing Crystal Reports are based directly off the data, but with sites that use BusinessObjects Edge and BusinessObjects Enterprise, universes are frequently the basis of all of their business intelligence content.

Crystal Reports is not really a tool for end-users to develop their own reports, as there is an expectation that they will need to know a bit about databases, how to write formulas, etc. In addition, there are a plethora of options within Crystal Reports that can leave a casual end-user scratching their heads.

So Crystal Reports is the perfect choice where you have large volumes of data, where you need a highly formatted report that is created by a technical resource, either directly from your data source, or through a BusinessObjects Universe.

WEB INTELLIGENCE

Web Intelligence is available both as a web-based tool and in a "rich client" that can be installed on a user's PC. Web Intelligence provides the ability to create ad-hoc reports, for answering one-off questions (i.e. "What were sales for Product X for last week?"), as well as formatted reports that are refreshed on demand, scheduled, etc. as well as providing analysis through the addition of filters, ranking, summaries, etc.

Web Intelligence also provides the ability for end-users to analyze their reports using an interactive viewer. Web Intelligence documents are based on Universes and can be created by anyone from end-users all the way to skilled report designers who may be part of an IT team.

For the casual end-user, Web Intelligence provides the tools to get the data they need quickly, without worrying about the underlying database structures or complexity involved in creating a Crystal Report.

So Web Intelligence is a great choice for ad-hoc querying, reporting and analysis where you want to push the report development out to end-users or non-technical users, but does require the development of a Universe across your data source.

WHAT ABOUT DESKTOP INTELLIGENCE?

Desktop Intelligence is the latest iteration of the old BusinessObjects "Full Client" and is still in use at organizations that have been using "classic" BusinessObjects for a number of years. There have been a number of attempts to move users off Desktop Intelligence, in favour of Web Intelligence. Web Intelligence has been designed with a robust, scalable web architecture and has a number of architectural advantages over Desktop Intelligence. Over the past few years, Web Intelligence has grown in features and functionality, to where it can now be used to replace Desktop Intelligence in most instances. There are still some areas where Desktop Intelligence features are being used, in particular by developers who use VBA macros in Desktop Intelligence. This can make some sites reluctant to part ways with the tool. It's important to note that Microsoft stopped supporting VBA a while ago and BusinessObjects has a feature-rich software development kit (SDK) that can be used to extend its "out of the box" capabilities.

SUMMARY

Now that you know a bit about what Crystal Reports can do, it's time to have a look at the most crucial part of the report development process — actually designing the report itself. The next chapter goes through some of the processes you may want to adopt for designing a report on paper before you actually crack open the Crystal Reports designer and start working. For some experienced developers, this may come as second nature — in addition, you may have your own processes in place for specifying and designing reports. If this is true, you can skip ahead to Chapter 3 to get right into the action.

Chapter 2
Report Design Overview

In This Chapter

- Report Design Lifecycle
- Gathering User Requirements
- Designing Your Report
- Report Testing Strategies
- Report Deployment Options
- Reviewing and Refining Reports

INTRODUCTION

Before we actually get into the technical report design, we need to spend a little time planning our report to alleviate some of the problems we might face later when we actually sit down and develop the report. Often as developers we will create reports that we think the user will want, using all of the latest features, bells, and whistles, only to find that the end-user would be happy with a simpler report that was easier to read and understand.

Another problem is that a developer will often sit down and crack open the Report Designer without really knowing what the end product will look like, or considering how the end user will use what has been created. What is called for is a bit of planning before we jump right into the technical aspect of designing reports. By planning our report before we get started, we can deliver a report that meets the users' needs and expectations.

If you already have a report design methodology and it works for you, please have a read over this chapter anyway — you may find a different way of doing things or be able to incorporate some small part of it into what you already do. In any case, there are a number of reports associated with this section available in the downloadable files for this chapter — feel free to modify these reports for your own use.

REPORT DESIGN METHODOLOGY

When we look at the methodology around business intelligence, there is a basic five step process that will help you, as the report developer, to ensure that you get all of the information you need to develop your report, as shown below in Figure 2.1.

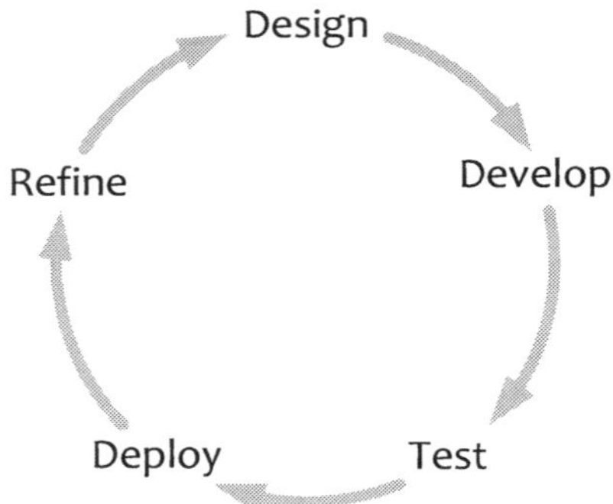

Figure 2.1 Report Design Methodology

In the following section, we will be looking at these different phases in depth, including some of the resources you will need to carry out each one.

DESIGN PHASE

In most applications, reports are usually tied to a specific function or area of an application. For example, if you have created an application that is used for entering telephone sales orders from customers, chances are there will be a suite of reports tied to that function, showing sales summaries, order totals, etc. The best way to determine what these reports should look like is to actually ask the people that will be using them on a daily basis.

If you are working in a large organization, you will probably have a business analyst who will interview the user, gather their requirements and then communicate these requirements back to you. If you are in a smaller organization or if, like the rest of us, you are forced to take on a number of roles, you may gather these requirements yourself.

In either case, end-user interviews are the key to targeting a report's content. Organize the interviews with the actual end-users (not their supervisors, personal assistants, or others), and ask them to bring along examples of reports they currently use, or would like to use. This is your chance to find out what information they need to better perform their job.

Be careful when interviewing, as users will sometimes come up with an arm's length long wish list for reports they would like to see.

In the interview, the user should be able to tell you how the report is used and what decisions are made based on this information. If they can't tell you either of these things, either you are interviewing the wrong person or they really don't need the information they have asked for.

Once you have interviewed the end-users, you should have a pretty good idea of what reports are required, and how they will be used. From this point, there are many different ways of documenting the user's requirements (such as user requirements statements, or user cases) but the most straightforward method is to create a formal Report Requirements document.

A Report Requirements document, like the one shown below in Figure 2.2, will outline what information needs to appear in the report, how the report is used (is it interactive or run in a batch?), what columns are required, etc.

Report Requirements Document

Completed By:	John Smith
Department	Finance
Completion Date:	Sep 05 2010

Overview (Type of report, what report will be used for, etc.)

This report will be used to track month-to-date and year-to-date spending across multiple cost centres. It will be distributed to a number of users across the organization to give visibility of their actual spend against the budget the provided at the start of the financial year.

Data Source

AccPac Accounting / General Ledger

Report Layout:

Column Based, Landscape

Required Columns

	1	2	3	4	5	6
Column Name	**Cost Centre**	**Cost Centre Name**	**MTD Actual**	**MTD Budget**	**YTD Actual**	**YTD Budget**
Description	*Cost Centre Code*	*Name of Cost Centre from short description*	*Actual values for Month to Date*	*Budget Values for Month to Date*	*Actual values for Year to Date*	*Budget Values for Year to Date*
Table/Field Name	GL.COST_CENTRE	GL.COST DESCRSHORT	GLDETAIL.ACTUALVAL	GLDETAIL.BUDVAL	GLDETAIL.ACTUALVAL	GLDETAIL.BUDVAL
Formatting	XXXX-XX					
Formula			Required to calculate MTD values	Required to calculate MTD values	Required to calculate YTD values	Required to calculate YTD values
Summary			SUM	SUM	SUM	SUM

Report Grouping

When the report is run, detail the fields that will be used for grouping the report contents

GL.CAT1	**General Ledger Category Code 1**
GL.CAT2	**General Ledger Category Code 2**
GL.CAT3	**General Ledger Category Code 3**
GL.CCPARENT	**General Ledger Cost Centre Parent**

Figure 2.2 A report requirements document

You may also create a mock-up of the report or a rough sketch of what it will look like. With your report requirements in hand, the next step in our method is to perform a technical review of the report's definition.

A good place to start with a technical review is to determine the data source for this report. Most likely the data source will be a relational database, but the data could also reside in spreadsheets, text files, OLAP (Online Analytical Processing) data structures, and even non-relational data sources (like Exchange or Outlook folders). Once you have found the data source, you will need to dig a little deeper to determine the exact tables and views that can provide the data required.

You may need to develop additional views or stored procedures to consolidate the data prior to developing a report (for speed and ease of use and re-use) and these will need to be documented as well. Again, all of this information is added to your Report Requirements document.

For the next step of the technical review, you will need to investigate whether or not the design of the report is feasible. The user may request twenty columns (when the page will only fit seven landscape) or may have based the design on an existing report or spreadsheet created by hand.

Once you are more experienced working with Crystal Reports, you will begin to understand how the tool works and the kind of output that can be achieved. In the meantime, browse through some of the sample reports that are available to get a feel for the types of reports Crystal can produce.

You can download the sample reports for Crystal Reports 2008 from the Start Page that appears when you open the Crystal Reports 2008 designer. There are links there for downloading sample reports and data sources from the SAP web site. You may also be able to find additional sample reports on the SAP Developer Network (http://sdn.sap.com).

Once you have completed the technical review and you understand where the data for the report resides and you are comfortable that Crystal Reports can deliver the required format, you have two choices. If you believe that your documentation clearly outlines what the user's requirements are, you can start developing the report based on that information.

However, if you think that you need to clearly communicate that back to the end-user, you may want to create a report prototype from your notes and preliminary sketches, like the one shown in Figure 2.3 so they can see what the finished report will look like.

Cost Center Summary

	MTD		YTD	
	Actual	Budget	Actual	Budget
XXXXXXXXXXXX	$9999.99	$9999.99	$9999.99	$9999.99
XXXXXX-XXXXXX	$9999.99	$9999.99	$9999.99	$9999.99
XXXXXXXXXXXXXX	$9999.99	$9999.99	$9999.99	$9999.99
XXXXXXXXXXXX	$9999.99	$9999.99	$9999.99	$9999.99
XXXXXXXXXXXX	$9999.99	$9999.99	$9999.99	$9999.99
XXXXXXX-XXXXX	$9999.99	$9999.99	$9999.99	$9999.99
XXXXXXXXXXXX	$9999.99	$9999.99	$9999.99	$9999.99
XXXXXXXXXXXX	$9999.99	$9999.99	$9999.99	$9999.99
XXXXXXXX-XXXXX	$9999.99	$9999.99	$9999.99	$9999.99
XXXXX-XXXXXXX	$9999.99	$9999.99	$9999.99	$9999.99
XXXXXXXXXXXX	$9999.99	$9999.99	$9999.99	$9999.99
XXXXXXXXXXX	$9999.99	$9999.99	$9999.99	$9999.99
XXXX-XXXXXXXX	$9999.99	$9999.99	$9999.99	$9999.99
XXXXXXXXXXXX	$9999.99	$9999.99	$9999.99	$9999.99
XXXXXXXXXXX	$9999.99	$9999.99	$9999.99	etc.
XXXXXXXXXXXX	$9999.99	$9999.99	etc.	
XXXXXXXXXXXX	$9999.99	etc.		
XXXXXX-XXXXXX	etc.			
etc.				

Figure 2.3 An example of a report prototype

This prototype can be created using Word, Excel, Visio, and so on, but should closely match the report's final layout and design. Again, another important check is to make sure that the layout and design you create can be created with the features and functionality that Crystal Reports has available.

The prototype, combined with a formal Report Requirements document will clearly communicate what the report should look like when it is finished. It also helps gain user acceptance for the design, as they can see what the finished product will look like (even before you have opened the Report Designer).

If you are working in a large team, this documentation will communicate the requirements to other report developers, so you don't actually have to brief them on every single report that needs to be developed.

If you are working as a business analyst, application developer, and report developer all-in-one, it can also help you keep on track with the user's requirements and make gaining user acceptance that much easier.

DEVELOPMENT

With your design in hand, it's time to get down to actually creating the report, as that is what this book is all about. During the development phase, you should refer back to your report specification document, as well as your prototype to make sure that your development is on track.

Also, during the report design process, check in with the end-user as you go along to ensure that what you are delivering will meet their needs. Often, users who are new to a formal specifications process will not be able to articulate what type of report they need.

This is often called "I will know it when I see it". This can be frustrating as it can cause the scope of your development to creep, but for the first few reports for a new user, it is worth letting them have a look at the report while it is in development. This "preview" approach can often catch problems before they arise, as well as help deliver exactly what the user wants.

During the development phase, you should also keep an eye out for issues that may affect the usability of the report — for example, if you find while you are developing the report that it takes 20–30 minutes to run, you probably need to look at some strategies during the development to improve performance or reduce the amount of data that is being returned.

TESTING

Once the report has been created, you are now ready to do some testing. Testing the report before you hand it over to end-users will save time and effort down the track, as you won't have to go back and make as many corrections later. There are three critical areas that need to be considered when testing your reports:

- Data accuracy
- Processing speed
- Formatting

In the following sections, we are going to look at these three areas in detail, as well as some strategies for testing your report. To facilitate your testing, you may want to create your own "system test" spreadsheet so you can tick off the areas you are testing. This can help document what testing you have done, as well as the results of the testing.

This spreadsheet could be as simple or complex as you like, but should reflect the key success criteria for the report. An example of this type of form is shown below in Figure 2.4.

Report System Testing

Date: 10-Oct-10

Report Name: MONTHLY SALES SUMMARY

	Pass	Fail	Comments
Data Accuracy			
Monthly Totals: Jan 2010	✓		
Monthly Totals: Feb 2010	✓		
Monthly Totals: Mar 2010	✓		
Monthly Totals: Apr 2010		✗	*Totals did not include 04/10 adjustment*
Processing Speed			
Report Refresh	✓		*30 seconds*
Open Document	✓		*96 milliseconds*
Format First Page	✓		*8 milliseconds*
Average Page Format	✓		*12 milliseconds*

Figure 2.4 Example of system testing form

DATA ACCURACY

To start off with, the most paramount issue around reporting is that the data is correct. So your first bit of testing should be around the accuracy of the data and numbers you are presented. You can check the data in a number of ways — first, if you are reporting off a transactional system (like an ERP or Accounting application) you can verify the numbers from the source system.

An easy way to do this is to filter your report for a specific time frame, customer, vendor, etc. to make the number checking a bit easier. This will enable you to perform multiple "spot checks" of the data, without waiting for the entire report to run each time. The theory is that if you check your totals against a few different months and they are accurate, then your report should be accurate for the rest of the year.

If you do experience problems with data accuracy, it can either be in the data you are returning, or alternately what manipulation you are doing in the report itself. To check the data that is being returned from the database, check the SQL that is being generated by Crystal Reports. From within the designer, select Database > Show SQL Query. You can then copy and paste this query into the SQL tool of your choice to determine what data is being returned and try and work things out from there.

If you suspect that it is your own formulas, totals, etc. that are incorrect, work your way through these formulas by restricting the data in your report. You may want to filter it so only a handful of records are returned, then work through the logic in your formulas, totals, etc. to find out where the issue is.

PROCESSING SPEED

Regardless of which deployment method you choose, the processing speed of the report is another key issue. While it is true that you may have to create reports that take a while to run, you want each report you create to be optimized so it can run in the least time possible.

There are a number of ways you can do this, from limiting the number of records that are coming back to the report, to removing unused formulas and fields. The list below represents only some of the tips you may use to optimize your reports:

- Create smaller reports, with smaller graphics, less data, etc.
- Don't overuse sub-reports
- Discard saved data when saving the report
- Remove unused fields and formulas from the report
- Watch "Page N of M" usage, as this can affect performance
- Keep in mind the 1:15 ratio, where a 1mb report can grow to 15mb with saved data
- "Right size" your reports and remove any unneeded features
- Use parameters to filter the report data to only retrieve the records required

To determine the effect these techniques (or your own) may have on the optimization of your report, you can view the report performance information in Crystal Reports 2008 by selecting Report > Performance Information to open the dialog shown in Figure 2.5.

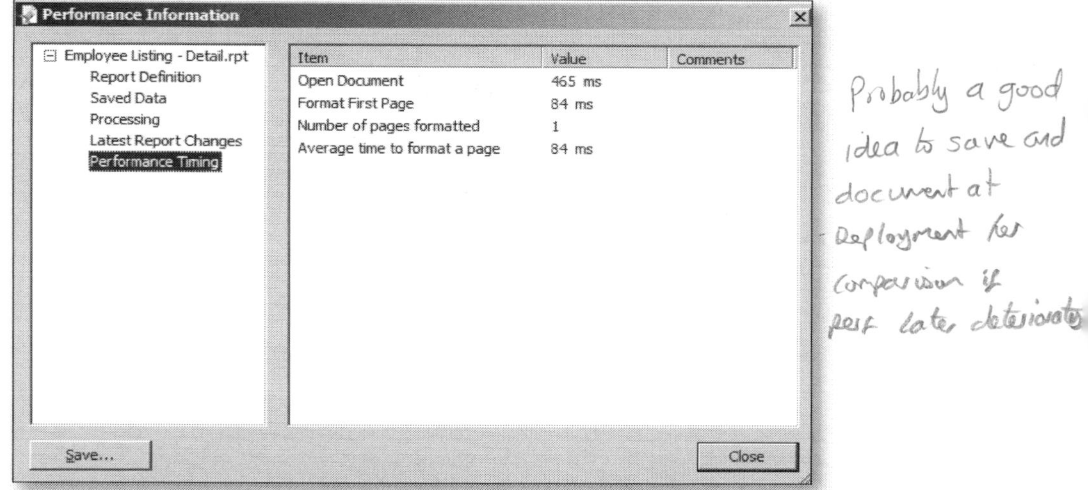

Figure 2.5 Report performance details

Using this dialog, you can check the affect your efforts has on the actual report performance — you will reach a point where you can't optimize the report any further, given the constraints of how many records you need to return, your database server or data source, etc. But it is always good to review your report to see where a few extra seconds of performance can be gained.

You can also save the performance information to a text file by clicking the "Save As" button on the Performance Information dialog. This is especially handy if you need to send this information to another report developer, database administrator, etc.

REPORT FORMATTING

Finally, once you have checked the report accuracy and performance, you need to lend a critical eye to the report formatting. Often this is the first thing that a user will pick up when they view the report. Common errors include not applying a consistent formatting style to fonts or numbers, or misalignment of fields or headings.

Often end-users will focus on the formatting over the data itself, so it's always a good idea to compare your report to the prototype that you created earlier, as well as any specific formatting options that were outlined in the report specification document.

DEPLOYMENT

Once the report is created and tested, you will need to deploy it somewhere for end users to view. For smaller organizations, this could mean that the report developer runs the report and then exports and sends the report in a format other users can access (like Microsoft Excel, Adobe Acrobat PDF, etc). Or you might choose to deploy the free Crystal Reports Viewer, so users can view the report in its native format and take advantage of all of the interactivity provided, including the ability to drill-down, change dynamic parameters on saved data, choose their own export format and more.

For mid-sized organizations, this could mean deploying your report to Crystal Reports Server or one of the other BusinessObjects server technologies (BusinessObjects Edge, BusinessObjects Enterprise).

How you deploy your report is up to you, but there are some factors to consider around how you deploy your report. Figure 2.6 below lists some of the deployment options, as well as some of the factors you may want to consider.

	Report Interactivity	Print-Quality Reporting	Excel Analysis	Refresh Report Data	Schedule Report	Web-based Delivery
Crystal Reports Designer	X	X		X		
Crystal Reports Designer Export - Data Only			X			
Crystal Reports Designer Export - WYSIWYG		X				
Crystal Reports Viewer	X	X				
Crystal Reports Server *	X	X		X	X	X
LiveOffice			X	X		

** Or alternately, BusinessObjects Edge or BusinessObjects Enterprise*

Figure 2.6 Report deployment options

REFINEMENT

In business, as in life, nothing stays the same forever. So you should also put a plan in place to periodically review your reports, for any modifications that need to be made or enhancements to be considered. Often end-users will drive this refinement process with their own additional requirements, based on their current needs.

Another key refinement area is tied back to the system testing you did earlier — reports that include date-based information will often grow in size and number of records as the year goes on and as your business grows. These reports may perform fine when first created, but in the months that pass, may become slower and slower as the volume of data increases. It is worth a periodic check of these reports to see if there are any refinements that could be made (parameters, record selection, etc.) to make the report process faster, or alternately find a different way of doing things.

To help yourself out, you should always keep a copy of the Report Specification for the report, as well as any other documentation you have prepared for the report. When a user does want to refine or "clone" this report for a new use, you already have a good starting point for the requirements gathering and design.

SUMMARY

Sometimes when I am teaching a training course, students will ask why I spend so much time making a point about the report development lifecycle and how important it is to your report development. The answer is always the same — with Crystal Reports (as in life,) a little bit of planning goes a long, long way!

By following a development methodology, you can gather the user's requirements and then communicate them back to the user, as well as the report developer to ensure that you always deliver exactly what is needed. With a technical review, you can be assured that you have the data to report on, before you actually start writing the report.

Once you have done all of the prep work, you can get down to the actual development of your report, and that is what Chapter 3 is all about — creating your very first report using the design you have created on paper.

Chapter 3
Your First Report

In This Chapter

- Accessing Data with Crystal Reports
- Creating Reports Using the Standard Report Wizard
- Saving Your Report
- Working with the Report Designer
- Report Design Environment
- Customizing the Design Environment

INTRODUCTION

At this point, you are ready to get down to business. You should have a report specification and a prototype of the report you want to create and should be well acquainted with the Standard Report Wizard and some of the other report wizards that are available to help you get started.

In this chapter, we take a look at the steps required to create your report using the wizard — keep in mind that everything you do with the wizard you can also do in the report designer user interface itself. The wizard just makes it a bit easier to get started, as all of the most commonly used report features have been included.

So even the most experienced report developers may use the report wizard to get started, in order to "jump start" their report development and get a report created quickly. As with most applications, you don't have to use the wizard — you can choose to just start with a blank report and build it up, feature-by-feature.

USING THE STANDARD REPORT DESIGN WIZARD

The easiest way to invoke one of the report wizards is from the Crystal Reports Start page, which appears whenever you open Crystal Reports, as shown below in Figure 3.1.

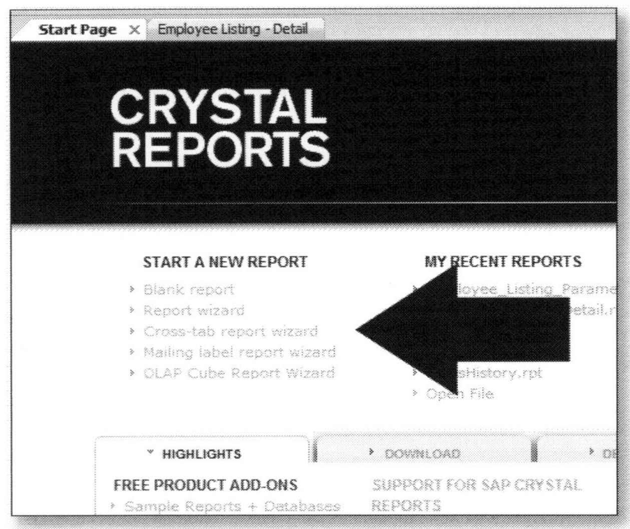

Figure 3.1 Crystal Reports wizards

Alternately, you can start one of the wizards by selecting File > New. Each of these wizards will guide you through the steps for creating a new report with the specified features. While there are a few to choose from, all of the wizards share some common steps, such as selection of the data source, fields, and so on. Once you are comfortable with the Standard Report wizard, you can apply your knowledge of how it works to other wizards. For now, we are going to step through creating a report using the Standard Report wizard.

SELECTING A DATA SOURCE

The first step in creating a report using the Standard Report Wizard is choosing the data source on which the report will be based.

For selecting your data source, the wizard will display a list of Available Data Sources, shown in Figure 3.2. This list is really just a number of different views of the data sources that you have available to use in your report:

My Connections — This section of the Available Data Sources shows any databases or sources that you may have connected to in the past, or that you are currently logged onto.

Create a New Connection — Used to create a new connection to your data source.

Since this is probably the first time you have used Crystal Reports, you will need to create a new connection. If you expand the node for "Create New Connection", you will see that there are a number of different data sources you can use in your report.

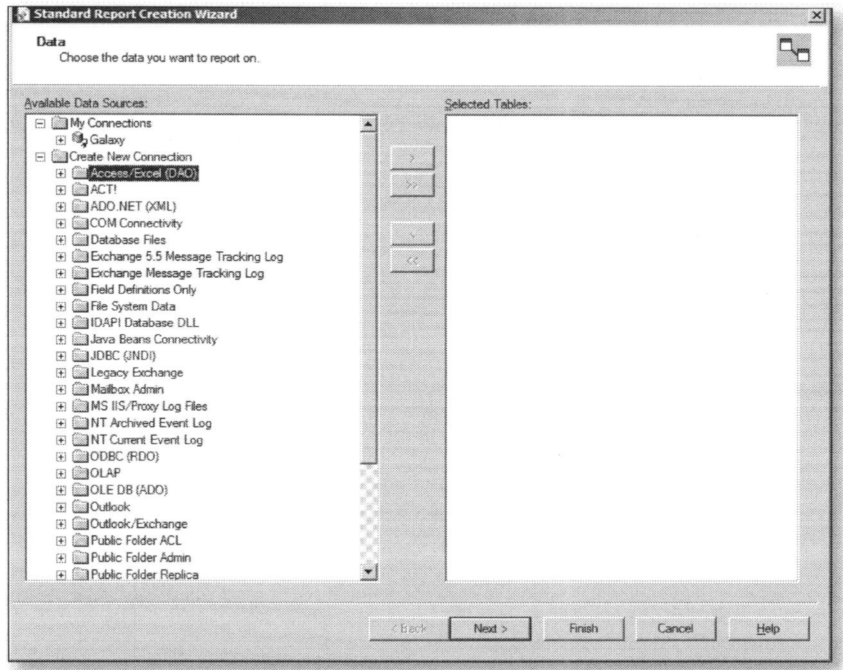

Figure 3.2 The Crystal Reports Data Explorer

The most common data source for Crystal Reports has to be ODBC. Most databases can be accessed through an ODBC driver and historically Crystal Reports has always shipped with a number of ODBC drivers for the most popular database formats, including Informix, Oracle, SQL Server, and so forth. You need to install and configure the appropriate ODBC driver for your database, as well as any database client software required to access the data source.

Running a close second is databases accessed using Crystal Reports' "native drivers". These drivers don't require ODBC and can make a direct connection to a particular data source, including SQL Server, Oracle and more. The advantage of these drivers is that they don't have a middle layer to go through (like ODBC) and can provide better performance.

Finally, rounding out the third most popular data source is local file-based data sources like Excel and Microsoft Access, as well XBase, DBase, Paradox and Btrieve that can also be accessed using native drivers available in Crystal Reports.

But that isn't the end of Crystal Reports' support for data sources — in addition to using standard relational databases, Crystal Reports can report from a number of other "non-traditional" data sources, including the local file system, message tracking logs, Internet Information Server (IIS)/Proxy logs, Windows NT event logs, Microsoft Exchange, Microsoft Outlook, and more.

For more information on working with different data sources, you may want to flip over to Chapter 16 where we look at these different data sources in detail.

In this chapter, we are just getting started so we are going to look at creating a report using an ODBC data source that has already been created for us. If you expand the ODBC data source node, a second dialog box appears asking you to select the data source you wish to use, as shown in Figure 3.3.

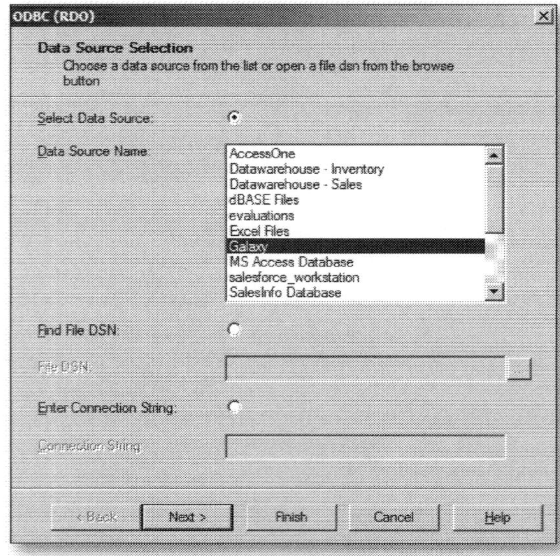

Figure 3.3 Selecting an ODBC data source

Throughout the book, we will be using a sample database named "Galaxy" which was designed for Galaxy Toys, a fictitious distributor of children's toys and games.

This database is included with the downloadable files that are available from www.kuiperpublishing.com. Also, check out "Setting up the Samples" section in the front of this book for instructions on how to configure the data sources, reports, etc.

For this example, select the "Galaxy" data source and click "Finish". A separate node will appear under "ODBC", showing the data source you have selected and all of the tables, views, stored procedures, etc. that are available for use.

For more information on working with Data Sources, including alternate data sources, tables, views, etc. check out Chapter 16.

To add a table to your report, expand the "Tables" node and double-click to add to the list of selected tables on the right. In this walk-through, we are going to add the Customer table to your report using this method, but you can also double-click the table name or highlight it and use the arrows to move it across.

LINKING DATABASE TABLES OR FILES

If you select one or more tables for your report, the Standard Report Wizard adds an additional step in the wizard titled Link, as shown in Figure 3.4 on the following page. Your database administrator should be able to provide you with an entity-relationship diagram that will show the relationships among the tables in your database. You need to re-create these relationships in Crystal Reports by drawing visual links between the tables.

By default, Crystal Reports will perform the links or joins for you based on the name of the field or keys that are present in the table. If you would like to learn about joining tables together, this is covered in detail in Chapter 16. For now, we need to select some fields to appear in our report.

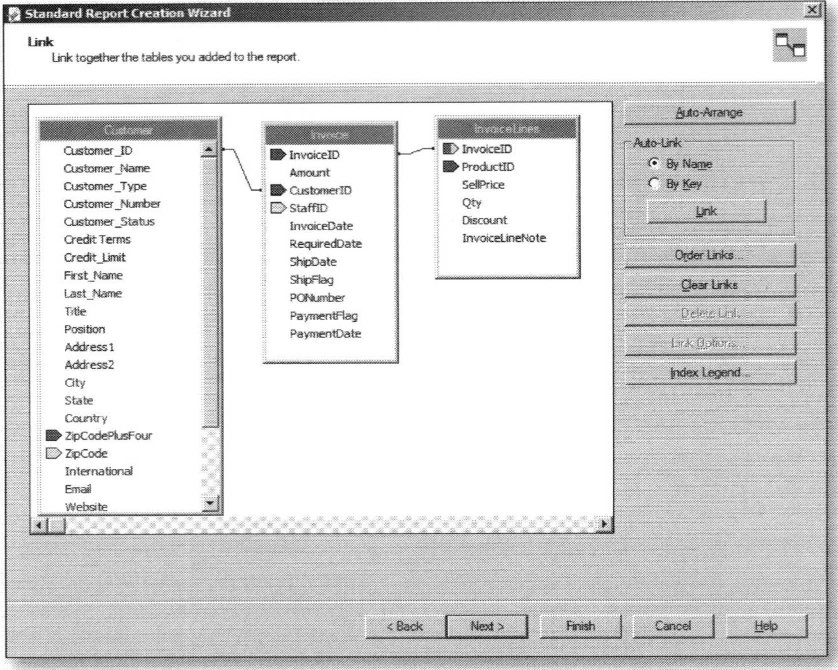

Figure 3.4 The Visual Linking tab helps you join two or more tables

CHOOSING FIELDS

The Fields page of the Standard Report Wizard, shown in Figure 3.5, is split into two sections. The left pane of the dialog box lists all of the fields that are available to be inserted into your report, grouped underneath their table name.

To add a field to your report, you need to move the field from the left pane to the right pane. You can accomplish this by double-clicking the field name or by highlighting the field and clicking the Add button. Additional buttons are also available to add or remove one or all fields. For our report, we are going to select the following fields: Customer Name, Region, Country and Last Year's Sales.

To select multiple fields, hold down the Control key while clicking.

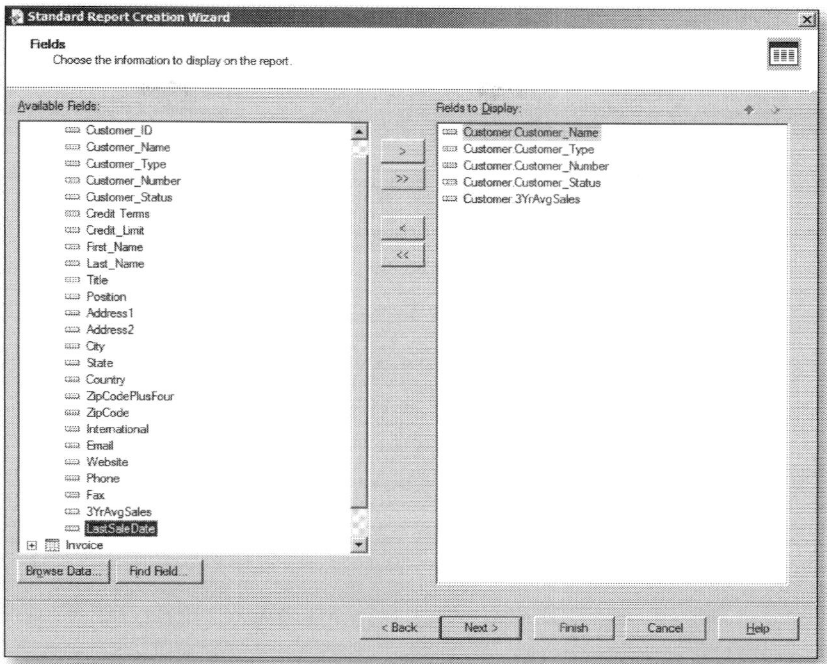

Figure 3.5 Using the Fields tab to select the fields for your report

If you are unsure of a field's definition or contents, you can use the Browse Data button to display a sample of the field's contents. Keep in mind that the sample returned is not based on the complete contents of the table; it is just a representative sample of up to (approximately) 200 records.

Another key feature of the Fields tab is the Find Field button, at the bottom left. This button allows you to search the selected tables for a field that matches your criteria.

Because we are creating a simple report to start with, we are not going to discuss the use of the Formula button to insert calculations and summaries at this time. If you can't wait to get started on using formulas, you can go straight to Chapter 11, where formulas are covered in depth.

To change the order of the fields you are inserting in your report, you can use the up and down arrows that appear in the upper-right corner of the dialog box to move fields up and down. At this point, you can also change the column heading associated with a field.

GROUPING & SORTING

The next step in creating a report using the Standard Report Wizard is selecting the sorting and grouping to use in your report, using the Grouping options shown in Figure 3.6. To select a field for grouping, you move it from the list on the left to the list on the right.

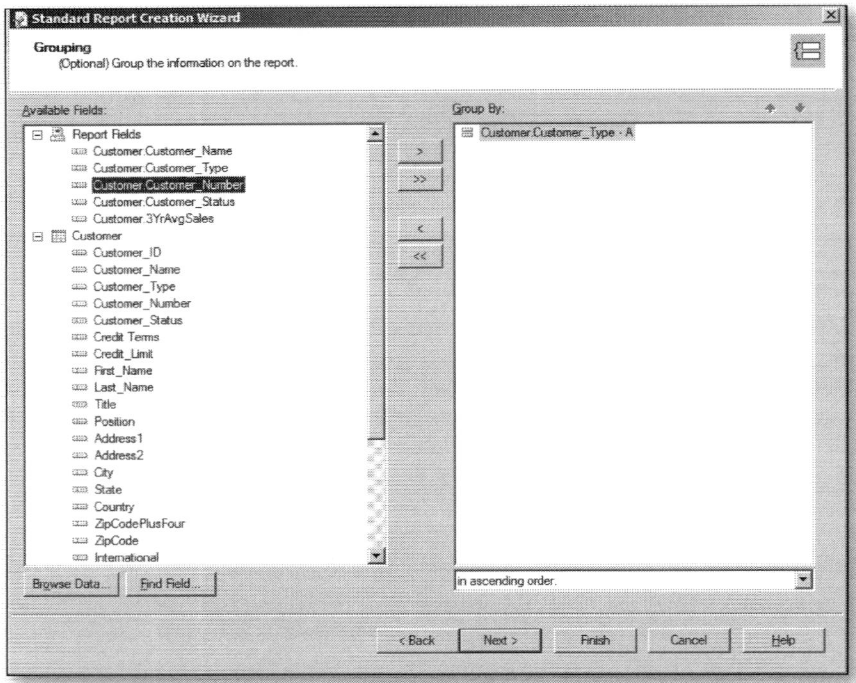

Figure 3.6 Grouping options

By specifying a field on the Group tab, you can add control breaks or groups to your report. For example, if you were to group on the State field, your report would be printed with all the records for each State together, with a break between each State. Another example is shown in Figure 3.7, where the report is being grouped by a Customer Type field — all of the Customers of a particular type will be grouped and shown together.

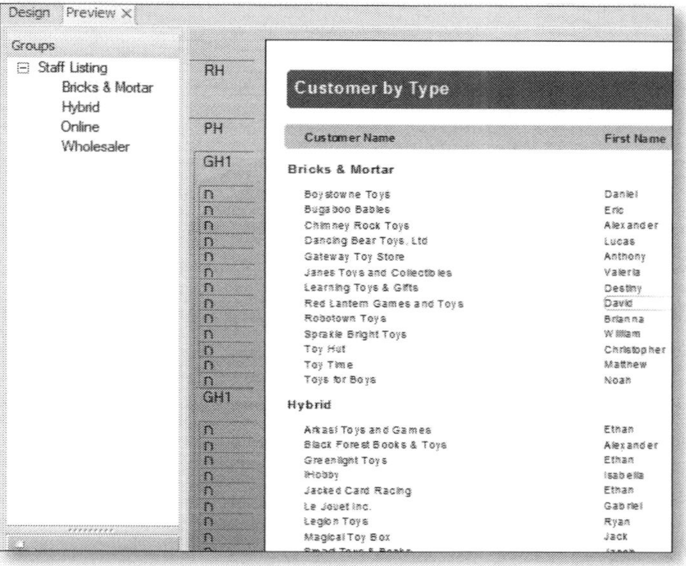

Figure 3.7 An example of a report grouped by Customer Type

For our purposes, we want to group by the Country field—after you select a field, notice that you have a choice of sort orders in the drop-down list below the list of selected fields:

In Ascending Order — This option groups the records by the field you have specified and orders those groups from A through Z, zero through nine, and so forth.

In Descending Order — This option groups the records by the field you have specified and orders those groups from Z through A, nine through zero, and so forth.

In Specified Order — Using this option, you can name and define your own grouping criteria. You might want to use this option, for example, if you want to group states into sales territories. You could create a group called Bob's Territory and set the criteria to North Carolina and South Carolina. When the report is printed, all of the records from North Carolina and South Carolina will be grouped together under the group name, Bob's Territory.

In Original Order — If your database has already performed some sorting on the data, this option leaves the records in their original order.

The Grouping step of the wizard also features a Browse Data button that allows you to search for a field to use for your group.

INSERTING SUMMARIES

The Summaries options, shown in Figure 3.8, are used to insert Crystal Reports summary fields into your report. Crystal Reports provides these summary fields so that you do not have to create a formula every time you want to insert a sum, average, etc.

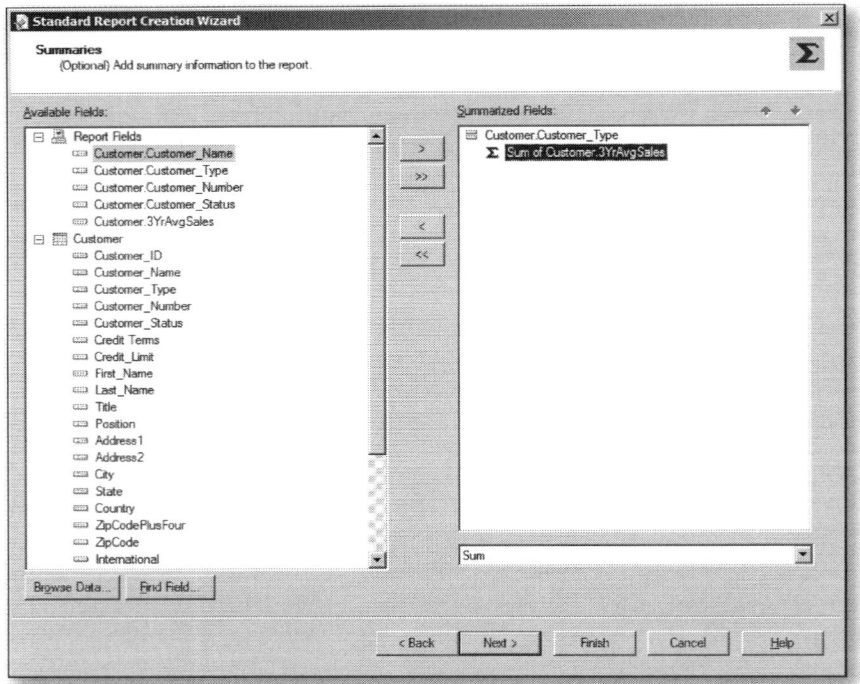

Figure 3.8 Inserting summary fields into your report

You can insert a number of summary fields into your report; the types of summaries vary based on the type of field, as shown on the next page in Table 3.1.

Table 3.1 Crystal Reports Summaries and Usage

Summary Type	With Numeric Fields	With Other Field Types
Sum	X	
Average	X	
Maximum	X	X
Minimum	X	X
Count	X	X
Distinct Count	X	X
Sample Variance	X	
Sample Standard Deviation	X	
Population Variance	X	
Population Standard Deviation	X	
Correlation	X	
Covariance	X	
Weighted Average	X	
Median	X	
Pth Percentile	X	
Nth Largest	X	X
Nth Smallest	X	X
Mode	X	X
Nth Most Frequent	X	X

For some of the statistical functions, you are asked to provide additional information, such as the value for N. For other functions, you specify that a summary is a certain percentage of a particular field.

When you add a summary to your report using the Standard Report Wizard, the summary appears immediately following each group, showing the summary for only that particular group. By default, if you have specified grouping for your report, Crystal Reports will add all of the numeric fields you have selected to be summarized.

Since we selected Last Year's Sales as one of the fields we want to appear in our report, this has been added to the list of fields to be summarized automatically.

USING GROUP SORTING

Group Sorting (sometimes called "TopN", "BottomN", etc.) shown in Figure 3.9, is a powerful analytical feature that allows you to order data based on subtotals or summaries and is an optional step when using most Report Wizards.

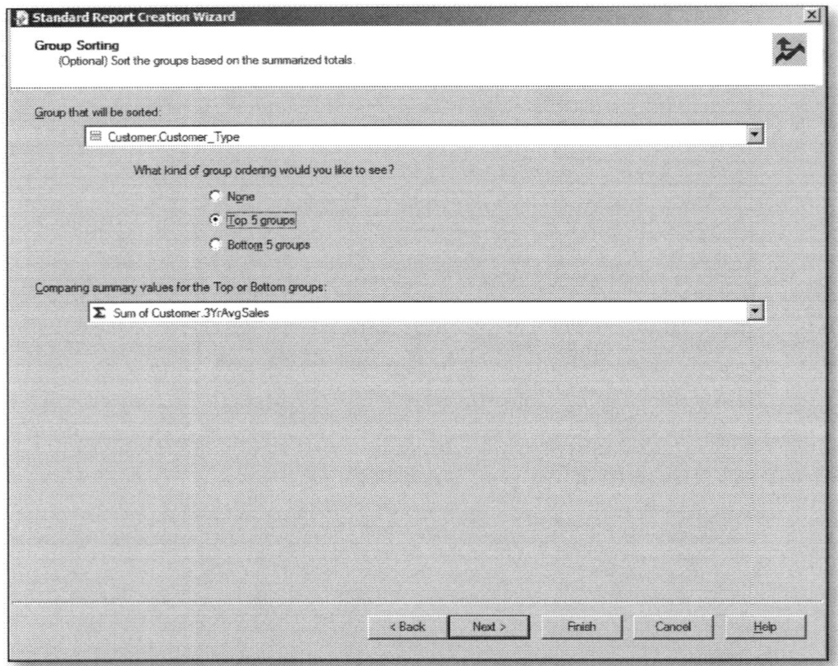

Figure 3.9 Group sorting and Top N analysis can be used to analyze report data to identify trends

For example, if you have a report totaling each customer's sales for the past year, you can use group sorting to determine your top 10 or top 20 customers by ranking their sales totals (and likewise, you can find your bottom 10 customers).

In addition to doing Top N and Bottom N analysis, this feature can be used to sort all of your customers, placing them in order from highest revenue to lowest (or vice-versa).

When working with Group Sorting and the Report Wizards, you need to keep two things in mind. The first is that to use Top N analysis, you need to have a group inserted into your report and a summary field summarizing some information within that group. The second is that Group Sorting analysis can be applied to multiple groups. You may want to show your top 10 customers, for instance, and, for each customer, the top 5 products purchased.

ADDING CHARTS

Crystal Reports uses a powerful third-party graphics engine to provide a wide range of graphs and charts, as shown in Figure 3.10. A number of standard chart types are available from within the wizard, including:

- Bar
- Line
- Pie

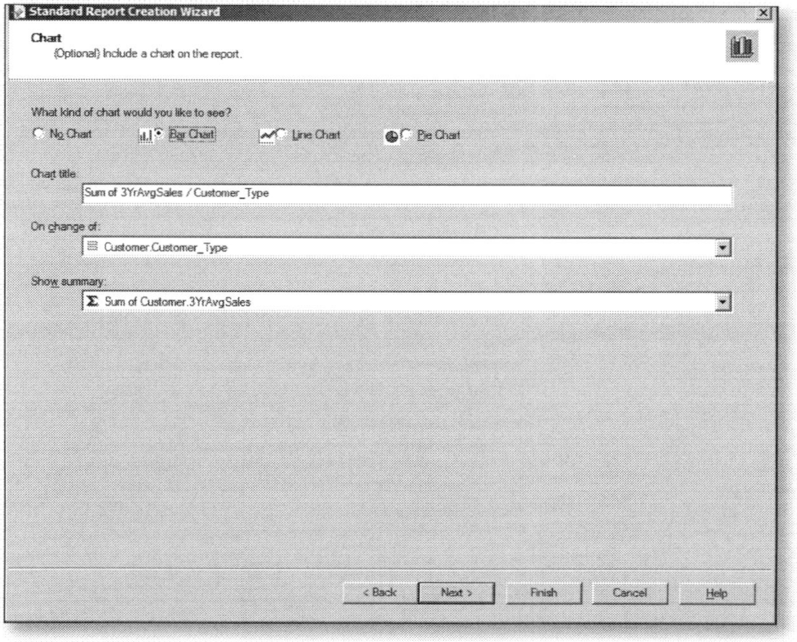

Figure 3.10 Charts and graphs available in the wizard

You also have the option of creating a custom chart type, based on your needs. In addition to the graph type, you can select the source of data for your graph. The charting function within the wizard is based on a group you have inserted and some summary field. These graphs will appear in the report header and represent some of the basic charting features available within Crystal Reports.

Crystal Reports has the ability to add complex charts and graphs after you are through with the Report Wizard, and this functionality is available within the Report Designer and detailed in Chapter 14: Charting and Mapping.

USING RECORD SELECTION

One of the last steps in creating a report using the Standard Report Wizard is setting the record selection for your report using the Record Selection options, shown in Figure 3.11.

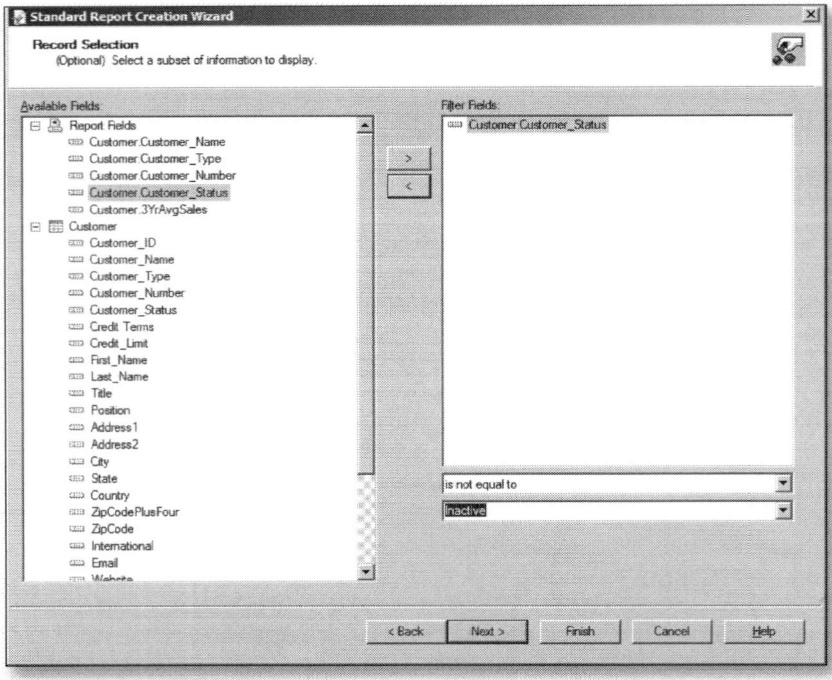

Figure 3.11 Choosing the record selection for your report

Record selection is important because you probably do not want to return every single record in the table for your report. You use record selection to narrow the data to get exactly the subset of information that you need. If you were creating a daily sales report, for example, you would probably want to return the sales records for only a single day.

Another key point about record selection is that you want to return only the records that you need for the report. It makes no sense running a report for 2 million customers when you are interested in only one particular customer. Using record selection, you can narrow in on that one customer and save on report processing time.

Using this step of the wizard, you can specify a field to use and then set your record selection criteria based on that field. If you are creating your report from a relatively small database, you can skip record selection and move to select a template and preview your report.

SELECTING A TEMPLATE

The final tab in the Standard Report Wizard, shown in Figure 3.12, is for applying a template to your report. A template is a set of formatting attributes that are applied to your

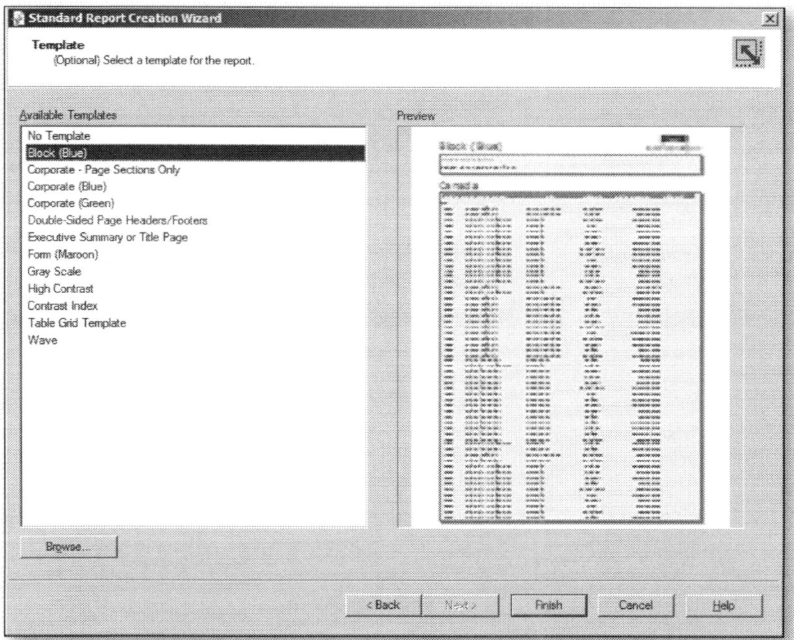

Figure 3.12 You can apply a preformatted report style to your report

report — if you select any of the predefined styles shown, you see a preview of what that particular style looks like on the right. You can also select a template that is not listed here by clicking the "Browse" button and selecting a template file.

Once you have selected a template (or none) you are almost finished. With the Report Wizard settings complete, click the "Finish" button to preview your new report.

If you are interested in creating your own report templates, we'll be looking at that a little later in Chapter 5.

SAVING YOUR REPORT

After you have finished with the Report Wizard, you are returned to the Preview tab of your report. At this point, you will probably want to save your report before we move on. You can save your report by selecting choose File > Save, or click the Save icon on the toolbar.

SAVE OPTIONS

On the File menu, the Save Data with Report option is checked by default. The only reason to leave this option checked would be if you were going to send the report to another Crystal Reports user who did not have access to your database. This user could open the report with saved data and view or print the results without having to go back to the original data source. Otherwise, this option will increase the report file size unnecessarily.

In addition, under File > Summary Info, shown in Figure 3.13, you can enter an author's name, keywords, and so forth. Although most people do not ever complete the summary information for any file that they save, you should complete this information. Crystal Reports treats these fields as special fields that can be inserted into your report. You can also use the check box at the bottom of the dialog box to generate a preview of the report. With this feature, when you are looking at a long list of reports, you can browse through a thumbnail picture of each report.

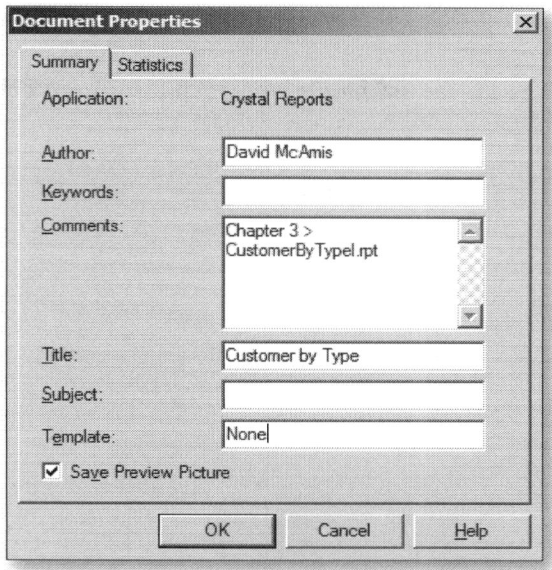

Figure 3.13 Summary information can be inserted into your report as a special field

If your report is based on a data source that is being updated regularly, you can refresh your report at any time by selecting Report > Refresh Report Data or by pressing the F5 function key. If you are working off a database that has live data entry during the day, you may notice that your report results change as new data is entered or deleted.

SAVING DATA WITH YOUR REPORT

Another issue to consider when saving your report is whether to save the data with the report. When a Crystal Reports report is run, a saved record set is written to a temporary file on the hard drive. (You may notice this feature when you make a change to your report or record selection and you are prompted with Use Saved Data? or Refresh?) When you save your report with the Save Data with Report option enabled, this saved record set becomes a part of the report (.RPT) file and increases the size of the file dramatically. To turn off this option, select File > Save Data with Report to remove the check mark.

SUMMARY

So now that you have your first report created, it is time to take a look at working with data sources in case you need to change the underlying data for your report or add additional tables. Every report you will create will be based on a data source, so this is a great place to get started — once we have finished looking at working with data, we will pick up in Chapter 4 with the report design environment itself and taking your report design to the next level.

Chapter 4
Working with the Report Designer

In This Chapter

- Working with the Report Designer
- Report Design Environment
- Design & Preview Tabs
- Navigation Methods
- Toolbars
- Report Explorer
- Customizing the Design Environment

INTRODUCTION

With your first simple report created, it is time to take a look at the place you will spend the most time as a report designer; the Report Design environment. In this section, you learn the basic skills that you will need throughout the report design process. You will learn how to navigate through reports, as well as the different parts of the design environment and the toolbars that are available.

We'll also be looking at the report explorer, to help you understand how reports are put together, as well as how to navigate through your reports. Finally, we'll finish the chapter with a look at how to customize the design environment to suit your own requirements and make your life as a report designer just a bit easier.

REPORT DESIGN ENVIRONMENT

The report design environment is divided into several areas, each with its own purpose and unique properties. Some areas of the environment are standard with every report that you create, such as the Design and Preview tabs, the navigation toolbar and other toolbars.

Other areas, like report sections, depend on your report's design and may appear multiple times, depending on your needs. Understanding what is happening in the report design environment is the key to understanding report design, and this section gives you an overview of the various areas of the design environment and their use.

REPORT TABS

There is now a tab shown for each report that you have open within the report designer. These tabs allow you to quickly switch between any reports that you have open, as well as the Crystal Reports Start Page.

You can close a report tab by clicking the small "X" that appears on the right-hand side of the tab. To switch between reports, click on the tab to open that particular report.

If you have a number of reports open at the same time, you will notice that there are "spinner" controls that appear at the end of the report tabs which allow you to scroll left and right through the open tabs, as shown below in Figure 4.1.

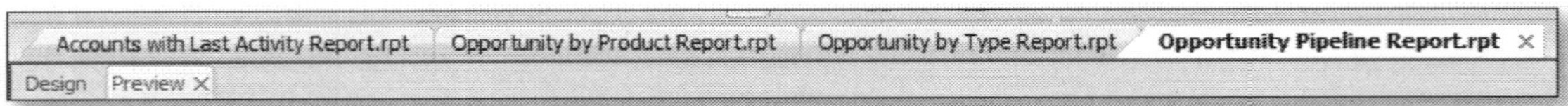

Figure 4.1 Report tabs in action

DESIGN & PREVIEW TABS

When working within a report, you can view that report in two modes: Design and Preview. Design mode, shown in Figure 4.2, offers a behind-the-scenes look at your report, and each section of the report is displayed once. Any changes you make to a section or objects in a section while in Design mode are reflected throughout the report. For example, if you change the title that appears in the Page Header section, that change is reflected in every page header, regardless of whether your report is 1 page or 100 pages.

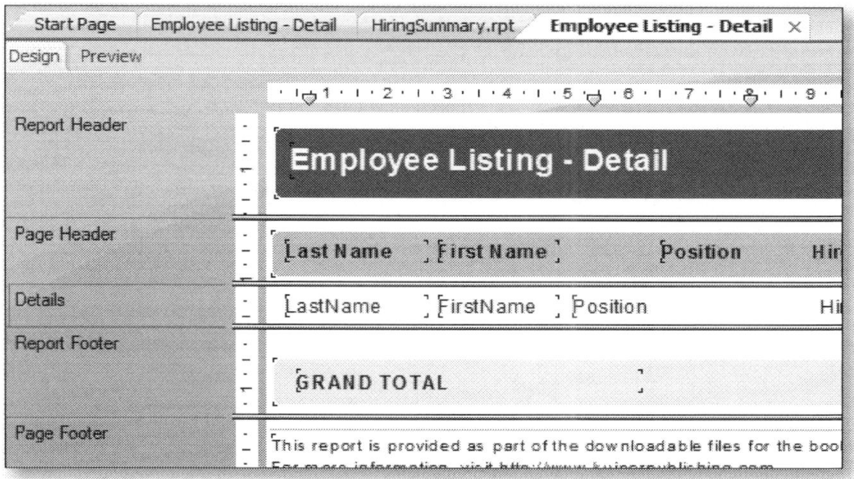

Figure 4.2 The Crystal Reports Design view

The Preview mode is a print preview that is prepared according to your default printer driver. It provides an accurate, multiple-page WYSIWYG (what you see is what you get) representation of your report. What you see in Preview mode is exactly what you see when you print your report.

For most operations, you can use either the Design or Preview mode, but it is sometimes easier to work exclusively in the Design mode because you can see precisely where you are placing objects. Also, when you are in the Design view of your report, you can view any objects that are suppressed or hidden in your report (and believe me, this comes in very handy sometimes!).

NAVIGATION METHODS

The navigation toolbar, shown on the Preview tab in Figure 4.3, is used to navigate between pages in Preview mode. The arrows move you a page at a time through the report, and the arrows with the vertical lines move you to the first or last page of your report. In addition, you can close preview windows and stop report processing using the same set of buttons.

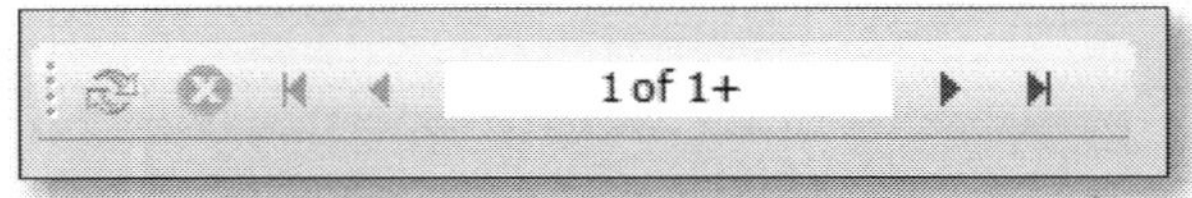

Figure 4.3 The navigation toolbar can be used to move backward and forward through your report

To navigate to a specific place in your report, you can also use the group tree that appears on the left side of the Preview window, shown in Figure 4.4. This tree will appear only if you have inserted one or more groups in your report. For quick access to information in your report, you can also search for a value using Crystal Reports' Find function (CTRL+F).

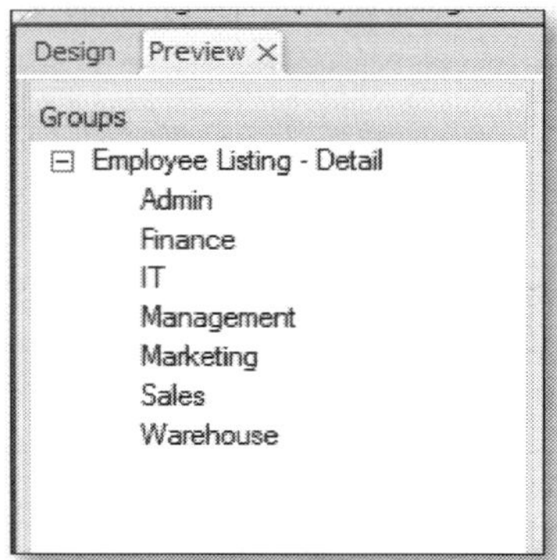

Figure 4.4 The group navigation tree

You can also jump to a specific page directly from the Report menu (CTRL+G).

Sometimes you may see the page numbers shown as "1 of 1+." This means that Crystal Reports knows that there is enough data to fill additional pages, but it doesn't know exactly how many. If you click the right arrow, Crystal Reports will advance to the next page. If you click the right arrow with the line (indicating to jump to the last page), Crystal Reports will show you the total page count.

In addition to navigating through pages, you can also zoom in and out of the report page using the zoom slider that appears on the status bar on the lower right-hand corner of the designer. If the status bar is not showing, you can select View > Status Bar to show it again. On the zoom slider, you have the ability to zoom to the whole page, just the page width or use the slider to select a custom zoom, as shown below in Figure 4.5.

Figure 4.5 Zoom slider in the status bar

TOOLBARS

A number of toolbars contain buttons or shortcuts to commonly used menu items. These graphics are also shown on the menus, to help you quickly locate the corresponding button on the toolbar.

You can display four different toolbars and the status bar by right-clicking in the toolbar area and selecting a toolbar from the list shown in Figure 4.6. The following is a list of the toolbars and their associated functions:

Standard — provides the standard Windows buttons for opening, saving, printing, and refreshing reports as well as buttons for cut, copy, paste, undo, redo, and basic Crystal Reports operations. This toolbar appears by default when you start Crystal Reports.

Formatting — Supplies shortcuts to common formatting options, including font, size, and alignment. This toolbar appears by default when you start Crystal Reports.

Insert — Includes tools inserting fields and other objects, including database fields, groups, and summary fields.

Experts — Provides shortcuts to commonly used experts, including the Running Total Expert and the Highlighting Expert.

Status Bar — Appears at the bottom of the report design page and shows object names, measurements, number of records, processing status, and so on. The status bar is shown by default when you start Crystal Reports.

Navigation — Provides navigation controls for navigating through the report, including next and previous page.

External Command — For displaying icons from any Crystal Reports add-ins you may have installed.

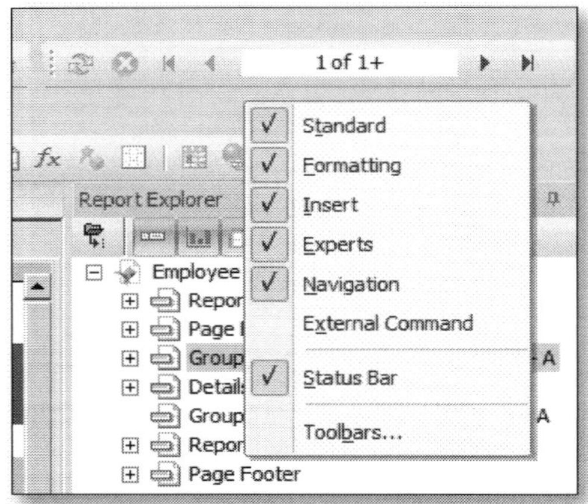

Figure 4.6 Crystal Reports toolbars

REPORT EXPLORER

Another important concept to understand in the creation of reports is that the report is broken into separate sections. The different sections that make up a report are shown on the left side of the screen, as well as in the Report Explorer, shown in Figure 4.7, which can be opened by clicking View > Report Explorer.

The Report Explorer provides a look at all of the elements of your report at a glance and can be helpful when working with complex reports. Regardless of how complex your reports may be, the following sections may appear in your report and are commonly used as described here:

Report Header/Footer — These appear at the top of the first page of the report and at the bottom of the last page. The report title appears most often in the report header, and a record count or end-of-report marker may appear in the report footer (that is, 10,000 records processed — end of report), in addition to any grand totals for your report.

Page Header/Footer — These appear at the top and bottom of every page. The page headers and footers are used to display information that is critical to understanding the data represented and may include field headings, page numbers, and the print date of the report.

Group Header/Footer — These appear immediately before and immediately after any groups you have inserted. The group header or footer usually contains the group name field, which provides a label for the group, and may also contain formulas, subtotals, and summaries based on the data in the group.

Details — This appears once for each record in your report and (unless you are creating a summary report) contains most of the report's data.

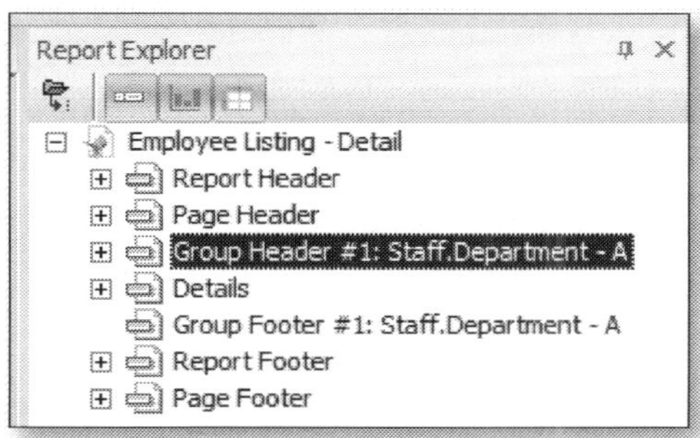

Figure 4.7 A number of sections make up a Crystal report

On the Design tab, each of these sections is represented once, but when the report is previewed or printed, these sections are repeated as many times as is needed, as shown in Figure 4.8.

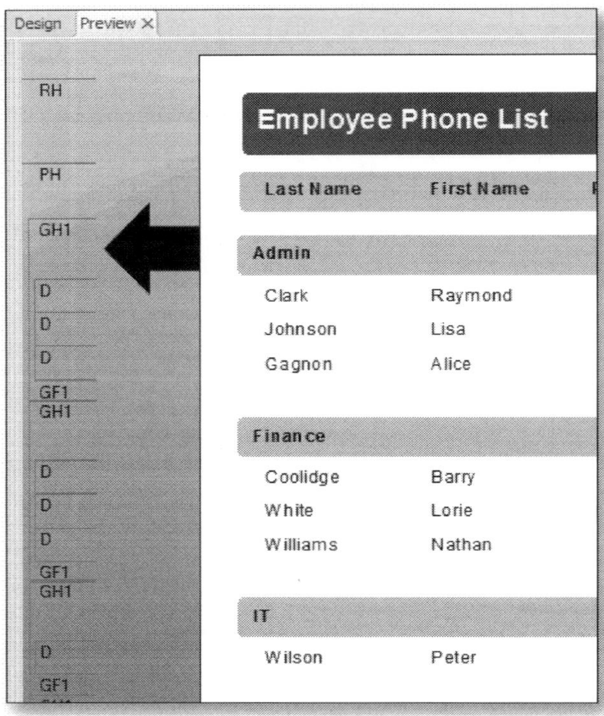

Figure 4.8 An example of a report preview

It is easy to become confused when switching between the Design and Preview modes, especially when you are looking at a preview of a report and wondering where that field came from. One trick to help you understand where different objects are placed is to click the object to select it. When you switch back to the Design mode, the object will still be selected and you can see in which section it appears.

Note, too, that you can split report sections into multiples, so you may see Report Header A and Report Header B to indicate that the report header section has been split into two (or more) segments. You can use this technique to create complex reports that may be impossible to create with just a single section.

For example, if you were creating a report for distribution in two different languages, you could have one report header set up with the English title, comments, and so on and a separate report header with the same information in Spanish. By looking at a database field (like a Country field), you could determine which header to display.

In the report shown below, the Report Header section was split into three parts in order to display the report title, chart title, and chart at the top of the report, as shown below in Figure 4.9.

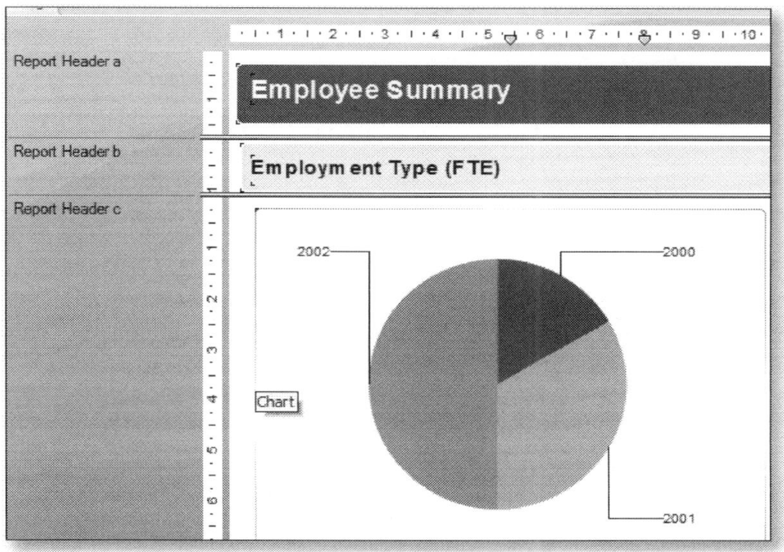

Figure 4.9 An example of a report with multiple sections

CUSTOMIZING THE DESIGN ENVIRONMENT

The report design environment can be customized to suit the way that you work and the reports that you need to design. Environment settings are established using two main option sets; global options and report options.

GLOBAL OPTIONS

Clicking File > Options provides global options that affect all reports you create using Crystal Reports. Using the options shown in Figure 4.10, you can customize the layout of the report design and preview windows, define where reports are stored, and set other options that apply to the report design environment, including which objects to show in the Data Explorer, etc.

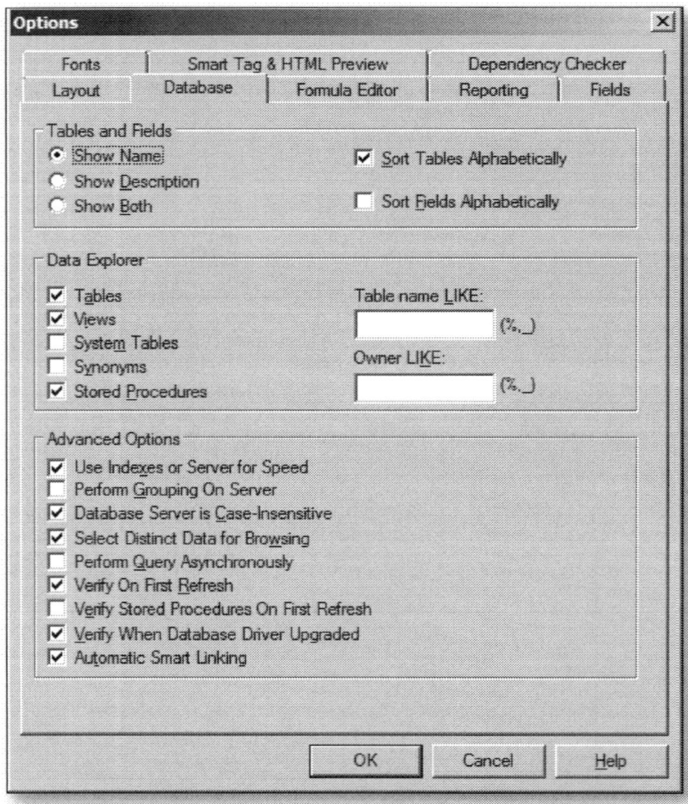

Figure 4.10 Options accessed using File > Options can be set globally and take effect with the next report you create

REPORT OPTIONS

Clicking File > Report Options, as shown in Figure 4.11, provides settings that apply to a specific report, and these settings take effect immediately. Report options specify how null fields are handled and whether data and/or summaries are saved with the report. Options also allow you to define preview page options and other properties.

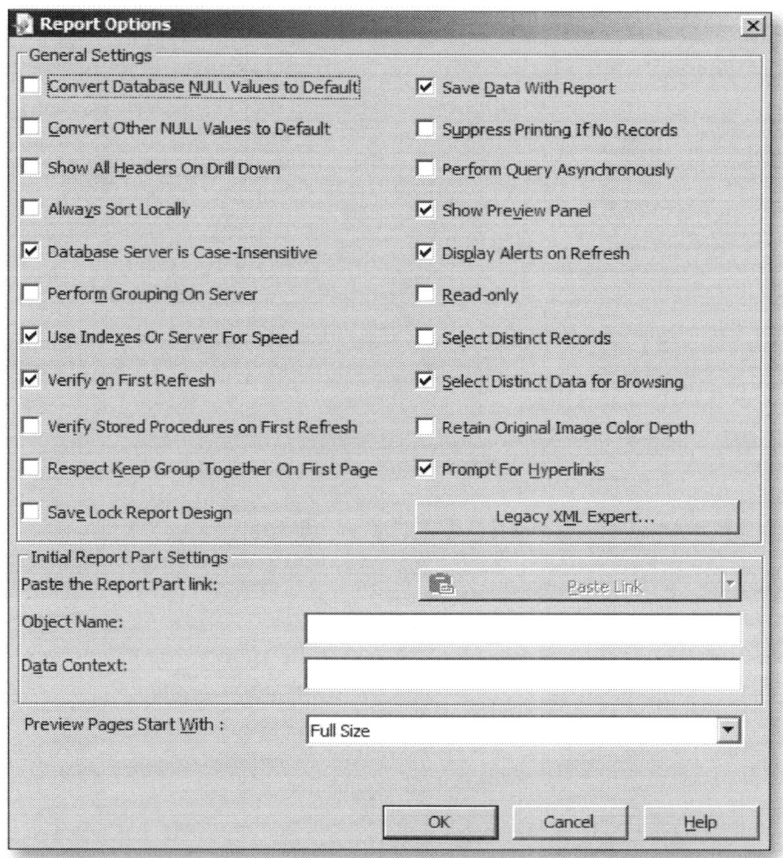

Figure 4.11 Options accessed using File > Report Options apply to a specific report only

LAYOUT OPTIONS

Everyone has his/her own work style and preferences when working with a software package, and Crystal Reports caters to your individuality. You can configure the design environment according to your preferences by clicking File > Options > Layout. You can control rulers that appear, guidelines (which we will talk about later), the display of section names, and more. Common layout options, shown in Figure 4.12, include options for both the Design or Preview view.

Which of these options you set is up to you. Some report designers prefer to view just the basics, without the rulers, guidelines, and so on to clutter up the design window, whereas others find that the rulers and guidelines help them get a feel for the report's dimensions and layout.

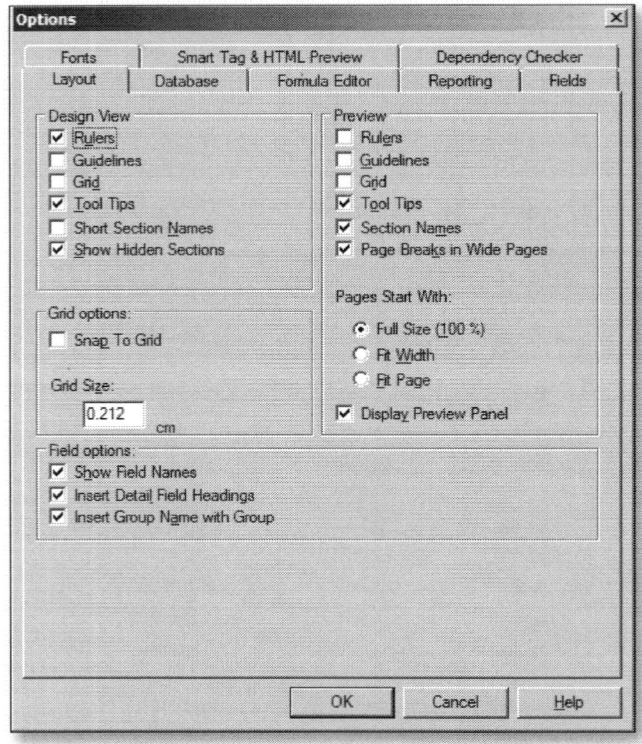

Figure 4.12 Options for the design environment

There are also options that apply to the objects shown on your report. Some of the options available include the following:

Show Field Names — This option displays the full field name instead of a placeholder such as XXXXXXX.

Insert Detail Field Titles — When a field is inserted into the Details section, a text object is inserted automatically into the page header with the field heading.

Insert Group Name with Group — A group tree can be inserted to aid in report navigation and organization.

You can also choose options for the preview page, starting with the initial image set to Full Size (100 percent), Fit Width, or Fit Page. You can use the Display Group Tree checkbox to control whether a group tree is generated from the contents of your report.

Selecting Insert Group Name with Group will cause a group tree to be generated and appear on the left side of the report preview window. Choosing not to generate the group tree aids in report performance, because there is one less item for Crystal Reports to display.

Finally, to help you keep your report design evenly spaced, Crystal Reports has an underlying grid that can be used to align objects. Options for this grid include the following:

Snap to Grid — Aligns all inserted or moved objects to the underlying grid.

Free-Form Placement — Disregards the grid's placement of an object and places the object where you indicate.

Grid Size — Provides the size of the underlying grid in inches or centimeters, depending on your regional settings.

Alternatively, you could use a shortcut to access some of these options, following these steps: Switch your report to the Design mode by clicking the Design tab in the top-left corner of the report design environment. In any white space (where no objects are placed), right-click. A menu headed by Snap to Grid should appear, showing most of the options. Use the checkboxes that appear on this menu to configure the underlying grid.

FORMATTING OPTIONS

Another time-saver is the ability to select the default field formats for any future reports using the Fields tab, shown below in Figure 4.13.

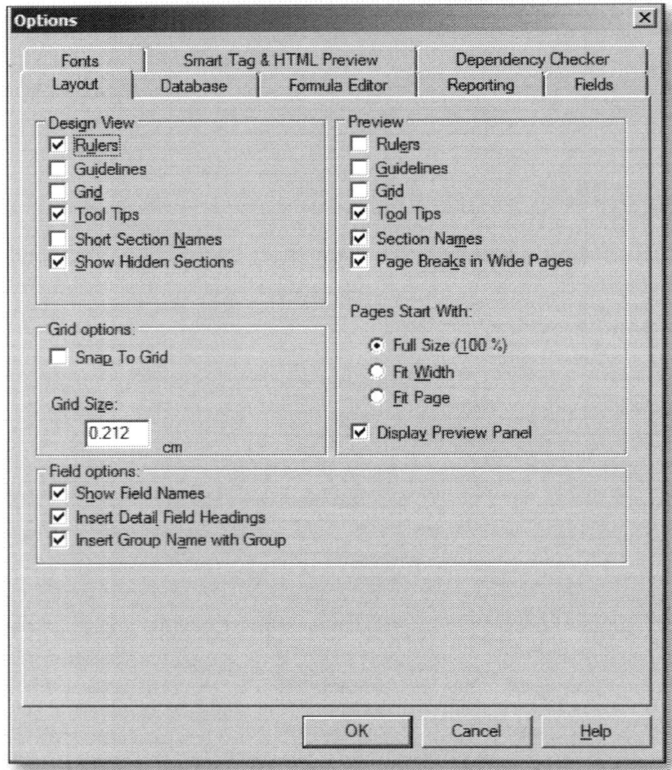

Figure 4.13 Default Field Format Options

To set the default field format, click on the field type then use the standard formatting dialog to select the field formatting attributes. We'll be looking at these formatting attributes in detail in the next chapter, so skip ahead if you want the details on how to format fields.

The formatting attributes you set here will be applied to the next report you create. Keep in mind that these settings are held locally on the user's PC and can't be transferred to another computer.

If you want to create a consistent look and feel for your reports, you may also want to consider using report templates, which we will be looking at in the next chapter.

SUMMARY

Now that you know a bit more about the design environment, we can get on with some of the formatting techniques you will need to create presentation-quality reports from your data. In the next chapter, we are going to start with looking at how to format the actual report page itself, as well as some of the types of fields you can add to your report. We'll also look at how to format these fields, including the ability to pick from a pre-defined format or customize it to your own needs.

And it doesn't stop there — we'll also look at how to use the new Sort Controls introduced with Crystal Reports 2008, as well as some of the tips, tricks and techniques you can use to quickly format your report canvas and get everything pixel perfect.

Chapter 5
Report Formatting

In This Chapter

- Working with Page Formatting
- Understanding Field Objects
- Inserting Field Objects
- Working with Field Objects
- Working with Sort Controls
- Formatting Field Objects

INTRODUCTION

Crystal Reports provides a powerful toolset for retrieving data from a variety of data sources, but just retrieving the data is not enough. We want to be able to format your reports to ensure that the information is presented in a logical manner. That is what this chapter is all about — the basics of report formatting, from setting up the actual report page to the different types of objects you can add to your report canvas.

To start this chapter off, we are going to look at page formatting and the type of fields you can add to your report, followed by how to format these fields. We're also going to be looking at sort controls, which are a new feature for Crystal Reports 2008 to help you create more interactive reports.

We'll also look at working with text fields and inserting large blocks of texts, as well as how to enhance the presentation of your report with lines, boxes, graphics and more. Whether you fancy yourself a plodder or a Picasso, it all starts here.

WORKING WITH PAGE FORMATTING

In older versions of Crystal Reports we were tied to a printer page size for our report layouts. With Crystal Reports 2008, we can choose to associate a page size with our reports (like Letter, A4, Legal, etc.) or alternately we can specify our own custom dimensions for the reports. In addition, no longer are we tied to a specific printer's format — we can design a report that will print nicely on any printer.

To change your printer selects, select File > Page Setup, which will open the dialog shown in Figure 5.1.

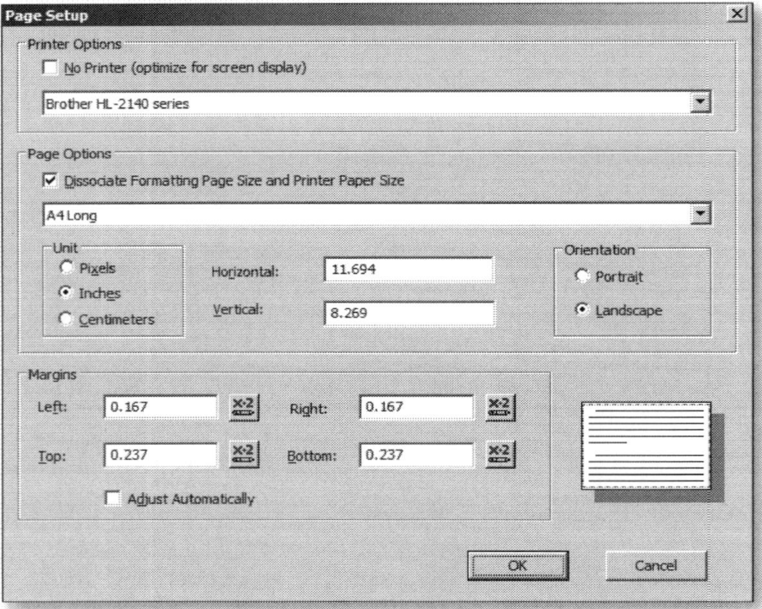

Figure 5.1 Crystal Reports page setup options

From this dialog, you can select a specific printer (or none) and you have the ability to disassociate the page formatting and paper size. You can also select your default units of measure, the orientation of the page and margins.

Keep in mind that in this release of Crystal Reports, you can also control the orientation of individual sections, choosing whether they are portrait or landscape. We'll look at this technique in Chapter 6.

For margins, the default is 0.25 inches but you can increase or decrease this as required. Keep in mind that when users are viewing and printing Crystal Reports, they may have a different printer with a different printable area, so it's best to keep a good margin around the edge of the report to ensure nothing is cut off the page.

UNDERSTANDING FIELD OBJECTS

Now that we have the basic structure of your report setup, we need to look at the basic building blocks for any report — field objects. Reports can include a number of different field objects from database fields, to parameter fields, to fields that hold text which you enter directly into your report. Chances are that you will need to use all of these types of objects to achieve the results that you desire. The following sections describe the common field objects, how to identify them, and their use within your reports.

DATABASE FIELDS

Database fields are drawn from the tables, views, or stored procedures used in your report. Database fields are represented with the designation `{TableName.FieldName}`. This information can be seen by clicking the field and looking in the bottom-left corner at the status bar. For example, if you were looking at a Phone Number field from a Customer table, the designation would be `{Customer.Phone}`. Database fields are used to display information from your database and are most commonly used in the Details section of your report, but they can also appear elsewhere.

FIELD HEADINGS

When you insert a database field into the details section of your report, a special type of text field is inserted to label the field. This field heading, which will appear in your page header, is derived from the field name stored in the database. When you move the database field, the field heading will move with the field itself.

FORMULA FIELDS

A formula field is a calculated field that can be inserted onto your report and is displayed the same way as any other field is displayed. Backed by a powerful formula language that looks like a cross between Pascal, Excel's formula language, and Visual Basic, formulas can incorporate database fields, parameter fields, and so on to perform complex calculations and string, date, and time manipulations.

Formula fields are always prefixed with an @ symbol and are enclosed in curly braces, for example, `{@commission}`. Formula fields, which can be inserted anywhere in your report, have a wide range of uses, including mathematical calculations, string manipulation, and the execution of complex logical statements and outcomes. A formula field can be used just about anywhere you need a calculated or derived field.

PARAMETER FIELDS

Parameter fields are used to prompt report users for information. Parameter fields are prefixed by a question mark and are enclosed in curly braces, for example, `{?EnterState}`. Parameter fields can be used with record selection, formulas, and so on and can be inserted anywhere in your report.

SPECIAL FIELDS

Special fields generated by Crystal Reports include page numbers and summary information fields. Special fields are designated only by their field name, and all of the field names are reserved words. Special fields contain system-generated information and can be inserted anywhere they are needed in your report.

RUNNING TOTAL FIELDS

A running total field is a specialized summary field that can be used to create running totals, averages, and so on and display this information on your report. A running total field is prefixed by a hash symbol and enclosed in curly braces, for example, `{#TotalSales}`. Running totals frequently appear in the page footer or in the Details section with the detail data, but they can be placed in any section of your report.

SQL EXPRESSION FIELDS

A structured query language (SQL) expression field is similar to a Crystal Reports formula field, in that an SQL expression field can be used for calculations. However, with an SQL expression field, these calculations occur on the database server itself and take advantage of the server's advanced processing power. An SQL expression field is prefixed by a percent sign and enclosed in curly brackets, for example, `{%CalcSummary}`. An SQL expression field can be inserted anywhere in your report.

SUMMARY FIELDS

A summary field can be used for calculations as simple as a subtotal or average or as complex as a standard or population deviation. At first glance, summary fields and formulas may appear to do the same thing, but the major difference is that a summary field does not require any coding. A summary field can be identified by the use of the summary type (Sum, Average, and so on), the word of, and the field that is being summarized, for example, `Sum of Sales`. A summary field is generally placed in the group, page, or report header or footer, but it can be placed anywhere in your report.

GROUP NAME FIELDS

A group name field is generated by Crystal Reports to label any group that you have inserted into your report. All group name fields can be identified by the same label — `Group #n Name` — where `n` is the number of the group with which you are working, such as `Group #3 Name`. Group names are generally inserted in their corresponding group header or footer. When a group name field is displayed on the Preview tab of your report, it will appear as the actual name of a group you have inserted. For example, if you inserted a grouping by state, the group name field might read Alaska, Alabama, and so on.

INSERTING FIELD OBJECTS

One of the most common formatting tasks when creating a report is the addition of fields to the report — in fact, sometimes it seems like a contest to see how many fields you can squeeze onto one page. To insert fields into your report, you use the Field Explorer, which can be opened by clicking View > Field Explorer. The Field Explorer is shown in Figure 5.2.

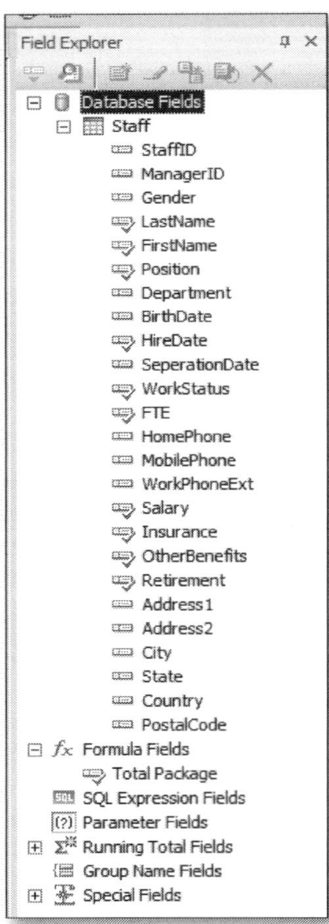

Figure 5.2 The Crystal Reports Field Explorer

You can resize the Field Explorer by dragging the bottom-right corner of the Field Explorer window.

The Field Explorer displays all of the different types of fields that you can insert into your report, broken down by the field type. You can see the fields that are available in each category by clicking the plus sign beside the category name to expand the group.

If you click the plus sign beside the field type Database Fields, a list of all of the database tables you have selected is displayed; to find the field you need, you may need to expand the contents of each table. For the rest of the field categories, you can simply expand the category to see all of the fields contained within it.

If you have a large number of tables and fields, you may want to consider sorting alphabetically by the table name, the field name, or both. You can find this option by clicking File > Options > Database.

Because most of your time will be spent adding and arranging fields in your report, Crystal Reports tries to make this process as intuitive as possible. Fields can be dragged directly from the Field Explorer onto your report, or you can highlight a field and press the ENTER key to attach the field to the tip of your mouse. Then you can position it where you want using your mouse — click once to release it; it's that simple.

When you insert fields into the Details section of your report, you will notice that a special Field Heading object appears in the page header to label the field you have inserted. This action works only with fields that are inserted into the Details section. This feature can be turned off from the dialog box that appears when you click File > Options.

WORKING WITH FIELD OBJECTS

Once a field has been inserted into your report, you can choose to control the way the field looks: its properties, size, font, and so on. One way you can control the way a field looks and behaves is by editing its properties.

FORMATTING FIELD OBJECTS

Every field object in Crystal Reports has properties associated with it. From the font that is used to display the field contents to the format of numbers contained within the field — you name it — there is a property to control it. To view the properties of a particular field, right-click the field and select Format Field from the shortcut menu that appears, shown in Figure 5.3.

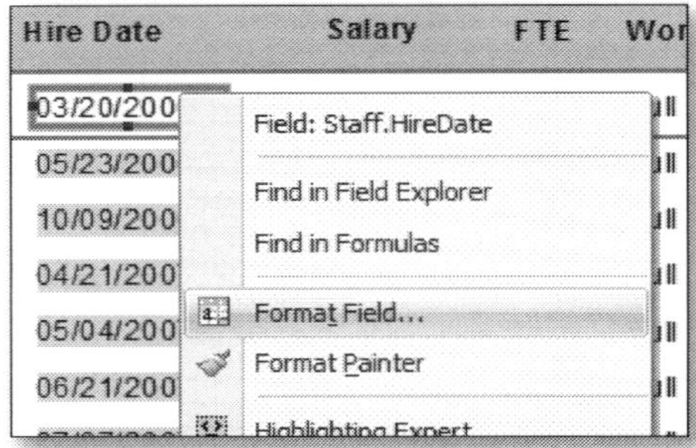

Figure 5.3 You can format a field by right-clicking the field and selecting Format Field

When you select Format Field from the right-click menu, the property pages for the object opens. In this example, we have selected a string field, so there are tabs for Common, Border, Font, and Hyperlink property pages. The property pages that appear depend on the type of field you select. For example, all field objects in Crystal Reports will display the property pages for Common, Border, Font, and Hyperlink, because all of these objects have these properties in common.

But when you compare a numeric field to a date field, you will notice some differences. A numeric field will have an additional tab with properties that relate only to numeric fields; likewise, a date field will have an extra tab with properties that can be set for date fields.

The following sections describe some of the most common object formatting properties.

Common Formatting Options

All of the different types of field objects in Crystal Reports have a common set of properties, as shown in Figure 5.4.

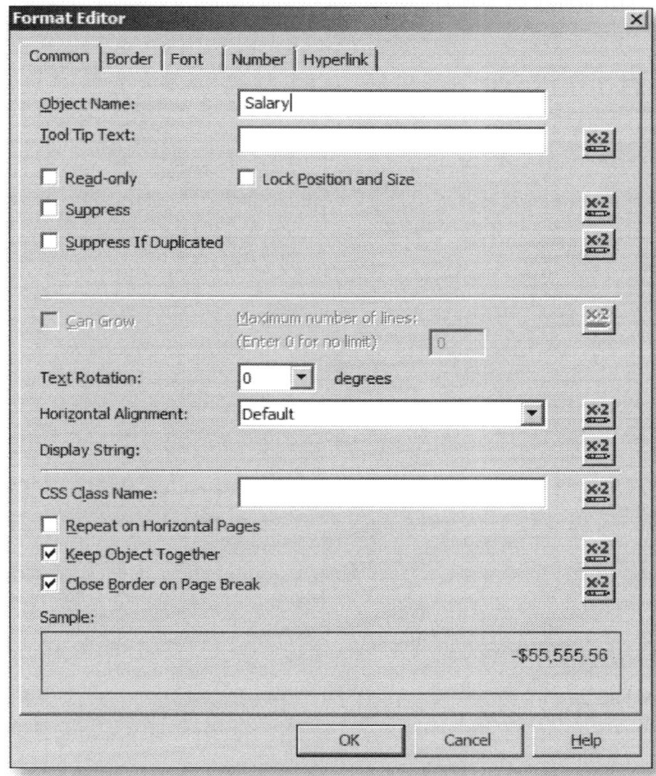

Figure 5.4 Common formatting properties

These properties include the following:

Object Name — The name of the object. Crystal Reports will fill this field in for you, but you may want to rename the objects in your report to something more meaningful.

Tool Tip Text — Enable tool tip text to appear when the user moves the mouse over a particular object. You can enter the text directly into the text box provided or enter the text or a formula using the X+2 button located on the right side of the dialog box. You need to enter the tool tip text in quotation marks (for example, "This is tool text") in the formula, then click Save and Close when finished with the Formula Editor. The X+2 icon turns from blue to red to indicate that tool tip text has been entered.

Read Only — This setting determines if the object's properties are able to be changed.

Lock Position and Size — This setting locks an object into place on your report.

Suppress — Click this setting to enable suppression so that the object does not appear on your report. (The object will remain in your report design but will not appear when the report is previewed or printed.)

Suppress If Duplicated — Use this option to suppress a field if the contents are duplicated exactly. The object still appears in your report design, but the data itself does not appear when previewing or printing.

Can Grow — For multiline objects, select this option to ensure that the object can grow as needed, whether 2 lines or 20 are used. To control the maximum size of any object, you can also set the maximum number of lines. By default, this is set to zero to indicate no limit.

Text Rotation — Use this setting to rotate the text in an object either 90 or 270 degrees.

Horizontal Alignment — Select from the drop-down list provided to left-align, center, right-align, or justify the contents of the object. The default setting varies by type of field.

Display String — This setting is for displaying different types of fields using custom formatting.

CSS Class Name — If you are going to be using this report on the Web, a Cascading Style Sheet (CSS) file can be associated with the report. CSS is used to apply consistent formatting across multiple Web pages (or in this case, reports).

Repeat on Horizontal Pages — Enable this option to repeat this object on any horizontal pages that are created (most often used with page numbers and cross-tabs that run horizontally).

Keep Object Together — Enable this option to attempt to keep large objects on the same page.

Close Border on Page Break — Set this option for objects that have a border to ensure that the border extends to the edge of the page and that the border closes before the next page begins.

Formatting Numbers and Currency

For formatting numbers and currency, Crystal Reports also offers a number of specific properties. To make things easier, you can set these properties and formatting options by example, as shown in Figure 5.5. Instead of actually setting all of the properties, you can just pick a format that looks similar to what you want.

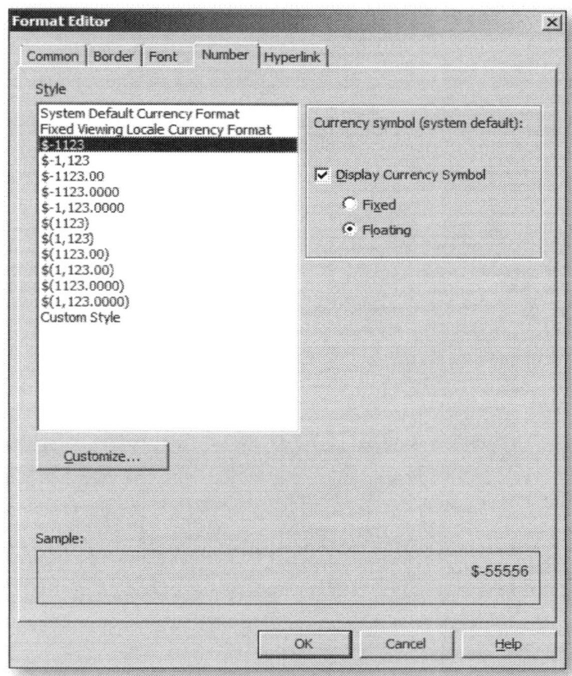

Figure 5.5 Selecting a numeric format by example

To add a currency symbol, use the options at the top-right corner of the dialog box. You can also specify whether the symbol should be fixed in one position or floating beside the numbers.

To specify a custom numeric format, click the Customize button at the bottom of the dialog box. The following options are available for customization on the Currency Symbol and the Number tabs:

Enable Currency Symbol — Specifies whether the symbol will be shown.

Fixed/Floating — Specifies whether the currency symbol is fixed in place in the left margin of the field or floats next to the first digit.

One Symbol per Page — Places one symbol at the top of each page.

Position — Determines where the currency symbol will be displayed.

Currency Symbol — Specifies the currency symbol that will be used with numeric fields.

Use Accounting Format — Fixes the currency symbol at the left and displays negative amounts as dashes.

Suppress if Zero — Suppresses the display of a field if that field's value is zero.

Decimals — Specifies the number of decimal places, by example, to 10 decimal places of accuracy.

Rounding — Specifies the number of places to round to, from 10 to 1,000,000 decimal places.

Negatives — Displays a negative symbol before or after the number or uses parentheses to indicate negative numbers.

Reverse Sign for Display — Reverses the negative sign that would normally appear beside a negative value (that is, it displays a negative sign for positive numbers).

Allow Field Clipping — Specifies whether field clipping is performed. Field clipping occurs when a field frame is not large enough to hold the entire contents of the field. In situations where this occurs, Crystal Reports clips the field by default and shows only part of the field. If you uncheck this option, Crystal Reports displays number signs (#####) to indicate that a field is longer than the space allotted to it. To get rid of the number signs, you need to drag the field frame so that it is large enough to accommodate the entire contents.

Decimal Separator — Changes the decimal separator. By default, Crystal Reports uses a period to mark the place between a whole number and the numbers after the decimal point.

Thousands Separator — Changes the thousands separator and symbol. By default, Crystal Reports uses a comma to indicate the thousands place in a number.

Leading Zero — Adds a leading zero to any numbers displayed as a decimal.

Show Zero Values As — Displays zero values in the default format, which is zeros or a dash (-).

Any custom formats that you create cannot be saved and must be re-created each time you want to use them.

Formatting Date Fields

Date fields can also be formatted by example. Crystal Reports gives you a number of predefined formats to serve as a starting point. If you locate the date or time field that you want to format, right-click the field, and select Format Field. You'll notice that the properties include a Date/Time tab, which allows you to choose a date format by example (just like you did for numbers).

To select a custom style, click Customize at the bottom of the dialog box and specify a custom numeric format. Options available for customization follow:

- Date/time order
- Separator
- Date type
- Calendar type
- Format (month, day, year)
- Era/period type
- Order
- Day of week type
- Enclosure
- Position
- 12/24 hour
- AM/PM breakdown
- Symbol position
- Format (hour, minute, second)

Bringing it all Together

With all of these formatting options, there are a number of ways we can improve the way a report looks. In the following walk-through, we are going to format the field objects of an existing Detailed Sales Report (shown in Figure 5.6) to remove the thousands separator in the Invoice Id field and to format the Print Date field and Invoice field on the report.

Detailed Sales Report

Print Date: 1/06/2010

Invoice ID	InvoiceDate	Customer Name	Amount
3,800	1/01/2009 12:00:00AM	Dancing Bear Toys, Ltd	$11,829.60
3,801	1/01/2009 12:00:00AM	Puzzle Master	$2,870.35
3,802	1/01/2009 12:00:00AM		$2,808.00
3,803	1/01/2009 12:00:00AM	Legion Toys	$125.00
3,804	1/01/2009 12:00:00AM		$1,800.00
3,805	4/01/2009 12:00:00AM	Turtle's Nest	$1,800.00
3,806	4/01/2009 12:00:00AM	Red Wagon Toys	$1,800.00
3,807	4/01/2009 12:00:00AM	abraKIDabra Toys Inc	$139.50

Figure 5.6 An Inventory Summary report

To apply these formatting changes, use the following steps:

1. Open Crystal Reports, and open the DETAILEDSALESREPORT.RPT report file from the book download files.
2. Click the Design tab to switch to the Design view of your report.
3. Locate the Invoice Id field in the Details section of your report, right-click the field, and select Format Field.
4. Click the Customize button, deselect the option for Thousands Separator, and then click OK twice.
5. Next, right-click the invoice date field in the Details section, right-click, and select Format Field.
6. Select the option for 03/01/1999, and then click OK.
7. Next, locate the Print Date field in the Page Header section, right-click, and select Format Field.
8. Select the option for Monday, March 1 1999 from the list, and then click OK.
9. Next, click the Preview tab to preview your report, which should now look like the report shown in Figure 5.7.

Detailed Sales Report

Print Date: Tuesday, June 1, 2010

Invoice ID	InvoiceDate	Customer Name	Amount
3800	01/01/2009	Dancing Bear Toys, Ltd	$11,829.60
3801	01/01/2009	Puzzle Master	$2,870.35
3802	01/01/2009		$2,808.00
3803	01/01/2009	Legion Toys	$125.00
3804	01/01/2009		$1,800.00
3805	01/04/2009	Turtle's Nest	$1,800.00
3806	01/04/2009	Red Wagon Toys	$1,800.00
3807	01/04/2009	abraKIDabra Toys Inc	$139.50
3808	01/04/2009	Toys and Stuff	$145.00
3809	01/04/2009	Hobby Time	$125.00
3810	01/07/2009		$145.00
3811	01/07/2009		$2,088.00
3812	01/07/2009	Kidz Stuff Toyz	$900.00
3813	01/07/2009	Black Forest Books & Toys	$3,750.00
3814	01/07/2009		$5,000.00
3814	01/07/2009		$198.00

Figure 5.7 The finished report with the formatting applied

WORKING WITH SORT CONTROLS

Sort controls were introduced in Crystal Reports 2008 and provide an easy way for end-users to sort the data in their reports. Using sort controls, you can create a report that can be manipulated by the end-user whenever the report is viewed. Sort controls are usually placed at the top of the page in the report or page header, as shown below in Figure 5.8.

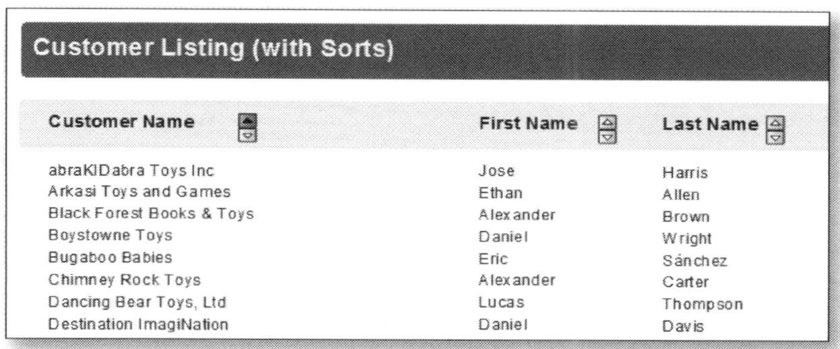

Figure 5.8 Sort controls in action

Creating a sort control is a two step process — first, you will need to have a group sort or record sort added to your report. From there, you can bind your control to one of the sorts you have created.

For example, if you wanted to create a sort control based on a "Product Name" field, you would first create a record-level sort on this field, then insert a sort control and bind it to the sort. To create a sort control like this, follow these steps:

1. From the book download files, open the CUSTOMERLISTING.RPT report
2. Within the report designer, select Report > Record Sort to open the Record Sort Expert, as shown on the next page in Figure 5.9.

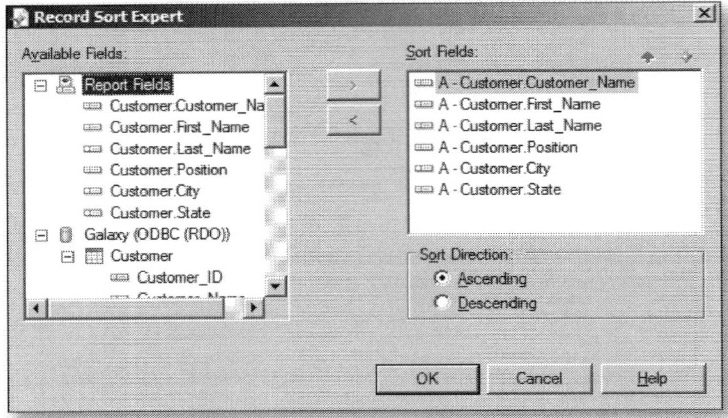

Figure 5.9 Record Sort Expert

3. Click on the Customer Name field and use the right arrow to move it to the list of selected fields, then click OK.
4. From the design or preview view of your report, select Insert > Sort Control to open the dialog shown below in Figure 5.10.

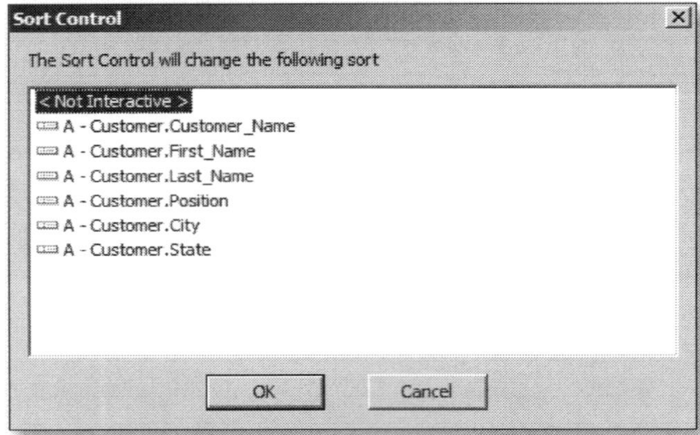

Figure 5.10 Sort control dialog

5. Select the Customer sort to bind your control to it and then click OK.
6. You can now draw your sort control on your page header. When you have finished drawing your sort control, you will be able to edit the label that appears alongside the control.

7. If you want to label your control, enter "Sort by Customer Name" and then click anywhere outside of the text object to get out of edit mode. Otherwise you can leave the field heading as is.

When you preview your report, you will now be able to sort your report by clicking on the sort control, as shown in Figure 5.11.

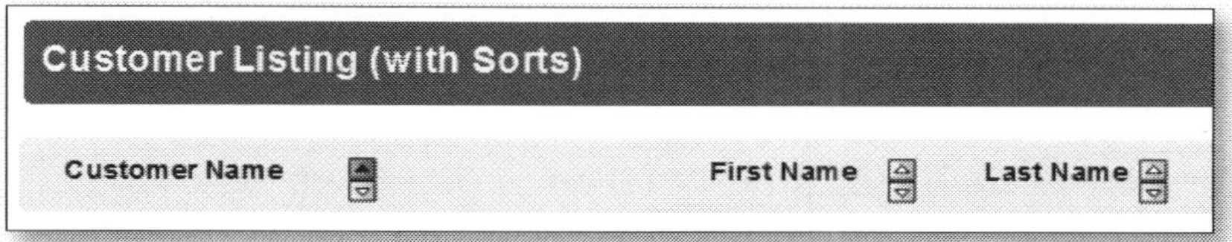

Figure 5.11 Finished report with sort controls

This technique can be repeated for as many group and record level sorts you have inserted into your report. Another quick way to add sort controls is to click on your existing column headings (they are text objects) and from the right-click menu, select "Bind Sort Control" as shown below in Figure 5.12.

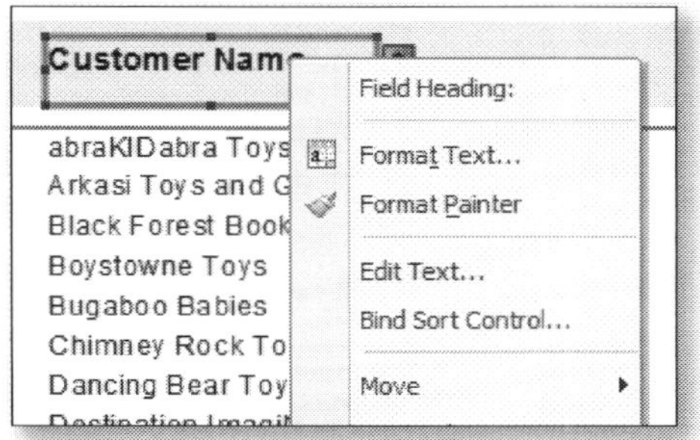

Figure 5.12 Converting text object to a sort control

This will turn any text field into a sort control and display the up and down arrow beside the field. This technique is a quick way to add more interactivity to your reports, as well as give it some of the features that end-users would normally export to another format to get.

FORMATTING FIELDS

There are a number of formatting techniques you can apply to fields in your reports — in the following sections we will outline some of these techniques you can use in your own reports.

RESIZING FIELDS

All of the fields you can insert onto your report can be resized. If you click a field object, you'll notice that four handles (or little blue boxes) appear on each side of the object. By moving these handles, you can resize the object.

Resizing objects in Crystal Reports works like resizing objects in Microsoft Office applications, such as Word and PowerPoint®.

As a time-saving feature, you can select multiple objects (even different types of objects), and when you resize one, all are resized in proportion. To resize multiple objects at the same time, use the following steps:

1. Locate the objects that you want to resize in your report, and multiple-select them. You perform a multiple-select operation by drawing a marquee box (sometimes called a stretch box) around the objects, or by clicking each while pressing SHIFT or CTRL.

You can also select all of the objects in a particular section by switching to the Design tab and right-clicking the section in the gray area on the left side of the screen. From the shortcut menu, select the Select all objects in section option.

2. Choose one of the objects, and resize its frame using the handles (or boxes) that appear on each side. Each object is resized proportionate to the object that you selected.
3. Click anywhere outside the selected fields to finish the operation.
4. If you need to resize a field with some precision, you can also specify the exact size and position of an object. Locate the object that you want to resize or position, and right-click it. From the shortcut menu that appears, select Size and Position. Using the dialog box shown in Figure 5.13, select the X and Y positions of your object and the object's height and width.

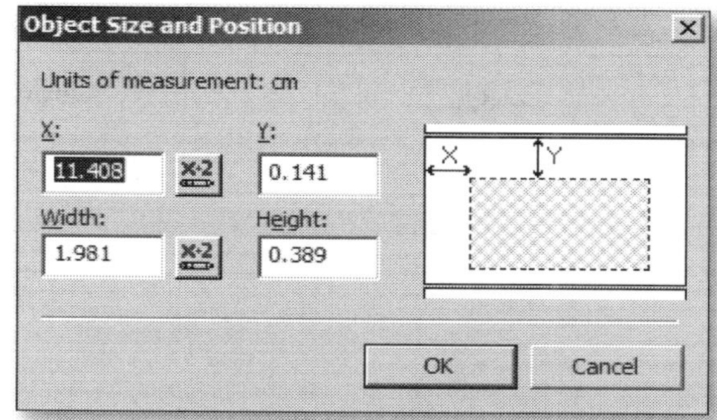

Figure 5.13 You can specify the exact size and location of objects in your report

---

The height and width settings use the measurement unit defined in your Windows setup.

MOVING FIELD OBJECTS

When moving field objects, you have a couple of choices. The first, dragging and dropping, is more of a Windows skill than a Crystal Reports technique. By clicking an object, holding the mouse button, and moving the mouse, you can drag field objects around and drop them where you like by releasing the mouse button. If you have used other Windows applications, chances are you have used this technique many times. Though it is the easiest way, it can also be the most time consuming, because you have to move each individual field, field heading, and so on. A much easier method is to use guidelines.

Another way to move field objects in the Design view of your report is to click the object to select it and then nudge the object around using the arrow keys.

Guidelines

Guidelines, shown in Figure 5.14, are invisible objects that can be used to align and move fields. The easiest way to think of a guideline is as a piece of string; you attach objects to that string, and when the string moves, everything attached to it moves as well.

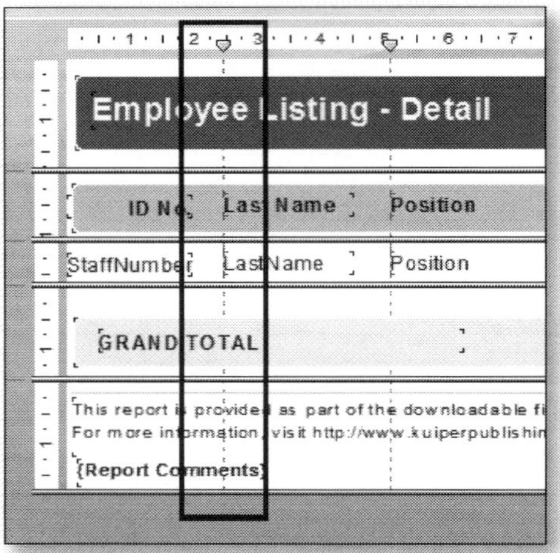

Figure 5.14 Guidelines can appear on both the Design and Preview tabs

Guidelines are invisible when you print your report. When working with the report design or preview, guidelines can appear as a dashed line in your report.

Guidelines can be added to your report by clicking anywhere in the ruler. A small icon (sometimes called a caret), shown in Figure 5.15 on the next page, will appear, indicating that you have created a guideline.

Guidelines are also created with each new field that you add to the Details section of your report.

Once you have created a guideline, you can then snap objects to it by moving them close to the guideline; you should see them jump a bit as you get closer. This appearance of jumping can be likened to the effect of a magnet; the object seems to want to stick to the guideline as you move the object closer.

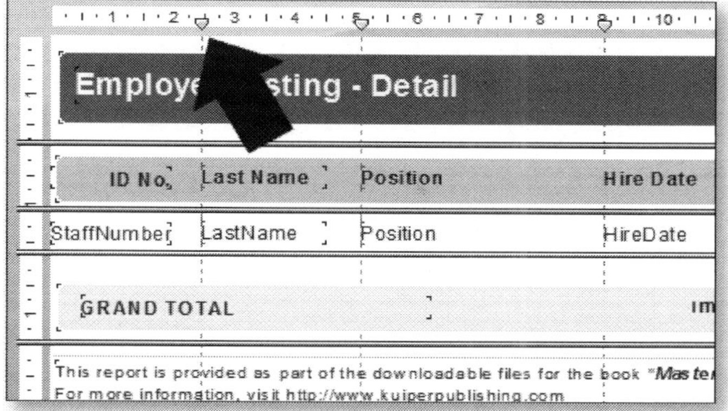

Figure 5.15 Guidelines are marked by a caret

When you move the guideline caret, the guideline and all of the objects attached to it move. To remove a guideline from your report, use the following steps:

1. Locate the guideline you want to remove, and drag the caret off the ruler.
2. To remove all guidelines, right-click the ruler in the toolbar, and select one of the Remove All options from the shortcut menu.

Moving or Aligning Multiple Objects

An alternative to using guidelines is to move or align multiple objects at the same time. To use this technique, use the following steps:

1. First, locate the objects that you want to align in your report, and then multiple-select them. You perform a multiple-select operation by drawing a marquee box around the objects or by clicking each while pressing SHIFT or CTRL.

A standard Windows shortcut is to use the SHIFT key to select contiguous items in a list and the CTRL key to select multiple items.

2. After you have selected all of the objects that you want to align, right-click one of the objects, and from the shortcut menu that appears, select Align, as shown in Figure 5.16 on the following page.
3. Select one of the options to align your fields, and then click anywhere outside the selected fields to finish.

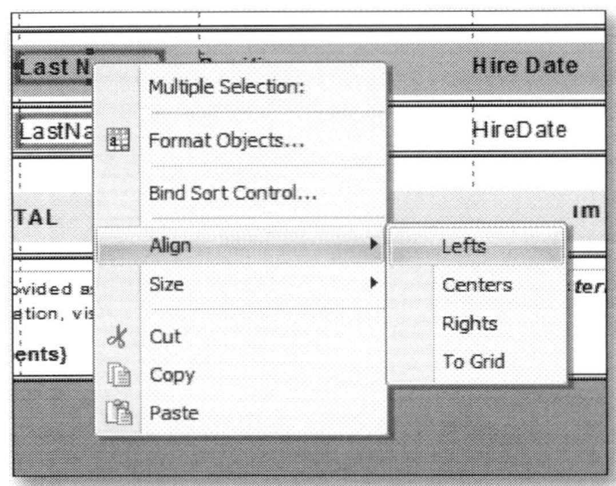

Figure 5.16 Alignment options can be found on the shortcut menu

Rulers, Guidelines, and Grids

Most of the user interface options for controlling the layout of your report are available by right-clicking a blank space in your report. A shortcut menu will open, allowing you to control the appearance of rulers, guidelines, and grids that may appear in the Design or Preview tabs.

Rulers are designed to show you the precise measurement of your report; guidelines and grids are used to control the position of elements on your report page.

If you are creating a report that mimics an existing report design or prototype, rules are the easiest way to ensure correct, precise alignment. You can measure the existing report or prototype to find the exact measurements and placement of fields.

Once you have established where all of the elements should be placed on the report, place guidelines using these measurements, so all of the fields on your report can be aligned.

Finally, if you have established a set gap between fields (that is, each field should be .05 inches apart), you can also change the underlying grid settings by clicking File > Options. These little tricks may not seem like much, but they can mean the difference between a few minutes and a few hours spent formatting your report.

Controlling Object Layering

Crystal Reports uses transparent object layering; that is, objects can be placed directly on top of one another. To control where objects sit within a layer, use the following steps:

1. Locate the object on your report that you want to use, and right-click it.
2. From the shortcut menu that appears (shown in Figure 5.17), select the layering option to control where the object is positioned.

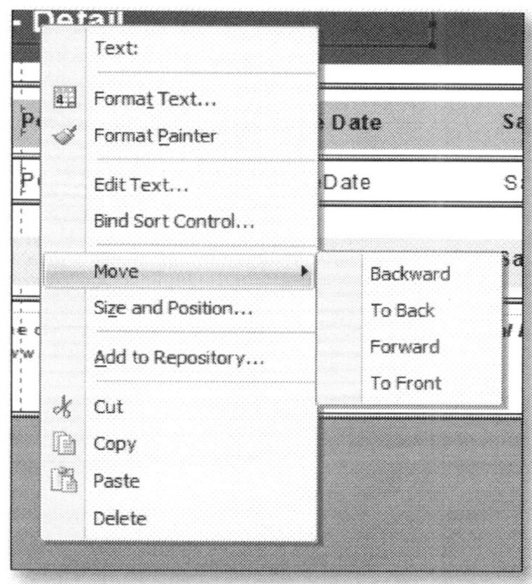

Figure 5.17 You can control object layering using this menu

TEXT OBJECTS

Text objects enable you to type text directly into your report. Text objects, simply labeled Text on the status bar, can be combined with database fields or formatted as paragraph text. Preformatted text, such as rich text format (RTF) and hypertext markup language (HTML), can be inserted directly into text objects. Text objects are used for report titles, field headings, and any text that needs to be inserted into your report.

Working with Text Objects

Text objects can be inserted anywhere in your report through the Insert menu. When you click Insert > Text Object, a crosshair icon will appear; use it to draw a text box in your report. Crystal Reports will immediately place the text object in Edit mode and place the tip of your mouse pointer inside the text object so you can start typing text. When you have finished editing the text, click anywhere outside the text object to leave Edit mode.

Once inserted on your report, text objects behave just like the field objects we saw earlier; they can be moved, resized, formatted, and so on. If you need to change the text you have entered in a text object, you can double-click the object to put it back into Edit mode, or you can right-click the object and select Edit Text from the shortcut menu.

For more control over a text object, you can also set paragraph formatting, including indentation, line spacing, character spacing, and tab stops.

Although text objects are most often used for report titles, column headings, and the like, they can also be combined with database and other fields where these formatting features come in handy.

Combining Text Objects and Other Fields

When working with Crystal Reports, you will reach the point where you are moving a field around in your report, and suddenly (and unexplainably) the field merges with a text object. Believe it or not, this is a feature. You can combine text objects with other fields to create form letters, statements, and so on. Imagine that you are writing a letter; you could merge the text Dear: with the database field containing the first name of your customer to create a personalized letter generated by Crystal Reports.

The mechanics of combining a text object with another field are simple — as mentioned earlier, you may have already done it by accident. In the walk-through that follows, we are going to use text objects to create a form letter like the one shown in Figure 5.18, integrating fields from a database with text you can enter or import.

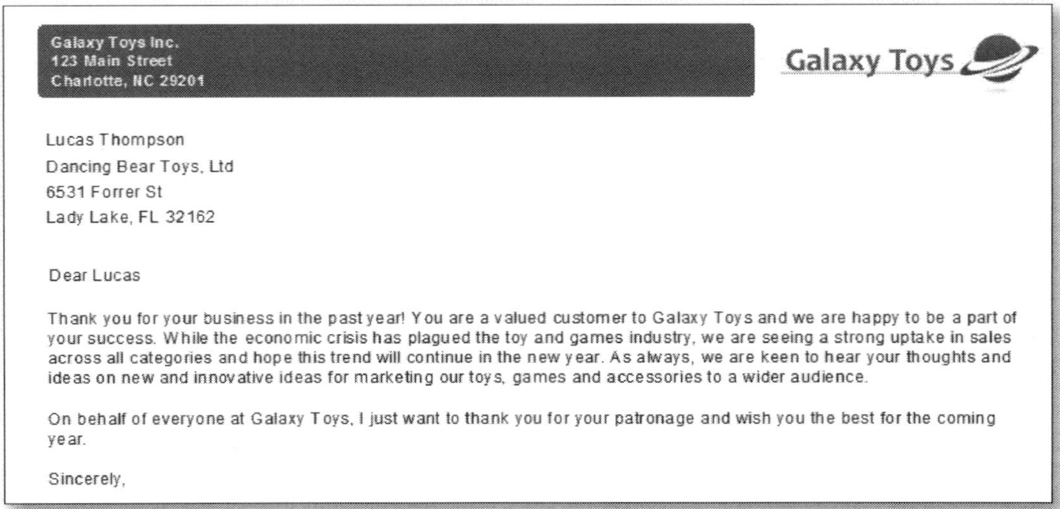

Galaxy Toys Inc.
123 Main Street
Charlotte, NC 29201

Galaxy Toys

Lucas Thompson
Dancing Bear Toys, Ltd
6531 Forrer St
Lady Lake, FL 32162

Dear Lucas

Thank you for your business in the past year! You are a valued customer to Galaxy Toys and we are happy to be a part of your success. While the economic crisis has plagued the toy and games industry, we are seeing a strong uptake in sales across all categories and hope this trend will continue in the new year. As always, we are keen to hear your thoughts and ideas on new and innovative ideas for marketing our toys, games and accessories to a wider audience.

On behalf of everyone at Galaxy Toys, I just want to thank you for your patronage and wish you the best for the coming year.

Sincerely,

Figure 5.18 A typical form letter created with Crystal Reports

To get started using this technique to create your own form letter, follow these steps:

1. Open Crystal Reports, and open the BLANKLETTER.RPT report file from the book download files.
2. Click the Design tab to switch to the Design view of your report.
3. To insert the text object for our salutation (Dear XXX:), click Insert > Text Object.
4. Use your mouse to draw a text object at the top of the Details section, and then enter the text Dear.
5. Next, click View > Field Explorer to open a list of available database fields.
6. Expand the Database Fields folder and the Customer table.
7. Drag the First_Name field from the Field Explorer to the text object you created. Your cursor will show your insertion point. Drop the field in the text object after the text.
8. Click the Preview tab to preview your report. As you go through the pages of the report, you will see that each page has its own salutation, as shown in Figure 5.19.

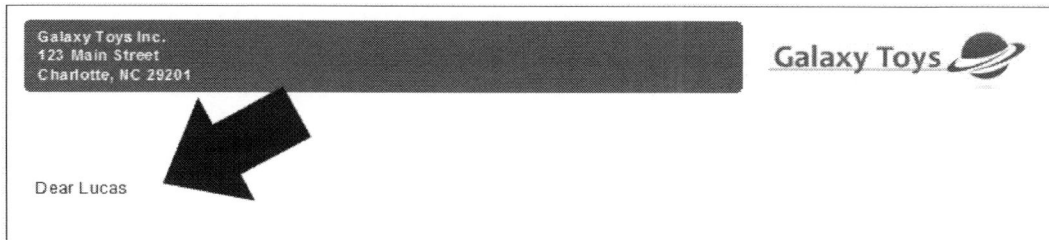

Figure 5.19 Text objects can be used to combine text you enter with database fields.

It may take some time to get the hang of the technique, but once you do, it is a handy trick to have up your sleeve. You can use this same technique to add an address block to the letter as well.

Inserting Preformatted Text

When working with text objects and form letters or statements, you probably don't want to have to enter all of the text directly into Crystal Reports. Crystal Reports does not have a spell check or grammar facility, and it is difficult to type and format large amounts of text directly into a text object. To help you, Crystal Reports allows you to use a word processing application to create form letters and text and then bring that preformatted text directly into Crystal Reports.

To add the body text for the form letter we have been creating, use the following steps:

1. Open Crystal Reports and the report we have been working on in this section.
2. Click the Design tab to switch to the Design view of your report.
3. To insert the text object for the body of our letter, click Insert > Text Object.
4. Use your mouse to draw a text object at the top of the details section below your salutation.

Right-click in the text object you just created, and select Insert from File (shown in Figure 5.20), which allows you to insert a text, RTF, or HTML file into your text object. The book download files contain an RTF file named LETTERTEXT.RTF — browse and insert this file into your text object.

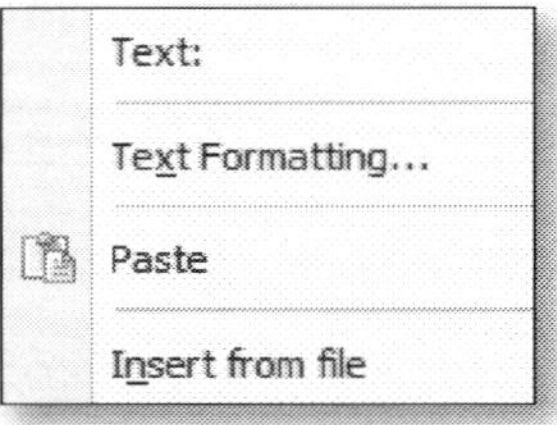

Figure 5.20 You can insert preformatted text into your text object

If you do need more room on your form letter report or if you want to change the page size, you can click File > Page Setup to access the settings for page size, orientation, margins, and other printer options.

WORKING WITH LINES AND BOXES

Another formatting technique is to use lines and boxes to define areas of your report, as well as highlight subtotals, group and page headers/footers and more. In the Formatting toolbar, there is a line drawing icon that will allow you to draw either horizontal or vertical lines on your report canvas. (You can also select Insert > Line)

You can then right-click on these lines to format the line style, width, color, etc. using the dialog shown in Figure 5.21.

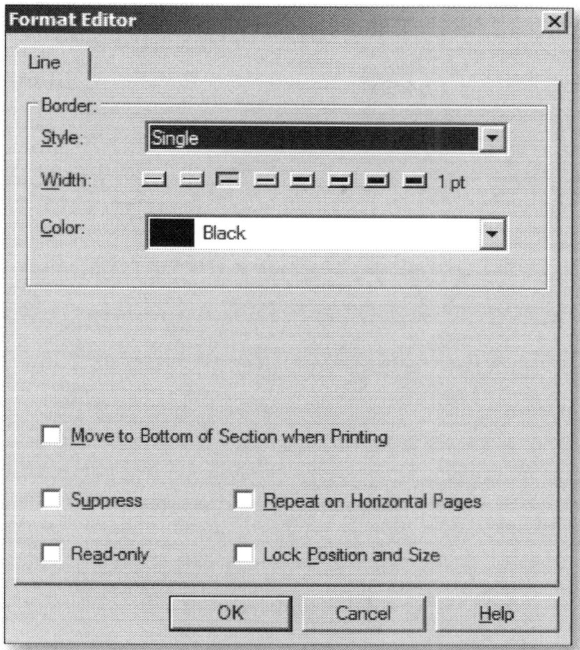

Figure 5.21 Line formatting options

In addition to lines, you can also add boxes to your report — this enables you, as the report developer, to create complex formats and separate columns, sections, etc. as required.

If you have trouble right-clicking or selecting your line, remember that you can use the Zoom control to make your report canvas larger so you can see the line more clearly. This technique also works well when you want to make sure your lines are placed precisely on the report canvas.

To draw a box, click on the Box icon and draw your box on the report canvas. If you right-click on the box, you will be able to format the box properties, as shown over the page in Figure 5.22.

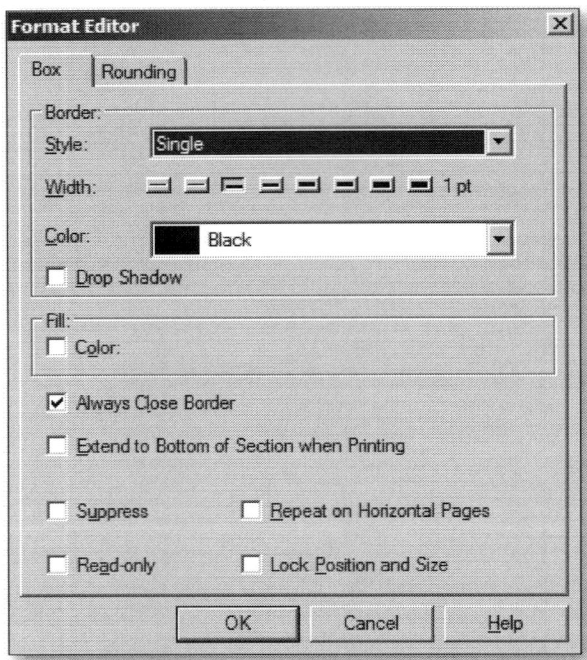

Figure 5.22 Box properties

You can control the line style, width, color and fill color of the box, as well as the rounding of the corners of the box. This technique has been used in the sample reports that are included with this book for most of the page and report headings. To change the rounding of the corners, click on the Rounding tab then use the slider shown in Figure 5.23 on the next page to change the corner rounding percentage.

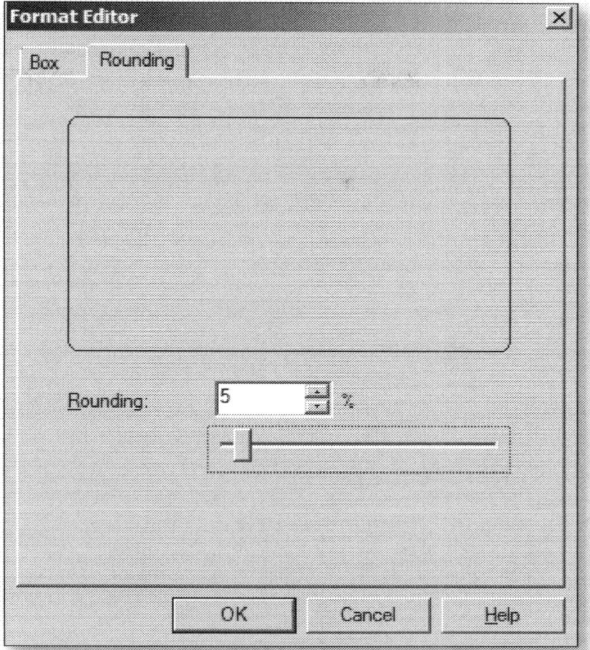

Figure 5.23 Rounding options

USING IMAGES

You can insert images into your reports to add corporate logos, watermarks, footers, etc. as required. Crystal Reports supports a wide range of image formats, including:

- Windows Metafile (.WMF)
- Windows Bitmaps (.BMP)
- Tagged Image File Format (.TIFF)
- JPEG files (.JPG)
- Portable Network Graphics (.PNG)

To insert an image, select Insert > Picture and then use the browser dialog to select an image file from your computer. This file will then be attached to your mouse pointer and you can click to insert the image on your report canvas.

Any images that are inserted onto your report can be moved, resized and stretched as required, although if you attempt to resize the image in Crystal Reports, you may end up skewing or stretching it, so just be careful when you are resizing.

In addition, large images can cause the report file itself to be extremely large, so it is best to make sure your image is resized prior to inserting into your report. Also, you should consider how your report is going to be used and then select the right image format for it. If your report is only ever going to be printed in black/white, then it may be worth inserting a black/white or grayscale version of your logo or image, which is usually a smaller file size.

CREATING AN IMAGE WATERMARK

Another technique used in report creation is the use of watermarks behind your report — a watermark could be anything from a company logo to a graphic with the word "unaudited" or for invoices, even a "past due" image, as shown below in Figure 5.24.

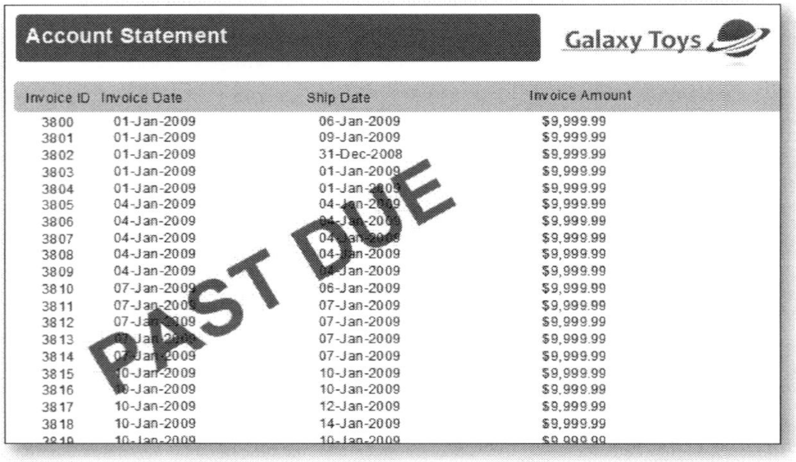
Account Statement

Galaxy Toys

Invoice ID	Invoice Date	Ship Date	Invoice Amount
3800	01-Jan-2009	06-Jan-2009	$9,999.99
3801	01-Jan-2009	09-Jan-2009	$9,999.99
3802	01-Jan-2009	31-Dec-2008	$9,999.99
3803	01-Jan-2009	01-Jan-2009	$9,999.99
3804	01-Jan-2009	01-Jan-2009	$9,999.99
3805	04-Jan-2009	04-Jan-2009	$9,999.99
3806	04-Jan-2009	04-Jan-2009	$9,999.99
3807	04-Jan-2009	04-Jan-2009	$9,999.99
3808	04-Jan-2009	04-Jan-2009	$9,999.99
3809	04-Jan-2009	04-Jan-2009	$9,999.99
3810	07-Jan-2009	06-Jan-2009	$9,999.99
3811	07-Jan-2009	07-Jan-2009	$9,999.99
3812	07-Jan-2009	07-Jan-2009	$9,999.99
3813	07-Jan-2009	07-Jan-2009	$9,999.99
3814	07-Jan-2009	07-Jan-2009	$9,999.99
3815	10-Jan-2009	10-Jan-2009	$9,999.99
3816	10-Jan-2009	10-Jan-2009	$9,999.99
3817	10-Jan-2009	12-Jan-2009	$9,999.99
3818	10-Jan-2009	14-Jan-2009	$9,999.99
3819	10-Jan-2009	10-Jan-2009	$9,999.99

Figure 5.24 An example of a watermark image

In the following example, we are going to look at how to create this report with the watermark image. To see how this technique works, follow these steps:

1. From the book download files, open the IMAGEWATERMARK.rpt report.
2. Switch to the Design tab of your report, just to make inserting the image into the Page Header a bit easier.
3. From the Insert menu, select Picture and browse to the book files and select the PASTDUE.JPG image from the Chapter 5 folder.

4. Click to insert this image into the "Page Header A" section of your report.
5. Next, right click on the Page Header section on the left-hand side of the design tab and select Section Expert.
6. From the section expert, select the tick box for "Underlay Following Sections" then click OK.
7. Switch to the Preview tab of your report and you should now see the image underlaid beneath the report text.

Using this technique, you may need to switch back and forth between the design and preview views to change the size and position of the image to get the desired results. But this technique can be used not only for images, but also for putting subreports and other report content side-by-side, which is a technique we'll look at a little later in Chapter 12.

WORKING WITH TOOLTIPS

Another formatting technique that can add value to your report is the use of Tooltips. A tooltip is a small text label that appears when the user mouses over a field in the report, as shown in Figure 5.25 below.

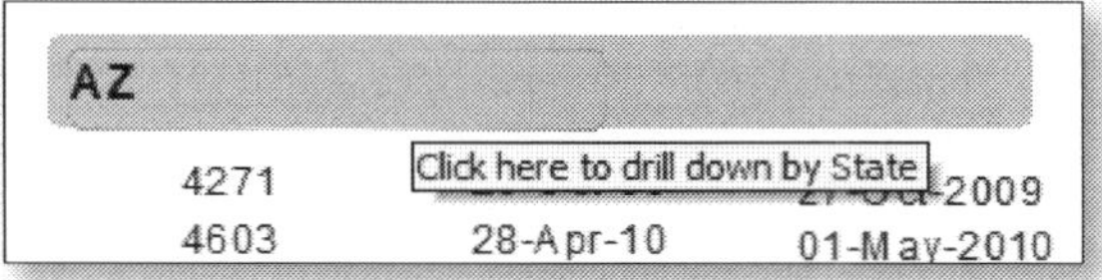

Figure 5.25 An example of a tooltip

Tooltips are a handy way to highlight information that may not otherwise appear on the report or even give instructions to users about report functionality. For example, you might have a tooltip on a Group Heading field of "Click here to drill down by Country".

If there is no tooltip created for a particular field, the tooltip will show the field name from the database, as well as the type of field (String, Number, etc).

To add a tooltip to a field in your report, right-click on the field and select Format Text from the right-click menu. This will open the dialog shown below in Figure 5.26, which has a field for your tooltip text.

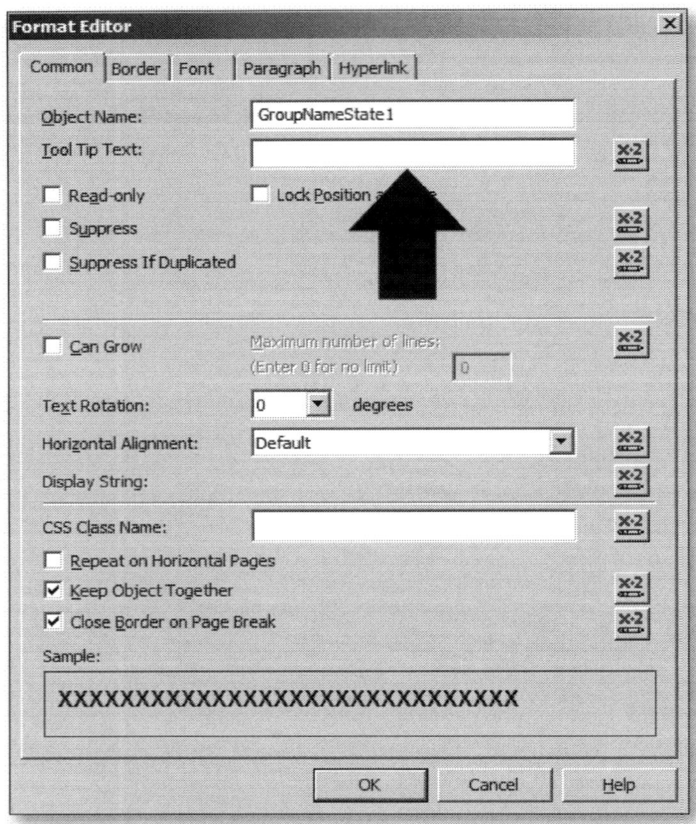

Figure 5.26 Tooltip field in the format dialog

You can enter your tooltip text or alternately use the "X+2" to create a formula to generate the tooltip text for you. This technique can be used to add logic to your tooltip text and the formula itself must return a value.

For example, if you wanted a tooltip to be based on a parameter field called "Show Additional Info", which would determine whether or not to show the tooltip, your formula would look something like this:

```
If {?Show Additional Info} = True then "This field is
derived from the INVOICE.AMOUNT field in the database
and is the total invoice amount" else ""
```

USING THE FORMAT PAINTER

Some report developers spend the majority of their time in the formatting of the report. Sometimes it is nice to have a few shortcuts to rely on to speed this process up, and one of these shortcuts is the Format Painter. The concept behind this functionality is that you can apply the formatting from one field to another by "painting" it on to the second field.

To use this technique, locate the field to serve as the source of your formatting options, then click the Format Painter icon from the Standard toolbar, as shown below in Figure 5.27.

Figure 5.27 Format Painter

Once you have clicked on the Format Painter icon, you can then click on your target field and the formatting properties will be copied from the Source to the Target field. You can repeat this process by clicking the Format Painter icon each time to copy the formatting attributes to each new target field.

CREATING BARCODES

One of the new features in Crystal Reports 2008 is the ability to quickly add barcodes to your reports through a right-click menu. To use this functionality, all you need to do is locate a field you wish to convert, then right-click the field and select "Change to Barcode" from the right-click menu, as shown on the following page in Figure 5.28.

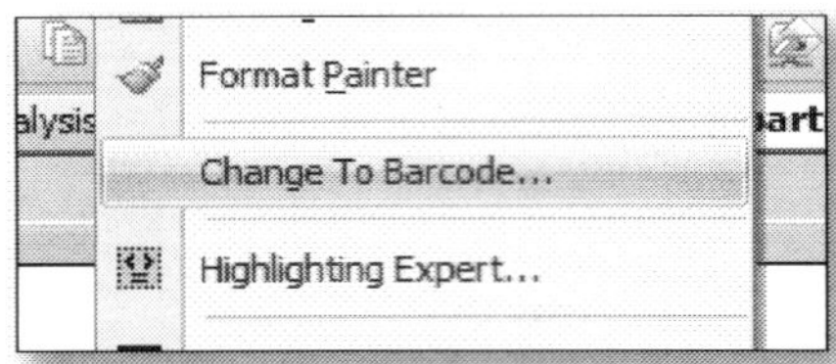

Figure 5.28 Barcode right-click menu

This right-click menu will open a dialog where you can select the barcode options for your report. To get some practice with this technique and add a barcode to your own reports, follow these steps:

1. From the book download fields, open the EMPLOYEESIGNIN.rpt report from the Chapter 5 files.
2. On the report canvas, locate the StaffNumber field in the details section and right-click on the field and select "Change to Barcode". This will open the dialog shown below in Figure 5.29.

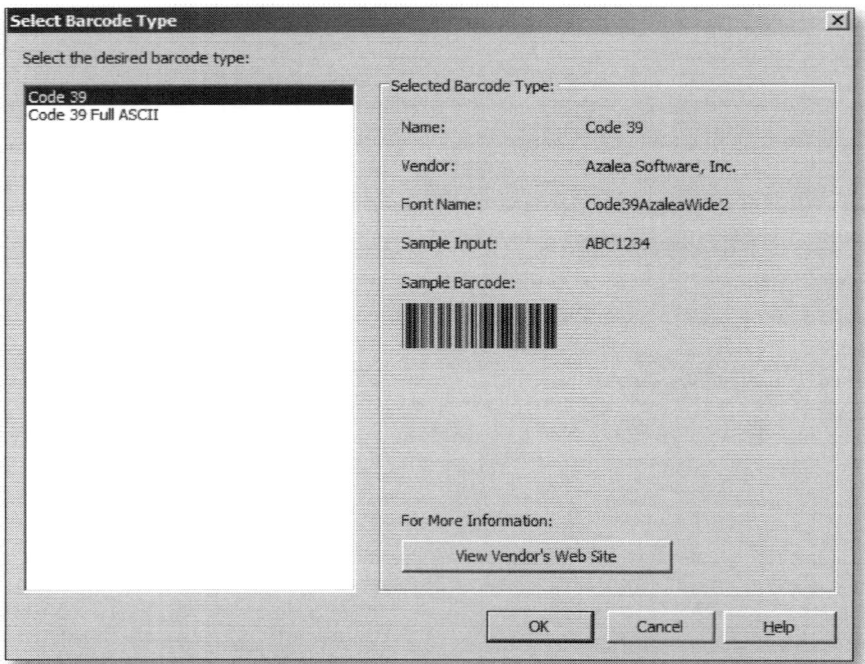

Figure 5.29 Barcode options

3. Select the Code 39 option for your barcode and then click “OK” to return to the report to see the field converted into a barcode.

You may notice in this dialog that there are only two barcode formats available — Code 39 and Code 39 Full Ascii. The fonts for these barcode formats were licensed from Azalea Software, which has long provided barcode fonts and software for use with Crystal Reports and other applications.

Code 39 barcodes can by created from any alphanumeric string and are an industry standard. You can use Code 39 barcodes with most barcode readers and they will interpret the text as if you typed it in. When creating a Code 39 barcode from Crystal Reports, you can have up to 254 characters using the numbers 0–9, and letters from A–Z in both uppercase and lowercase.

If you would like to print additional types of barcodes (UPC, Code93, etc.) you can purchase the appropriate barcode fonts directly from Azalea Software (http://www.azaleasoftware.com). In addition, they have a number of User Function Libraries (UFL) and formula examples on their web site of how to format the data appropriately for different barcode formats.

WORKING WITH REPORT TEMPLATES

An easy way to quickly apply formatting to your reports is through the use of report templates. You may remember that one of the last steps in the Standard Report Wizard gives you the ability to apply a template to your report. But you don't need to use the report wizard to apply a template — if you select Report > Template Expert, you can use the dialog shown in Figure 5.30.

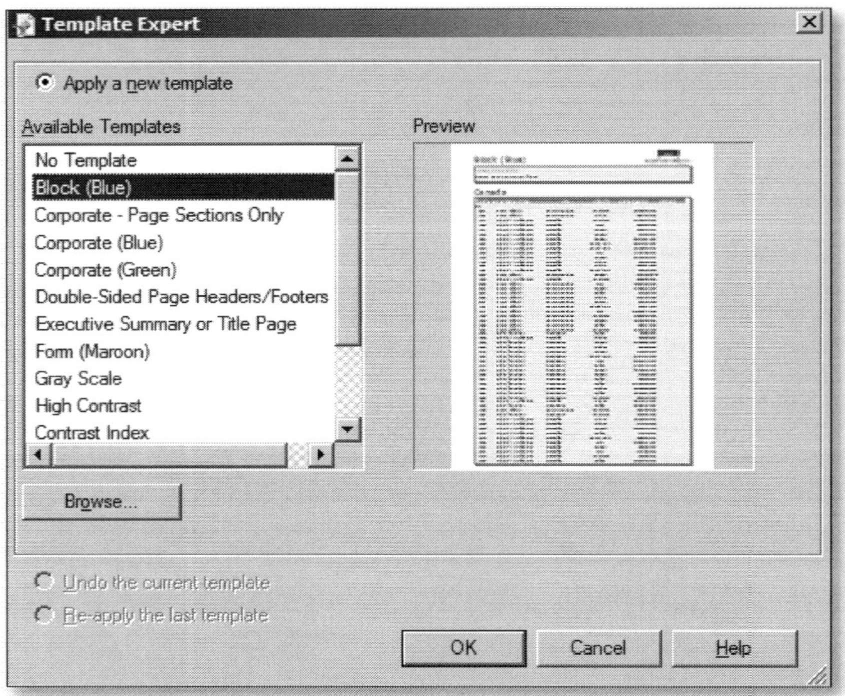

Figure 5.30 Template Expert

There are 12 different standard templates that can be applied to your report, with corporate color schemes, gray scale, etc. While these templates provide a good starting point, chances are you will want to create your own, using your own corporate color scheme, company logo and more.

In the following section, we are going to look at how to create report templates and how to apply them to your own reports.

CREATING REPORT TEMPLATES

The easiest way to think of a report template is as a Crystal Report that has no data, but has all of the report formatting attributes you would like to see in the final report. To apply these formatting attributes, we will add "Template Fields" to your blank report and then format these template fields. When the template is applied to a report, these fields will be used to represent the data in your report.

To create a basic report template, follow these steps:

1. From within Crystal Reports, select File > New > Blank Report. When the Database Expert appears, click Cancel, as you don't actually want to connect your report template to a specific data source.
2. To create your template, change the page orientation, setup, etc. using File > Page Setup.
3. Once you have the page setup complete, select Insert > Template Field Object to insert fields on to your report canvas. You can then format these fields as you normally would, changing the font, background color, etc.
4. You can also add Special Fields to your report, including the Report Title, Comments, Print Date and Time, etc.
5. Before you save your report template, select File > Summary Info and enter a Title for your report template. This will appear when you go to select the template later.
6. Save your template as SIMPLETEMPLATE.RPT
7. Open one of your existing reports, or one of the sample reports from the book download files and then select Report > Template Expert.
8. Use the Browse button to browse to the location where you have saved your report template and click OK to select the file, then OK to apply the template to your report.

When you create your first report template, you will spend some time going back and forth to get the formatting exactly as you want. You may notice that when you create a report template, Crystal Reports creates a formula field for each of the Template Fields that you add to your report, as shown in Figure 5.31 on the following page.

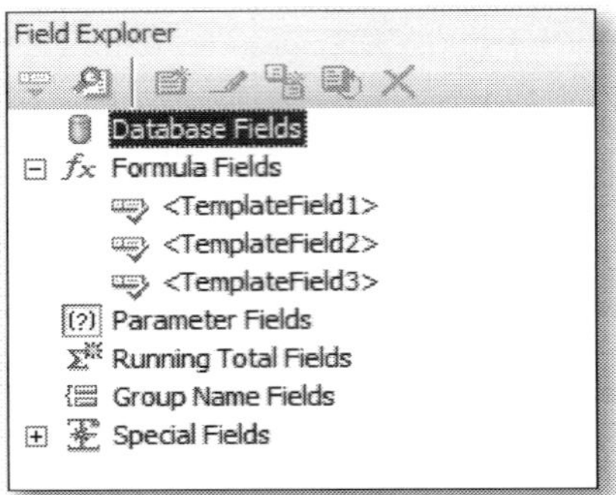

Figure 5.31 Template formula fields

If you want to actually add some sample data to your template, to give you a better idea of what the report will look like, you can edit this formula, which looks like this:

```
WhilePrintingRecords;
Space(10);
```

You can replace the text Space(10) with some sample data, shown in quotes for strings or alternate for numeric data, just the numbers, etc. This will help give you some idea of what the report will look like when formatted.

When working with a template, the following elements can be added to the template and will be applied to your target report:

- Bitmaps
- Fields
- Group Charts
- Groups
- Hyperlinks
- Lines, Boxes, Borders
- Static OLE Objects
- Summary Fields

It is important to note that when you apply a template, you may be overwriting your existing report formatting, so make sure you save a copy of the report before applying any changes (just in case).

In the Template Expert dialog, you also have the ability to undo the current template, or re-apply the last template, as shown below in Figure 5.32.

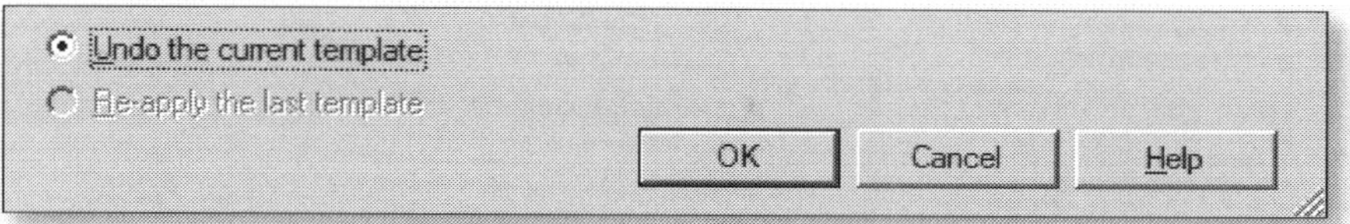

Figure 5.32 Template application options

If you would like to add your own custom report templates to the list, first make sure that they have a report title in the Summary Info of the report, then save the template .RPT file to the following location:

C:\Program Files\BusinessObjects\Crystal Reports 12.0\Templates\en

where C:\Program Files is where you have installed Crystal Reports. If you selected a custom directory, this path may be different but will follow the same structure.

In terms of some general recommendations for report templates, you may want to create both a "portrait" and "landscape" version of the template to suit both types of reports, as well as a few templates that utilise different levels of grouping.

Remember not to get frustrated with the template process — you may create a template that has 10 fields for example, but your report actually has 12 fields appearing before you apply the template. The template expert does the best job it can in applying a template, but you may still have some manual formatting to do.

WORKING WITH REPORT ELEMENTS

There are several "tricks of the trade" when formatting your reports and in the following sections we are going to look at a few of these. You may use some of these techniques in your own reports, or develop your own formatting shortcuts, but either way it is up to you.

"BUMPING" OBJECTS

Whenever you are in the Design view of your report, you can click on a report object and "bump" it into position using the arrow keys. This technique works especially well when you are trying to just nudge an object over a little bit at a time. You can also hold down the arrow key and the object will move in the direction you have selected until you release it.

SIZING OBJECTS TOGETHER

To get a number of fields (including fields and field headings) the same size, use CTRL-CLICK to select both objects and then select Format > Make Same Size and choose either Width, Height or Both.

ALIGNING OBJECTS

Using the menu available under Format > Align as shown below in Figure 5.33, you can align objects easily without having to move each object individually.

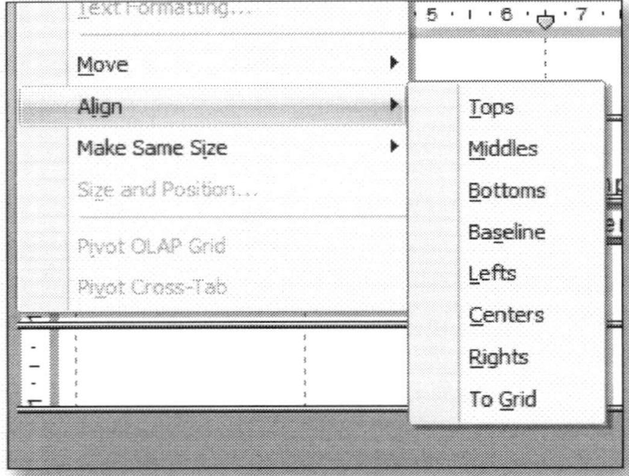

Figure 5.33 Alignment options

LOCKING OBJECTS IN PLACE

One of the problems with a crowded report canvas is that you may end up moving an object you didn't mean to. To alleviate this problem, you can tick on the option for "Lock Size and Position" from the Format Editor.

To turn this option on, click on a field and from the right-click menu select Format Field and then tick on the box shown on the following page in Figure 5.34.

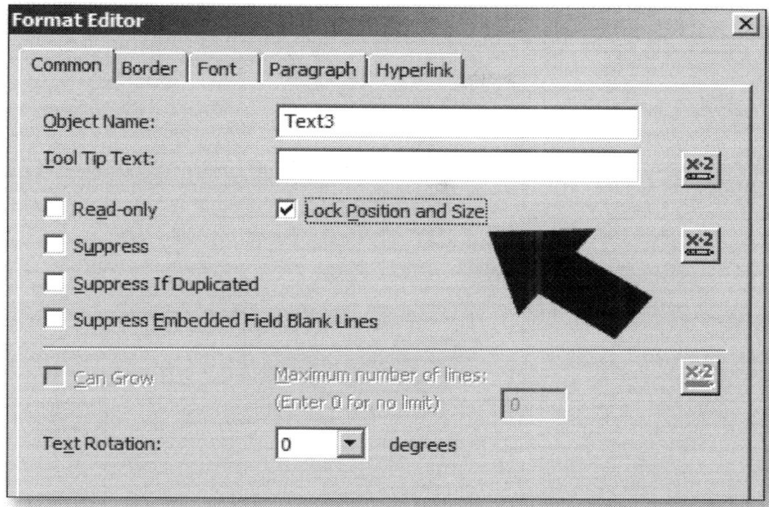

Figure 5.34 You can lock the size and position of an object in your report

SUMMARY

So now you are familiar with how to do some basic formatting, we are going to take it up a notch in the next chapter, when we look at how to format the different sections of your report. Sections form the basic building blocks of any report and you'll learn how to insert and delete sections, as well as apply section formatting and advanced techniques to create multi-column reports and more.

Chapter 6
Working with Sections

In This Chapter

- Working with Sections
- Formatting Sections
- Working with Drill Down Reports
- Advanced Section Formatting
- Creating Multi-Column Reports

INTRODUCTION

Behind the scenes, your Crystal Reports are broken down into sections — we looked at sections a bit earlier when you were creating your first report. But now, it's time to drill-down a bit deeper into sections and some of the advanced formatting you can use to create complex report layouts.

Each Crystal Report you create will have a Report Header and Footer, a Page Header and Footer, and a Details section, as shown in Figure 6.1 on the following page. You may also have a Group Header and Footer if you have any groups inserted into your report.

For basic reporting, chances are you won't need more than one occurrence of the sections listed here, but Crystal Reports allows you to create multiple sections and set a number of section-specific properties to assist with tricky formatting problems you may encounter with complex reports.

An example of where the "multiple section" concept could come into play is if you were creating form letters for your company and wanted to show two different return addresses on the letter (one for your head office and one for a regional office, where appropriate). Using multiple sections, you could create two page headers and use conditional formatting to show the correct header for each page, based on the customer's address.

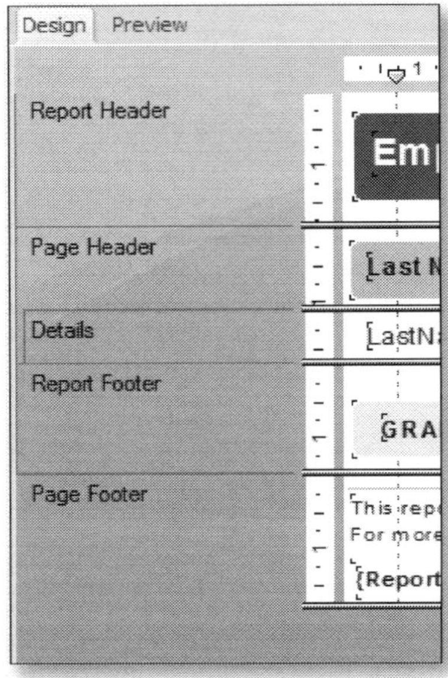

Figure 6.1 A look at report sections

ADDING SECTIONS TO YOUR REPORT

To accomplish this type of report, we need to insert two page headers (Page Header "A" and Page Header "B") and apply a little bit of conditional formatting to make this happen. Conditional formatting is something new, but if you have worked with Crystal Reports formulas before, this experience will come in handy. Conditional formatting allows you to create a formula, and if that formula is true, then something will happen.

If you have never worked with Crystal Reports formulas before, they are covered in-depth in Chapter 11, so you may want to flip ahead to get some background before working with conditional formatting.

Using the example of the two page headers, we could create two formulas and where the State was equal to "CA," we could show the header with the California return address; where the State was any other value, we could show the New York return address.

This sample Form Letter report is included in the download files for this chapter.

This is just one example of how multiple sections can be used to solve common formatting problems. Before we can get into the specifics of how sections can be used, we need to understand how to perform some basic operations, like inserting and removing sections and merging sections.

When working in the Design tab of Crystal Reports, each section has its own full-sized area on the left side of the design environment. If you were to right-click on this area to the left, a menu would appear, similar to the one shown in Figure 6.2.

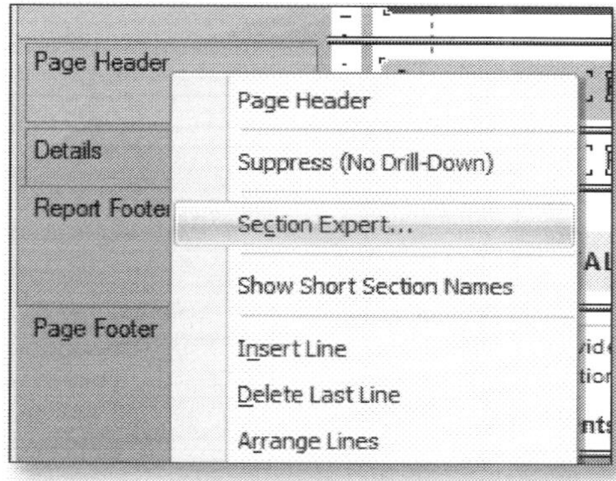

Figure 6.2 Each section can be edited through the Design tab

This menu provides a quick and easy way to work with the different sections in your report, and it is where you will find the basic section functions (insert, merge, and so on).

The Preview tab also shows the different section names, but by default these are shown in abbreviated form (PH, GH1, and so forth).

If you would like to follow along with the step-by-step instructions in this chapter, this book includes downloadable files that are available from www.kuiperpublishing.com. Also, check out "Setting up the Samples" section in the front of this book for instructions on how to configure the data sources, reports, etc.

INSERTING A NEW SECTION

In the earlier example, we looked at a situation where you might want different page headers for your report to display two different return addresses for a form letter report you create. To insert a new page header section into your report, use the following steps:

1. Open Crystal Reports and from the book download files, open the INTERNATIONALFORMLETTER.RPT report.
2. Switch to the Design tab of your report.
3. Locate the page header section, right-click it, and select Insert Section Below.

Crystal Reports inserts a section immediately below the page header, as shown in Figure 6.3.

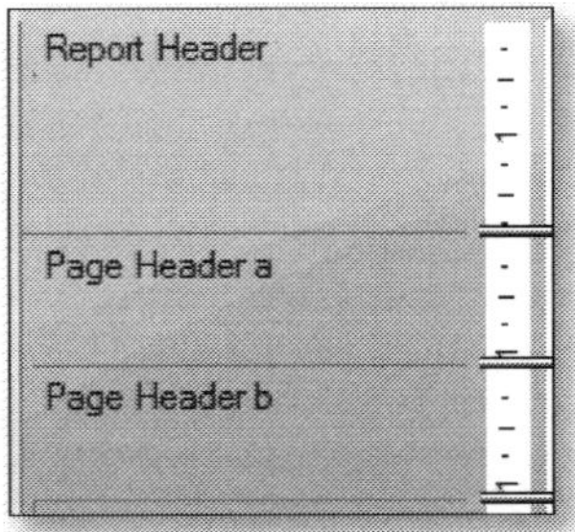

Figure 6.3 Sections you have inserted appear below the original section

You'll notice that Crystal Reports has named this section Page Header B and your original page header Page Header A. Crystal Reports follows this naming convention throughout the different sections, so if you were to create a section below Page Header B, it would be labeled Page Header C.

DELETING A SECTION

When working with sections, you can also delete any unused sections you may have inserted by right-clicking the section and selecting Delete Section from the shortcut menu. Any objects that you have placed in that section will also be deleted, so if you want to keep any of them, make sure you have moved them out of the section before you delete it.

You will be unable to remove the following sections from your report using this method: Report Header/Footer, Page Header/Footer, Group Header/Footer, and the Details section. You can delete a Group Header/ Footer by deleting the group to which they relate.

RESIZING A SECTION

Another handy trick when working with sections is resizing. Sections can be resized to accommodate whatever information you need to insert, but they may not be larger than the page itself. To resize a section, use the following steps:

1. In the Form Letter report, locate the Page Header B section in the Design view of your report.
2. At the bottom of each section is a divider line. Move your mouse pointer over the divider line until the pointer changes to a double-headed arrow.
3. Using your mouse, you can drag the bottom border of the section up or down to resize. It helps if you select an area that is free of objects; otherwise, you may end up moving the object instead of the section border.

Remember, you can always use Ctrl+Z to undo if you accidentally move an object instead of a border line.

SPLITTING A SECTION

Often, when resizing a section you will "split" that section by mistake. Splitting a section will just separate one section out into two (that is, the Details section becomes Details A and Details B). Using this method is very tricky and takes a little practice, but it comes in handy when you don't want to move a lot of objects around to get two separate sections.

To split a section of your report, use the following steps:

1. Using your own report, locate the section you wish to split, and move your mouse toward the left, along the bottom of the section, until you reach the intersection of the ruler line and the section's divider line.

2. Your mouse pointer should turn into the Split icon with a single line and one up-and-down arrow.
3. Use the Split icon at the intersection of the ruler line and bottom of the section, dragging your mouse down to split the section into two.

> **This technique takes a little practice. Remember, a double line with an up-and-down arrow indicates that you can resize the section.**

After spending a half-hour trying to get the technique down, a lot of people find it is just easier to insert another section and drag all of the objects down from the original section. Either way, it is up to you.

MERGING REPORT SECTIONS TOGETHER

When working with multiple sections, you may need to occasionally merge sections together to clean up or simplify the report's design. To merge two sections together, right-click the section above the one you wish to merge, and select Merge Section Below from the shortcut menu. When you merge two sections, all of the objects in those sections are retained.

One thing to note is that you can merge two sections only of the same type. For example, if you have Page Header A and Page Header B, you can merge them together, but if you try to merge Report Header A with Page Header A, it won't work.

CHANGING THE ORDER OF SECTIONS

Finally, along with all of the other skills you have picked up for working with sections, you can also change the order of sections that appear in your report (without having to delete and recreate the same). To change the order of sections, simply drag-and-drop the section to its new location. When you first hold down the mouse button to drag, your cursor should change to the hand icon. Once you have positioned your section where you want it, release the mouse button to drop the section into place.

FORMATTING SECTIONS

The Section Expert, which can be opened by clicking Report > Section Expert, is key to understanding how the different sections of a report work together. All of the sections of your report are listed, as shown in Figure 6.4, and all of their formatting options are available from this dialog box.

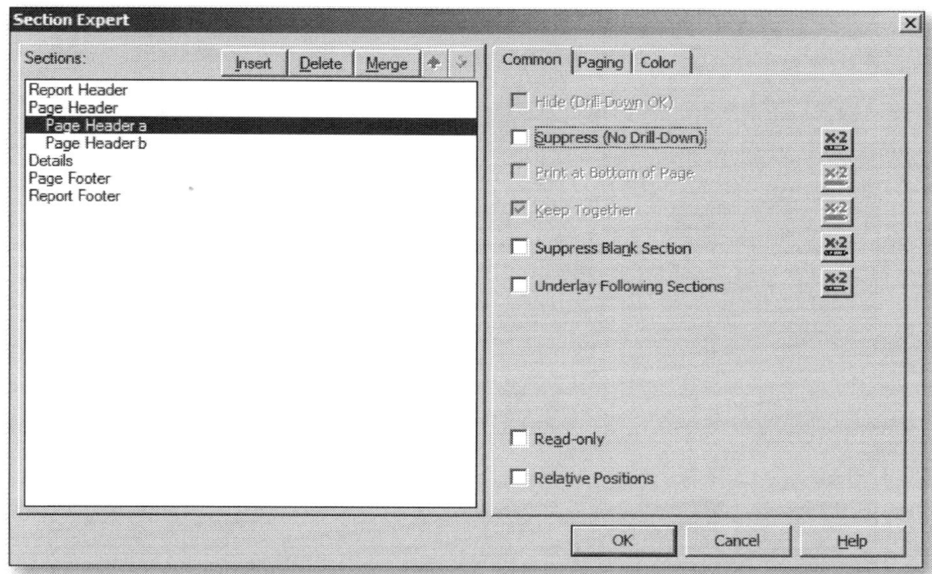

Figure 6.4 The Section Expert contains all of the formatting options for sections contained in your report

A number of options deal specifically with creating, rearranging, and deleting sections of your report. Although some of these options are also available from the shortcut menu we used earlier, it may be easier to use the Section Expert to get the big picture. These options can be found at the top of the Section Expert dialog box and are described in the following list:

Insert — Inserts a new section into your report.

Delete — Deletes a section that you have inserted into your report.

Merge — Takes all of the objects out of two sections and merges them into one section.

Move Up/Move Down — Changes the order of multiple sections you have inserted into your report.

Each section in your report also has a number of specific options associated with it that control the section's behavior and appearance. A list of available options follows:

Free-Form Placement — Allows you to place objects anywhere in a section, disregarding the underlying grid and/or guidelines.

Hide (Drill-Down OK) — Hides a section of your report but still allows for drill-down, to show this section when required.

Suppress (No Drill-Down) — Completely suppresses a section of your report — you will not be able to drill down to show this section.

Print at Bottom of Page — Prints an entire section of your report at the bottom of the page.

New Page Before — Creates a page break immediately preceding a section.

New Page After — Creates a page break immediately following a section.

Reset Page Number After — Resets the page number immediately following a section.

Keep Together — Attempts to keep a particular section on one page to eliminate orphaning, or sections split between multiple pages.

Suppress Blank Section — Suppresses any sections that do not display any data, text fields, and so on.

Underlay Following Sections — Makes a section transparent and places it underneath the section immediately following.

Format with Multiple Columns — (Details Section Only) Creates multicolumn reports for mailing labels and the like.

Reserve a Minimum Page Footer — (Page Footer) Maintains the minimum page footer required by your printer or report design.

Color — Enables and sets the background color for a particular section.

We will be going through the most commonly used options a little later in this chapter.

Another option at the bottom of this list of properties makes the section read only, which would prevent users from making any changes to the section formatting.

HIDING OR SUPPRESSING A SECTION OF YOUR REPORT

Earlier in this book, you learned how to create drill-down reports by hiding different sections of your report and summary reports by suppressing the details of your report. The Section Expert provides the same functionality and provides a quicker method for hiding or suppressing multiple sections. From the Section Expert, you will need to click to highlight a particular section name and then select the Hide or Suppress property from the options shown on the right side of the page. This method has the same effect as right-clicking the section name in Design view and selecting Hide or Suppress from the shortcut menu.

SHOWING HIDDEN SECTIONS IN DESIGN

When you return to your report's design, it is sometimes difficult to determine what sections are present, hidden, and so forth, but you do have some options to help you out. To show all of the hidden sections in the Design tab, use the following steps:

1. Click File > Options, and click the Layout tab.
2. From there, locate the Design View section, and click the option for Show Hidden Sections.
3. Click OK to accept your changes.

When you return to your report's Design view, the hidden sections in your report will now appear in the Design tab, but they will be grayed out.

PRINTING A SECTION AT THE BOTTOM OF THE PAGE

Another handy feature is the ability to print a section at the bottom of the page. This technique can be used with invoices to print a remittance slip or with form letters to include a "return comments" form. To use this technique, use the following steps:

1. Open Crystal Reports and your own report.
2. Switch to the Design view of your report to make things a bit easier, and identify the section you wish to print at the bottom of each page.

This section will be shown above the bottom page margin when the report is previewed or printed, but it will appear in the Design tab in its correct place.

3. Right-click the section, and select Section Expert from the shortcut menu.
4. In the Section Expert, select the option of Print at the bottom of the page.

This section will be printed at the bottom of the page. Where it appears in your report is subject to which section you select. If you select the report header, for example, the section will be printed on the bottom of the very first page and nowhere else. (Likewise, setting this option on the report footer would print the section on the bottom of the very last page.)

CREATING A PAGE BREAK BEFORE OR AFTER A SECTION

Often you want to create a page break before or after a section. This technique can be used with invoices (throwing a page break between invoice numbers), form letters (a separate page for each letter), or anywhere else you need to add a break. To create a page break before or after a section, you can set the properties within the Paging tab of the Section Expert of New Page Before or New Page After. When your report is previewed or printed, a page break will occur in the location you have specified.

If you are using New Page Before, your report may show a blank page for the very first page. This is due to the report header appearing and then a page break is thrown. To eliminate this problem, suppress the report header section.

RESETTING PAGE NUMBERING AFTER A SECTION

When working with statements, invoices, form letters, or reports created for distribution to a number of different parties, you can reset the page number after a specific section to print pages that can be distributed, with each showing the correct page number. In this example, we are going to combine the New Page After option and Reset Page Numbering to throw a page break and reset the page numbering after a customer statement. To use these techniques, use the following steps:

1. Open Crystal Reports, and open the CUSTOMERSTATEMENT.RPT report file from the book download files.

To reset page numbering after a section, always first make sure you have a Page Number field inserted in your report. If you don't, there won't be any way to tell if this option has actually worked.

2. Using the Design view of your report, click View > Field Explorer. Expand the section of the Field Explorer marked Special Fields. From the list select either the Page Number or Page N of M field, and drag it into your report in the Group Footer #1 section.
3. Right-click the Group Footer #1 section, and select Section Expert from the shortcut menu.
4. Using the properties on the right side of the Section Expert, select New Page After and Reset Page Number After.
5. Click OK to return to your report's Design or Preview.

When you preview your report, you should see that either the Page Number or Page N of M field you have inserted will reset after the section you have specified. In addition, your report will now run over multiple pages, with one (or more) pages for each statement, with the page numbering correct for each customer.

SUPPRESSING A BLANK SECTION

You can suppress a blank section to tighten up your report's design and get rid of any unwanted white space. This technique is frequently used when working with names and addresses. You can create two different sections for the address lines (here, Address1 and Address2, as shown in Figure 6.5) and enable the option for Suppress a Blank Section where there is no Address2 field. That way, when your report is printed, it won't appear as if a line is missing.

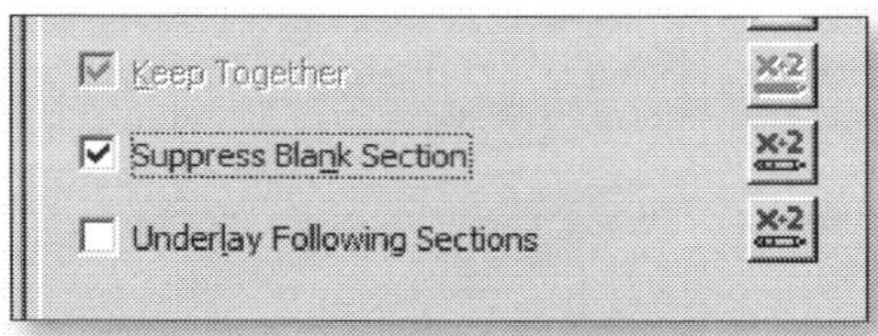

Figure 6.5 Using the Section Expert you can suppress blank sections in your report

To suppress a blank section, use the Section Expert and from the properties on the right side of the page, select Suppress if blank.

You may recognize the Underlay option from where we looked at underlaying images in your report in Chapter 5.

CONTROLLING SECTION PAGE ORIENTATION

Another new feature in Crystal Reports 2008 is the ability to control the page orientation at the section level. This technique allows you to create a report that can feature both portrait and landscape orientation in the same report. This formatting may suit a report where one section has a particularly large amount of information that needs to be shown in landscape, while the rest of the report is best formatted in portrait.

To see how you can change the page orientation in your reports, follow these steps:

1. Open Crystal Reports and open the PAGEORIENTATION.rpt report from the book download files.
2. Right-click on the Group Header #1A section and select Section Expert.
3. Click on the Paging tab of the Section Expert and select the option under Orientation for "Landscape."
4. Tick on the setting for "New Page After" and then click OK to return to your report.

Now when the report is previewed, the Group Header #1A section with the chart will be shown as a landscape page, with a page break thrown each time it is printed.

This technique is especially handy when you are trying to format large reports to server as "report packs" which has a bulk of summary information to display on a single page, but where there are also detail records to be viewed.

CREATING A MULTICOLUMN REPORT

Until now, all of the report designs we have seen have been single-column layouts; in other words, all of the fields and elements of your report were simply listed down the page. Through some special section formatting, reports can be created with multiple columns, allowing you to create flexible reports for phone lists, contact lists, and any other format that requires a large amount of information within a set area.

To create a multiple column layout, create a report as you normally would using the Standard Report Expert, inserting any fields you want to appear in your report as well as any groups or summary fields. Once you have a preview of your report, it is time for some multicolumn magic. In this example, we are going to create a multicolumn report to display an inventory stock list, which will have a list of products running across three columns with text boxes to enter the current inventory count, as shown in Figure 6.6.

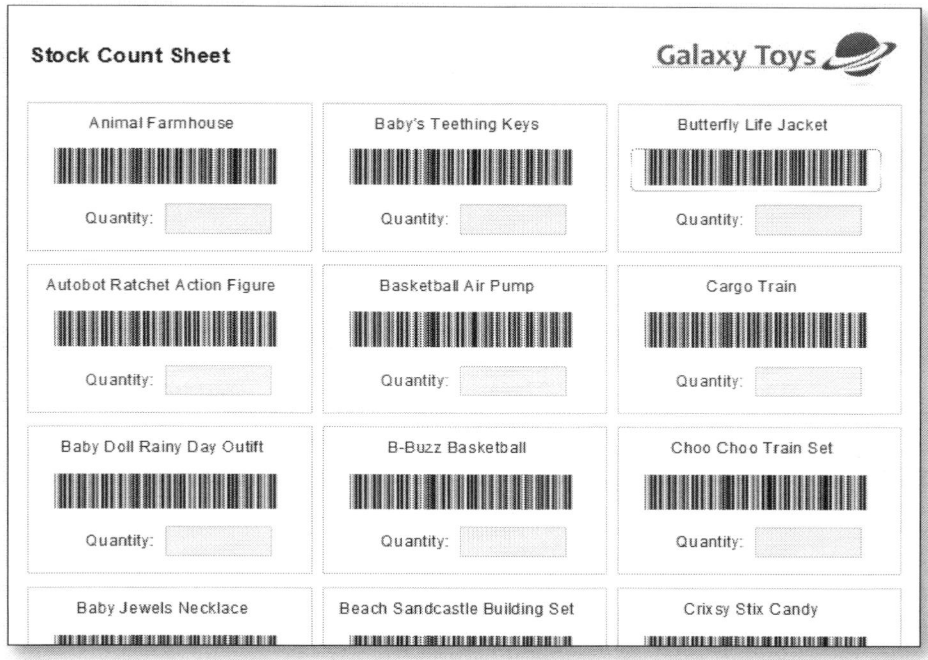

Figure 6.6 You can format your report with multiple columns to create complex report designs

To create this multicolumn report, use the following steps:

1. Open Crystal Reports, and open the STOCKLIST.RPT report file from the book download files.
2. Switch to the Design view of your report.
3. Right-click the Details section of the report, and select Section Expert from the shortcut menu.
4. From the list of options on the right side, select Format with Multiple Columns.
5. Once you have selected this option, the Layout tab will appear at the top of the list of options. Click this tab to open the dialog box shown in Figure 6.7.

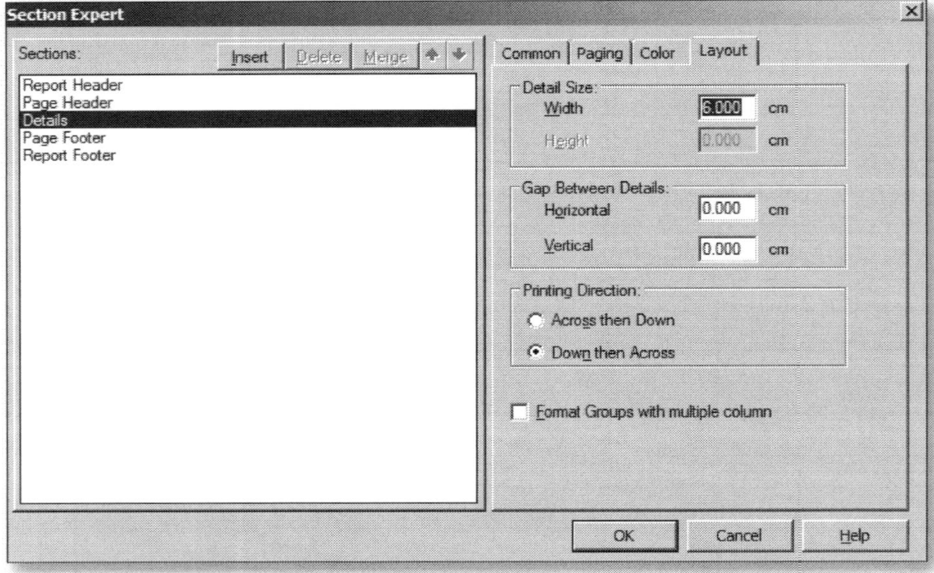

Figure 6.7 The Layout tab will not appear in the Section Expert unless you have specified you want the Details section formatted with multiple columns

6. The first thing you will need to do is choose the width for your columns, as well as the horizontal and vertical gaps between each. For this report, enter a width of 2 or so inches.
7. In this dialog box, you will also need to select a print direction using the radio buttons shown, either Across then Down or Down then Across (the default). For this report, the default is fine.

If you want to also format any groups you have inserted with multiple columns, click the checkbox at the bottom of the dialog box.

---

8. To finish, click OK to accept your changes and return to your report's Design or Preview.
9. Click Insert > Box to draw the boxes shown and adjust your fields and boxes as required.

The design view of your report will show a gray section line (see Figure 6.8) indicating the size of the column you have specified. Use this line as a guide to rearrange your report fields to fit in the columns you have created.

Figure 6.8 The gray section lines indicate the size of your column

If you need to resize your column or change the horizontal or vertical spacing, you must return to the Section Expert by clicking Report > Section Expert and selecting the Details section.

SUMMARY

Using sections, you can create quite complex report layouts that support a variety of uses. When combined with other techniques, like conditional formatting and hiding/showing sections, you can create flexible reports that suit many different types of users.

In the next chapter, we are going to be looking at groups, which provide a method of logically organizing your report data to make the report easier to read, as well as easier to find information in. We'll also look at some advanced techniques for creating summary and drill-down reports, which allow users to view data at a top level then drill down to the details.

Chapter 7
Working with Groups

In This Chapter

- Group Formatting Options
- Changing Groups
- Specified Grouping
- Grouping on a Date or Date-Time Field
- Record-Level Sorting
- Drill-Down & Summary Reports

INTRODUCTION

To this point, we have been working with some very simple reports, with columns and rows of data, and occasionally (as in the form letter example we used earlier) you may have noticed that we used groups. There is nothing really complicated about groups — a group, simply put, is a collection of related records.

When used in a report, groups allow you to put records together, in order, to analyze the information that they contain. If you were creating a sales report, for example, you may want to group your customer records by the states where the customers reside, as shown over the page in Figure 7.1. Alternatively, for an analysis of orders you have received, you may want to group the orders by the customers who placed them.

Figure 7.1 A sales report grouped by state

For each group that you create in Crystal Reports, sections are created for the group header and footer, as shown in Figure 7.2. The group header or footer is usually where you put the name of the group as well as any summaries that are created from the group's data.

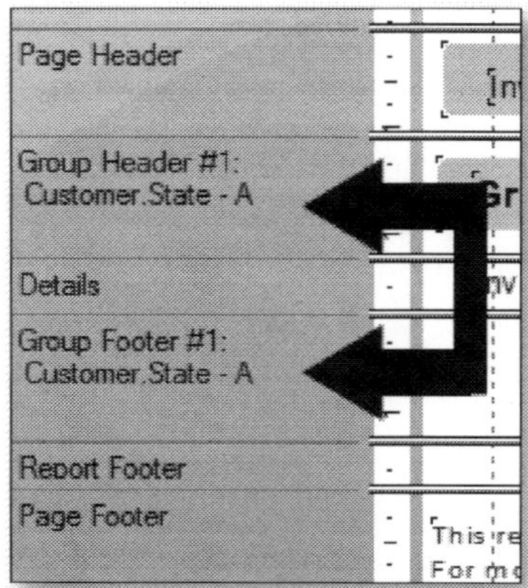

Figure 7.2 A group header and footer

If you previously have worked with different types of reports, you may notice that groups in Crystal Reports closely resemble control breaks in other reporting tools and platforms.

Both groups and control breaks share the same concept of putting like items together and placing space between the like items to indicate where one group ends and the next begins.

If you would like to follow along with the step-by-step instructions in this chapter, this book includes downloadable files that are available from www.kuiperpublishing.com. Also, check out "Setting up the Samples" section in the front of this book for instructions on how to configure the data sources, reports, etc.

Groups can be based on any of the database fields, parameter fields, formula fields, or SQL expressions that appear in your report. To insert a group into your report, use the following steps:

1. Open Crystal Reports, and open the CUSTOMERLISTING.RPT report from the book download files.
2. Select Insert > Group. Using the dialog box shown in Figure 7.3, select a field to be used to sort and group the records in your report as well as the sort order for that particular group.

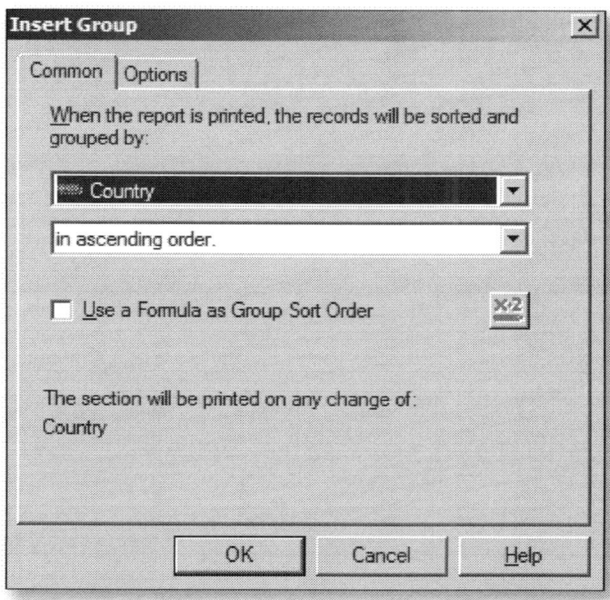

Figure 7.3 Basic group settings

3. Select the Country field from the drop-down list, and select In ascending order for the sort order.

With grouping in Crystal Reports, the groups are not only separated by the field criteria that you specify but are also arranged in the order that you specify. You can choose to sort groups in ascending or descending order (A through Z, or Z through A), use the original sort order from the database, or specify your own order and groupings, which we will look at a little later in this chapter.

4. When you have finished setting your group options, click OK to insert the group into your report.
5. Save this report as CUSTOMERCOUNTRY.RPT (we will be using it a little later in the chapter).

The groups should now appear in the group tree on the left side of the page, and a group header and footer and a group name have been added to your report design.

To delete a group, switch to the Design tab, right-click either the group header or footer, and select Delete Group from the shortcut menu.

GROUP FORMATTING OPTIONS

With any group you insert into your report, you want to control how that group looks and the formatting options applied. Using the tips and tricks in the next section, you should be able to make a group do just about anything you need it to do.

INSERTING GROUP NAMES

Group names, generated by Crystal Reports, are used to label the groups you create. You have already seen them in action — Crystal Reports automatically inserts them whenever you insert a group, and they usually appear in both the group header and footer.

There will be instances where you want to insert group names manually, and Crystal Reports lets you do this as well. To insert a group name, use the following steps:

1. To make it easier to see where you are going to place the field, switch the Design view of your report by clicking the Design tab in the top-left corner of the screen.
2. Then click View > Field Explorer. This step opens Field Explorer, shown in Figure 7.4.

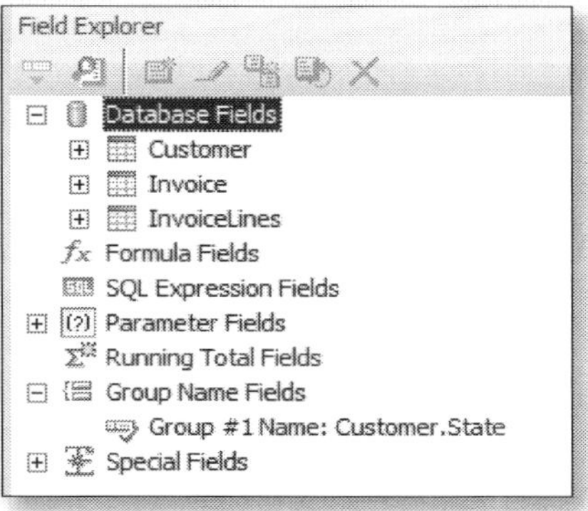

Figure 7.4 The Field Explorer can be used to insert group name fields onto your report

3. Click the plus sign to open to the section for Group Name fields. A list of all of the group names in your report appears.
4. Select the field that you want to insert into your report, and press the ENTER key. This step attaches the field to the tip of your mouse, and as you move your mouse around the page, you should see the outline of the field follow.

5. Position your mouse in the area where you want to place the Group Name field, and click once to insert the field in your report. When your report is previewed, this field will be replaced with the name of the group it represents.

CUSTOMIZING THE GROUP NAME FIELD

Group Name fields can be customized in a number of different ways. You can access the group options to customize a group name when you first insert a new group or when you are changing a group's properties.

To change an existing group, you can use the Group Expert by clicking Report > Group Expert.

As shown in Figure 7.5, you can choose a group name for an existing field by selecting Choose from Existing Field and then selecting a field name. A common example of when you would use this is when you have grouped on a company ID and want to display the company name.

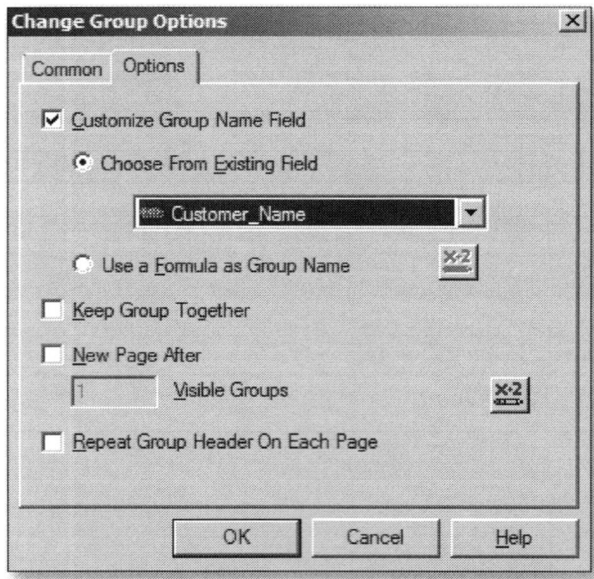

Figure 7.5 You can customize the Group Name field based on an existing field, formula, and so forth

You can also choose a group name based on a formula by clicking Use a Formula as Group Name and entering a formula using the X+2 button. This step opens the Crystal Reports Formula Editor and allows you to enter a formula that returns a group name. An example of when you would use a formula-based group name is when you have grouped by a sales rep code and want to display the sales rep's name. The formula in this situation follows:

```
If (Customer.RepNo} = 112 then "Nathan's Customers"
else
If (Customer.RepNo} = 234 then "Kelly's Customers"
else
If (Customer.RepNo} = 258 then "Aleigha's Customers"
else
"Other Customers"
```

After you have entered a custom formula for your group name and exited the Editor, you'll notice that the X+2 button changes from blue to red and the pencil icon is moved from horizontal to slanted to indicate that you have entered formula text (see Figure 7.6).

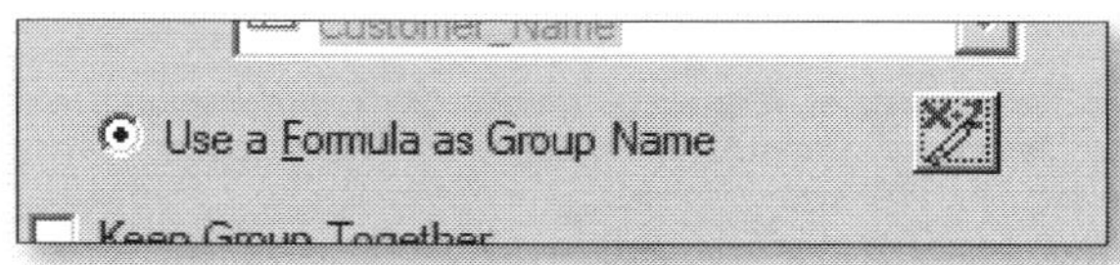

Figure 7.6 The X+2 button changes from blue to red and the pencil icon changes its position to indicate that a formula has been entered

CHANGING GROUP CRITERIA

After a group has been inserted, you may need to change the group criteria — there is a handy little trick to help you out. In Design mode, locate the group header or footer for the group that you want to change. After you have located the group that you want to change, right-click the group header or footer that appears in the gray area. From the menu shown on the following page in Figure 7.7, select Change Group.

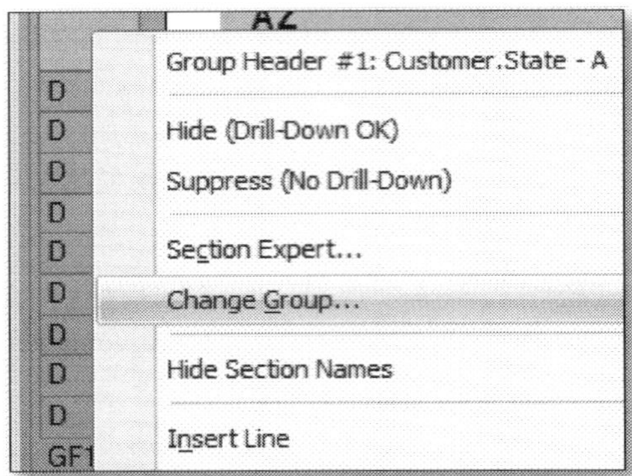

Figure 7.7 The Change Group Options dialog box

You can then make any changes to the group using the Change Group Options dialog box. Click OK to accept your changes. The changes should be reflected immediately in the Report Design or Preview window.

KEEPING A GROUP TOGETHER ACROSS MULTIPLE PAGES

The option for keeping a group together attempts to prevent one section of the group from being orphaned on a separate page. Where possible, Crystal Reports tries to display the complete group on the same page. You can access the options to keep a group together when you first insert a new group or when you are changing a group's properties.

From either the Insert Group dialog box or the Change Group Options dialog box, select Keep Group Together. Click OK to accept your changes. When your report is previewed or printed, Crystal Reports attempts to move all of the group records to a single page.

In any case where Crystal Reports is unable to fit all of the records on the same page, it places the records on separate pages, even with this setting turned on.

REPEATING A GROUP HEADER ON EACH PAGE

For long reports, a group header may be required on each page to identify each group because the group header and footer may be 10 or even 20 pages apart. To repeat a group header on each page, choose the group option Repeat a Group Header when you first insert a new group or when you are changing a group's properties. When your report is previewed or printed, the group header section is printed at the top of each page immediately under the Page Header section, as shown in Figure 7.8.

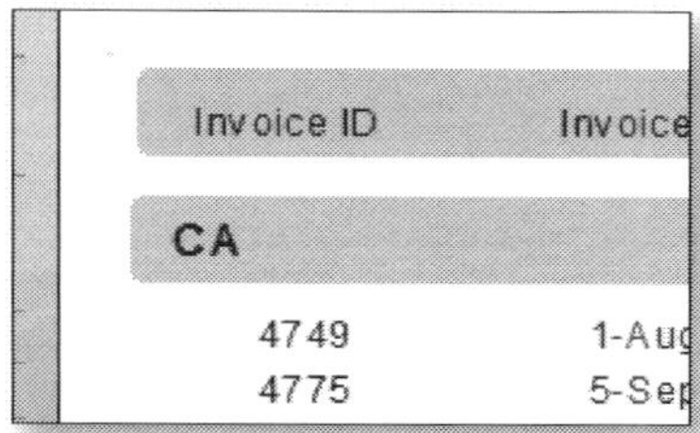

Figure 7.8 The group header can be printed at the top of each page

CREATING A PAGE BREAK BETWEEN GROUPS

For readability, you may want to consider inserting a page break between the groups that appear in your report. To insert a page break, use the following steps:

1. Click Format > Section to open the Section Expert.
2. Locate the group footer of the group that you want to use as a page break. Highlight the group by clicking it.
3. From the options that appear on the right side of the page, select New Page After.
4. Click OK to return to your report.

Alternatively, you can select the group header and select New Page Before.

Your report should now have a page break at the end of each group, making it a bit easier to read.

CHANGING GROUPS

Once you have a group inserted onto your report, you can change the options for the group without having to remove and add the group back again. You may want to change the group field or sort order or just have a look at the options you have set. In any case, you will want to get to know the Change Group Expert.

USING THE CHANGE GROUP EXPERT

The Change Group Expert can be used for changing groups and group options. To invoke the expert, click Report > Change Group Expert. As shown in Figure 7.9, select the group that you want to change, and then select Options.

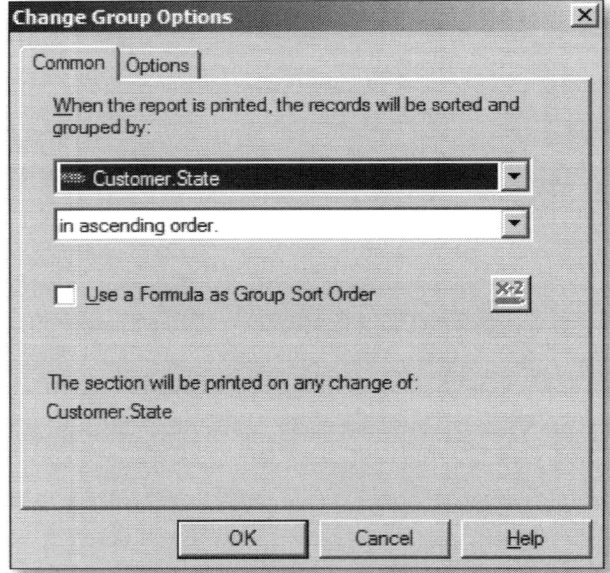

Figure 7.9 You can use the Change Group Expert to change group criteria

The standard Change Group Options dialog box appears, allowing you to make changes to the group field, sort order, and so forth. Click OK to accept your changes. The changes should be reflected in the report design immediately.

REORDERING GROUPS

When working with groups, it is easy to get the hierarchy out of order (for example, by inserting a group for the country inside a group for the state). To reorder the groups that appear in your report, you can simply drag and drop the sections using the following instructions:

1. In Design mode, locate the group headers and footers in the gray area on the left side of the page.
2. Locate the groups that you want to reorder, and move these up or down by dragging the appropriate section of the report.
3. When you have a group selected, your mouse pointer appears as a hand cursor, as shown in Figure 7.10, which indicates that you have attached to a group.

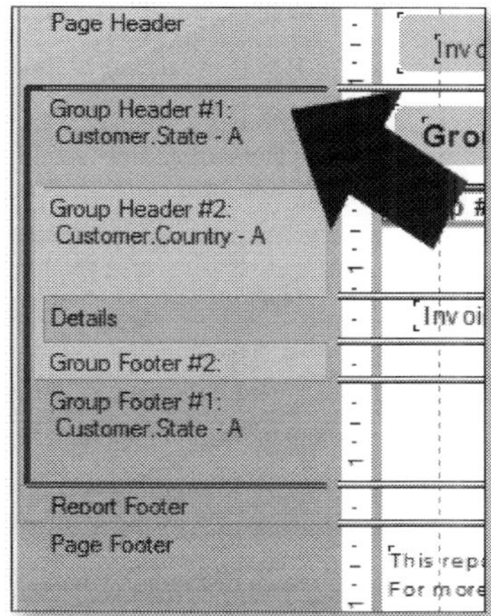

Figure 7.10 The cursor changes to a hand icon to indicate that you are dragging an entire group or section of the report

4. While you are moving the group up and down, a thick bold line indicates where in the group hierarchy your group will appear.
5. After you are satisfied with the new location, drop the group into position by releasing the mouse button.

CHANGING THE GROUP SELECTION FORMULA

When a group is created, Crystal Reports automatically creates a group selection formula. Editing this formula gives you more control of the data that appears in each group. You can edit the group selection formula by clicking Report > Selection Formulas > Group. The Crystal Reports Formula Editor opens. You can then use the Formula Editor to enter your group selection formula.

If you are interested in writing your own formulas, you may want to skip ahead to Chapter 11.

HIERARCHICAL GROUPING

When working with groups and sorting, you sometimes make assumptions about the design of your database. For a report on international sales, for instance, you might assume that separate fields for the customer's country, region, and city are in the database. To create a hierarchical group based on these fields, you would then simply insert three groups: one for the country, one for the region, and one for the city.

What about the instance where the data that makes up the hierarchy is all stored in one table in one single field? That is when you need to use hierarchical grouping. A common example of where you need to use this feature is when working with employee data. Imagine that you have an employee table that contains all of the details for the employees that work at your company. In this table, a field indicates the manager of each employee, using the manager's employee ID.

How would you show this relationship on a report? If you attempted to use a simple group, you could produce a report that has a separate group for each manager, listing his or her employees. Unfortunately, that method would not show the hierarchical relationship among all of the employees (that is, Justin works for Morgan who works for Amy, and so forth).

If you use hierarchical grouping in this situation, you can specify a parent field for your group, and the report produced will display the complete hierarchical view of your employee data. Although it sounds complex, setting up a hierarchical group requires only one extra step.

To create a hierarchical group, use the following steps:

1. Open Crystal Reports, and open the HIERARCHY.RPT report from the book download files.
2. Insert a group as you normally would, by clicking Insert > Group and choosing the field that represents the link to the next step in the hierarchy. In this example, we are working with an employee table that has a Supervisor ID field. You want to show the hierarchy of which employees work for whom, so select the Supervisor ID field for your group, and click OK.
3. Click Report > Hierarchical Grouping Options to open the dialog box, shown in Figure 7.11.

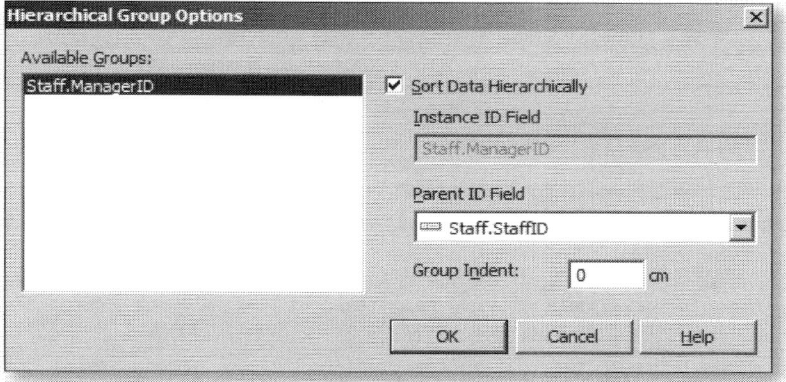

Figure 7.11 The Hierarchical Options dialog box

4. Select the Sort Data Hierarchically option. From the Available Groups list, you would select the group that you want to group hierarchically.
5. Select the parent ID field on which this hierarchy is based. In the example here, the field is the Employee ID field.
6. Finally, set the indentation size for when your group is displayed. Click OK to accept your changes. The group that you have inserted should now reflect the hierarchy that you created.

SPECIFIED GROUPING

Specified grouping is a powerful feature that allows you to regroup data based on criteria that you establish. For example, suppose you have sales territories comprising of a number of states. You can use a specified group to create a separate group for each sales territory and to establish your own criteria (for example, North Carolina plus South Carolina is Bob's territory). To create a specified group, use the following steps:

1. Open Crystal Reports, and open the SALESTEAM.RPT report from the book download files.
2. Click Insert > Group, and select the Employee Last Name field.
3. In the same Insert Group dialog box, change the sort order to In Specified Order. A second tab, labeled Specified Order, should appear in the Insert Group dialog box, as shown in Figure 7.12.

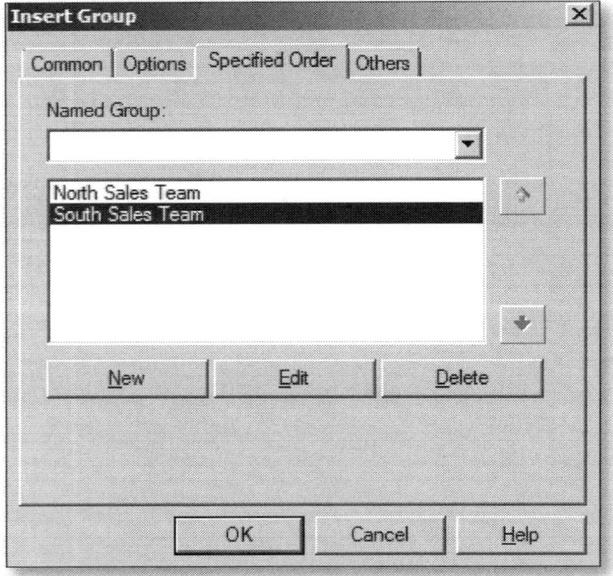

Figure 7.12 The Specified Order option page

For each specified group we want to create, we need to define a group name and specify the group criteria.

4. Type all of the group names you want to create, pressing the ENTER key after each, to build a list of group names.
5. Once you have all of the group names defined, highlight each one, and click the Edit button to specify the criteria.
6. To establish the criteria for records to be added to your group, use the drop-down menu to select an operator and value or values.

These are the same operators used with record selection.

7. You can add other criteria by clicking the New tab and using the operators to specify additional selection criteria, which are evaluated with an `or` statement between the criteria that you specify.
8. After you have entered a single group, another tab appears with options for records that fall outside of the criteria that you specify. By default, all of the leftover records are placed in their own group, labeled Others. You can change the name of this group by simply editing the name on the Others tab, which is shown in Figure 7.13. You can also choose to discard all of the other records or to leave them in their own groups.

Figure 7.13 Options for dealing with other records

9. After you have defined your specified groups and criteria and have reviewed the settings for other records, click OK to accept the changes to the group.

   Your specified grouping is now reflected in your report.

GROUPING ON A DATE OR DATE-TIME FIELD

Grouping on a Date field requires you to specify how the dates are grouped. To create a group based on a Date field, you would simply insert a group by clicking Insert > Group, and then selecting a date or date-time field (Database, Formula, or Parameter) to be used to sort and group the records in your report.

You will need to select a sort order for your group (ascending, descending, specified, or original). Notice that Crystal Reports has added an additional option box for selecting the interval at which the group should be printed. The interval options are slightly different for dates and time fields, as listed below.

For dates:

- For Each Day
- For Each Week
- For Each Two Weeks
- For Each Half Month
- For Each Month
- For Each Quarter
- For Each Half Year
- For Each Year

For times:

- For Each Second
- For Each Minute
- For Each Hour
- For Each AM/PM

You can set any group options for customizing the group name, keeping the group together, and so forth. Once you have clicked OK and your group has been created, check the Design tab. You should see the group that you inserted represented by a group header and footer that appear in the gray area on the left side of the page.

When you preview your report, the group name is generated from the interval that you picked when creating your group. You can format this group name just like any other time or date-time-type field.

Crystal Reports usually uses the last date in the interval to create a group name. If you select a grouping by week, it will display the last day of each week as the group name.

REPEATING GROUP HEADERS

When working with groups, you are probably going to have groups that run across multiple pages. For users who are viewing your reports, this can be a little confusing as it will appear that the group stops at the bottom of the page. We can, however, use a bit of logic to print a message at the bottom of the page to indicate that the group is continued on the following page, as shown below in Figure 7.14.

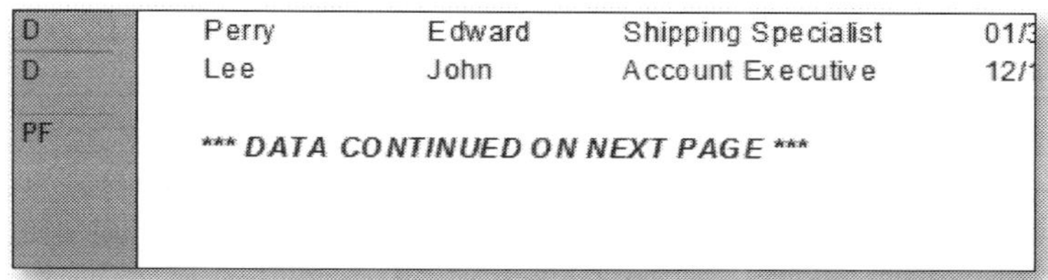

Figure 7.14 Continued group message

To use this technique in your report, you will need to create some formulas using variables — we will be covering formulas a bit later in Chapter 11, but the technique below can be adapted without having to know a lot about the formula language at this point.

To use this technique in your report, follow these steps:

1. From the book download files, open the CUSTOMERBYTYPE.RPT report from the Chapter 7 files.
2. You will need to create 3 formula fields. To create a formula field, select View > Field Explorer then right-click on the Formula Field node and select "New Formula". The names and text for these formulas are as described on the following page.

@ContinuedGroupHeader

```
WhilePrintingRecords;
BooleanVar ContinuedNextPage:= True
```

@ContinuedGroupFooter

```
WhilePrintingRecords;
BooleanVar ContinuedNextPage:= False
```

@ContinuedPageFooter

```
WhilePrintingRecords;
BooleanVar ContinuedNextPage;
if ContinuedNextPage:= True
then "Data continued on next page..."
else ""
```

3. Once you have created all three formulas, switch over to the Design view of your report and drag the ContinuedGroupHeader formula into the Group Header #1 section.
4. Right-click on the formula you have created and select Format Field, then on the Common object properties tab, select "Suppress" (as you don't want this field to appear when the report is run).
5. Repeat these two steps for the ContinuedGroupFooter formula, copying it into the Group Footer #2 section and suppressing the field.
6. Finally, drag the ContinuedPageFooter formula into your report's page footer.

When you preview the report, the continued message should appear, letting users know there is additional data on the next page. The message that is printed out can be customized to your own requirements and should give some clarity to users who are not familiar with viewing long reports.

RECORD-LEVEL SORTING

In addition to sorting records into groups, you can also use record-level sorting to sort records without separating them. For example, you may have a simple report that lists all of the invoices for a particular day. Because this is a simple list, you probably don't want to use grouping, but you do want to put the invoice in order on the report. This is where record-level sorting comes into play, because you can specify the sort order at the most basic level.

This method also works with groups inserted into a report, because you can sort the contents of a group using record-level sorting.

To add record-level sorting to your report, click Report > Record Sort Expert to open the dialog box shown in Figure 7.15.

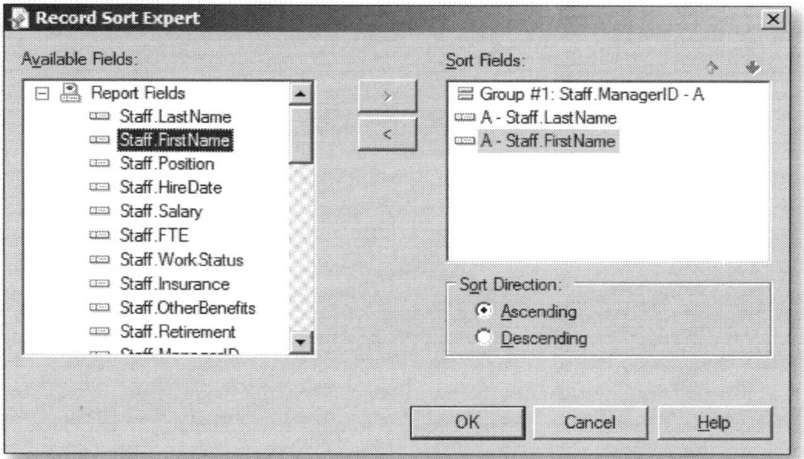

Figure 7.15 Options for record-level sorting

In this dialog box, you move the fields you want to use in sorting from the left list to the right using the arrows in the middle of the dialog box. Highlight a field on the left, then click the right arrow to move the field to the list on the right. You can use the radio buttons below the list to select the sort order. The arrows on the top-right corner of the dialog box are used to specify the sort-order precedence.

You will see any groups you have inserted into your report shown here. Group sorting always takes precedence over record-level sorting. You cannot remove groups using this dialog box.

Don't forget that you can use record-level sorting in conjunction with Sort Controls to give users most control over how the report is sorted when they are viewing it. If you missed this technique, flip back to Chapter 5 to check out how to add sort controls to your report.

DRILL-DOWN & SUMMARY REPORTS

Creating meaningful information from large volumes of data can be difficult. If a report is too long, report consumers generally tune out before they get to the section of data that is relevant to them. One trick for the concise presentation of information is to use a drill-down or summary report.

Drill-down and summary reports are similar because they contain a summary of information from your data. For a sales report, you may want to summarize a particular salesperson's sales for a given month, showing the total sales figure but not the details that it comprises, as shown below in Figure 7.16.

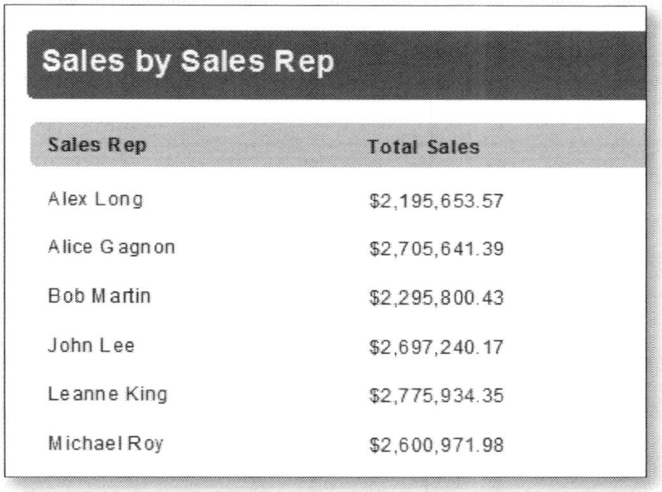

Sales by Sales Rep

Sales Rep	Total Sales
Alex Long	$2,195,653.57
Alice Gagnon	$2,705,641.39
Bob Martin	$2,295,800.43
John Lee	$2,697,240.17
Leanne King	$2,775,934.35
Michael Roy	$2,600,971.98

Figure 7.16 A typical Sales Summary report

Although drill-down and summary reports of this information look similar, one difference does exist. A drill-down report allows you to double-click the salesperson's name or any of the summary fields to display all of the details that make up that summary. In a drill-down report, the details are simply hidden from view, but they are still available when a user needs them. With a summary report, the details are suppressed from view and are not available to be seen by the user.

Each type has specific uses. If the report that you have created is likely to raise questions, such as "Why are John's sales up this month?" or "Why is that number negative?" a drill-down report can provide report consumers with the information they need to answer these questions without having to have another report created.

Drill-down reports can be used to drill down into the data as many times as required, and with each additional query, users open a separate Preview tab. This information can be printed independently of the main report; it can be exported as well. For report consumers, this flexibility provides a significant ad hoc capability. Instead of having to ask for a new report to be created for each request, they may be able to navigate through an existing comprehensive report and extract the information that they need. For this situation, a drill-down report is ideal.

On the other hand, if you are distributing a report that is a summary of expenses, including payroll figures that are confidential, you may not want to give users the ability to drill down into that information.

Regardless of which type of report you choose, both drill-down and summary reports can be used to add real value to the information that you present in your report.

CREATING A DRILL DOWN REPORT

Expanding on a report we looked at earlier, where a group had been inserted on the Country field, we could easily create a drill-down report that would show only the countries. The user could then double-click to drill down into the details of the report.

To create a summary report, use the following steps:

1. Open Crystal Reports, and open the CUSTOMERCOUNTRY.RPT report we were working on earlier; you can also open a copy from the book download files.
2. Switch to the Design view of your report, and locate the group header and footer for the group you are working with — in this case, the Country field.
3. Right-click the header in the gray area on the left side of the page, and select Hide from the shortcut menu.

Your report will now show only the summary information. To drill down into your report, move your mouse across a group header or footer field and double-click where your mouse pointer turns into the hourglass icon. An additional tab (beside Design and Preview) opens to display your drill-down data.

If you wanted to show the headings when you drilled down, you would need to click File > Report Options and select Show All Headers on Drill-Down. Click OK to accept your changes.

You could then hide any other headers or footers by repeating the same process. Likewise, if you wanted to suppress the details (not show the details), you can right-click on the section and select the option to Suppress.

A separate tab appears showing you the drill-down report you have selected. You can use the red X that appears on the navigation bar to close any drill-down tabs that you have opened.

Drill-down reports are the best means of quickly displaying an overview of the information contained within your report. Drill-down tabs can be printed individually, allowing a user to use one report for many sets of information. When combined with summary fields, drill-down reports offer a powerful analysis function, allowing users to see the details that make up a particular sum or total.

You can also find additional examples of different drill-down reports and techniques in the book download files in the Chapter 7 folder.

FORMATTING DRILL-DOWN REPORTS

When working with drill-down reports one of the most common complaints is that when you drill down to the details in your report, you lose the page headings, as shown in Figure 7.17.

GH1	AZ				
D	4271	26-Oct-09	27-Oct-2009	26-Oct-2009	Black Forest Books & Toys
D	4603	28-Apr-10	01-May-2010	28-Apr-2010	Black Forest Books & Toys
D	4538	20-Mar-10	27-Mar-2010	24-Mar-2010	Black Forest Books & Toys
D	4303	30-Oct-09	02-Nov-2009	30-Oct-2009	Black Forest Books & Toys
D	4170	27-Aug-09	29-Aug-2009	26-Aug-2009	Black Forest Books & Toys
D	4023	13-May-09	22-May-2009	13-May-2009	Black Forest Books & Toys
D	3868	9-Feb-09	11-Feb-2009	09-Feb-2009	Black Forest Books & Toys
D	3813	7-Jan-09	16-Jan-2009	07-Jan-2009	Black Forest Books & Toys

Figure 7.17 Drill Down without page headings

There is a formatting technique that you can use that will allow you to simulate page headings on the drill-down tabs using the Crystal Reports formula language and the DrillDownGroupLevel function.

This function will return a number that corresponds to where you have drilled in your report. For example, if you have not yet drilled down into the report, it will return 0. But if you have a report that has groupings for Customer Type, Country and State and drill down into Customer Type, it will return a 1. Likewise, when you drill down into Country it will return 2 and so forth.

To use this function to show a page header for each drill-down level, follow these steps:

1. From the book download files, open the DRILLDOWNHEADER.RPT report from the Chapter 7 folder.
2. Switch to the Design tab of your report and right-click on the Group Header #1 section and from the right-click menu, select Insert Section Below.
3. Drag the section you have just created to swap Group Header #1a and Group Header #1b. This section is going to be our simulated page header when the user drills down into the report.
4. Right-click on the Page Header and from the right-click menu choose "Select All Section Objects" and then right-click on the objects and select "Copy".
5. Right-click in the Group Header #1a section and select "Paste". The objects will be attached to the tip of your mouse pointer and you can click to place them in this section.
6. Right-click on the Group Header #1a section and select Section Expert.
7. Find the Suppress property and click the conditional formatting icon (X+2) next to it.
8. Enter the following formula:

```
DrillDownGroupLevel = 0
```

9. Click OK to return to your report.

When you preview your report and drill down into the Customer Type group, you will now see that every time you drill down, you now have a page header, as shown in Figure 7.18 on the following page.

	Invoice ID	Invoice Date	Required Date	Ship Date	Customer Name
GH1a	Invoice ID	Invoice Date	Required Date	Ship Date	Customer Name
GH1b	AZ				
D	4271	26-Oct-09	27-Oct-2009	26-Oct-2009	Black Forest Books & Toys
D	4603	28-Apr-10	01-May-2010	28-Apr-2010	Black Forest Books & Toys
D	4538	20-Mar-10	27-Mar-2010	24-Mar-2010	Black Forest Books & Toys
D	4303	30-Oct-09	02-Nov-2009	30-Oct-2009	Black Forest Books & Toys
D	4170	27-Aug-09	29-Aug-2009	26-Aug-2009	Black Forest Books & Toys
D	4023	13-May-09	22-May-2009	13-May-2009	Black Forest Books & Toys
D	3868	9-Feb-09	11-Feb-2009	09-Feb-2009	Black Forest Books & Toys
D	3813	7-Jan-09	16-Jan-2009	07-Jan-2009	Black Forest Books & Toys
D	4316	22-Nov-09	29-Nov-2009	22-Nov-2009	Black Forest Books & Toys

Figure 7.18 Report with page headings for drill downs

If you have multiple groups, you would need to duplicate each group's Group Header section and put another formula behind the suppression of that section. This technique is a must-have for anyone who has ever heard users complain "But where are my headings?!"

SUMMARY

A large part of report design is formatting your reports and taking large amounts of data and condensing it into very concise pieces of information. In this chapter, we looked at some of the techniques to help you format and organize your reports to make the information more meaningful and easier to read. In the next chapter, we are going to look at another way of adding value to reports by analyzing the data they contain with summaries, running totals and more.

Chapter 8
Summarizing Report Data

In This Chapter

- Summarizing Report Data
- Summaries vs Formulas
- Adding Summary Fields to Your Report
- Working with Statistical Summaries
- Inserting Grand Totals
- Using Running Totals
- Conditional Highlighting

INTRODUCTION

Reports can contain large amounts of data, and report users often find it difficult to sort and shift through hundreds of thousands of rows to find the information they need. Luckily, Crystal Reports provides the ability to summarize and filter report data to find only the relevant information the user needs. This chapter is dedicated to these techniques.

The first part of this chapter looks at the ways you can summarize report data, including some of the differences between Crystal Report's built-in summary operators and formula languages. If you need to summarize large amounts of data and present it in a concise, meaningful way, this chapter is for you — so let's get started.

SUMMARIZING REPORT DATA

One of the most common ways to analyze the data in your own reports is to display summaries of the underlying data. For example, you could have a report with groups that display the different salespeople in your organization and then show a summary at the bottom of each column to indicate their average sale, total sales, and so on, as shown in Figure 8.1.

Employee Listing - Detail

Last Name	First Name	Position	Hire Date	Salary	FTE
Admin Department					
Clark	Raymond	General Admin	10/18/2009	$68,500.00	1.00
Johnson	Lisa	Admin Assistant	10/19/2010	$80,500.00	1.00
Gagnon	Alice	Office Manager	12/17/2010	$77,000.00	1.00
	Average Salary:	$75,333.33	Total Salary:	$226,000.00	
Finance Department					

Figure 8.1 A typical summary report

You can display these summaries in a couple of ways. The following section will help you decide which is best for your report.

SUMMARIES VERSUS FORMULAS

When it comes to adding calculations to your report, you have two choices. You can either insert a summary field or create your own calculation, either using formulas or SQL Expressions. Summary fields are designed to eliminate the need to write formulas or SQL Expressions for common calculations, including sums, averages, counts, and others.

However, despite their ease of use, summary fields do have limitations. Summary fields are not as flexible as formulas that you write, and they are limited to the 19 summary operators currently available. Summaries are also tied to a particular group or the grand total for your report.

If you need full control of how a calculated field is derived or where it is placed, you probably need to create a formula field. Likewise, if you are familiar with writing SQL statements and want the calculation to be performed on the database server, you can create an SQL Expression.

If you are looking for a quick and easy standard calculation based on some grouping you have inserted into your report, summary fields may be for you. A list follows of the most popular summary fields and a description of how they are used:

Sum — Provides a sum of the contents of a numeric or currency field.

Average — Provides a simple average of a numeric or currency field (that is, the values in the field are all added together and divided by the total number).

Minimum — Determines the smallest value present in a database field. This field is for use with number, currency, string, and date fields.

Maximum — Determines the largest value present in a database field. This field is for use with number, currency, string, and date fields.

Count — Counts the values present in a database field. This field is for use with all types of fields.

Distinct Count — Counts the values present in a database field, but counts any duplicate values only once.

A number of statistical summary functions are also available for use, including:

- Correlation
- Covariance
- Weighted average
- Median
- Pth percentile
- Nth largest
- Nth smallest
- Mode
- Nth most frequent
- Sample variance
- Sample standard deviation
- Population variance
- Population standard deviation

If this all looks like Greek to you, you may want to check out the Internet Glossary of Statistical Terms, available at www.animatedsoftware.com/statglos/statglos.htm. If you are not familiar with statistics, you can use this glossary as a guide to determine if any of these summary operators may be of use in your organization.

ADDING SUMMARY FIELDS TO YOUR REPORT

Simple summary fields include sum, average, minimum, maximum, and other calculations that do not require any additional criteria. In the following example, we are going to add a summary field to an existing sales report to make it easier to read and interpret the information that is contained within the report.

If you would like to follow along with the step-by-step instructions in this chapter, this book includes downloadable files that are available from www.kuiperpublishing.com. Also, check out "Setting up the Samples" section in the front of this book for instructions on how to configure the data sources, reports, etc.

To create a simple summary field, use the following steps:

1. Open Crystal Reports, and open the STAFFPAYROLL.RPT report from the book download files.
2. Locate and click the field that you want to summarize. In this case, it is the Staff.Salary field.

Make sure that you click the Order Amount field that appears in your report's Details section. You may want to switch to the Design view of your report before clicking to make things a bit easier.

3. Click Insert > Summary. In the dialog box, shown in Figure 8.2 on the next page, select the summary operation you want to use from the Calculate this summary box, and select a location from the drop-down box at the bottom of the dialog box. For this example, select Sum and choose the option to "Add to all group levels".
4. Click OK to accept your changes. Your summary field is inserted in the location you have selected, as shown in Figure 8.3.

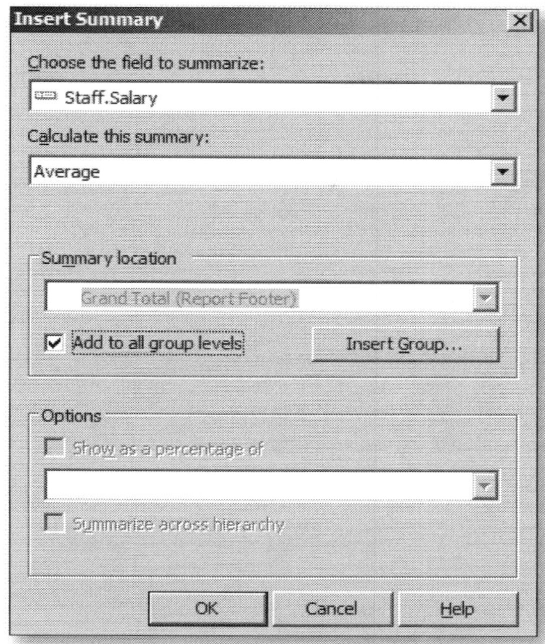

Figure 8.2 Options for inserting simple summary fields into your report

As you scroll through the report, you will notice that underneath the Order Amount field column is now a Sum field that appears with each group. This field represents the total sales for a particular salesperson. Summaries are most often used with groups and appear on the group footer if a group has been selected. If you selected the option to create a Grand Total, these sums will always appear in the report footer on the very last page of your report, showing a total for the entire report.

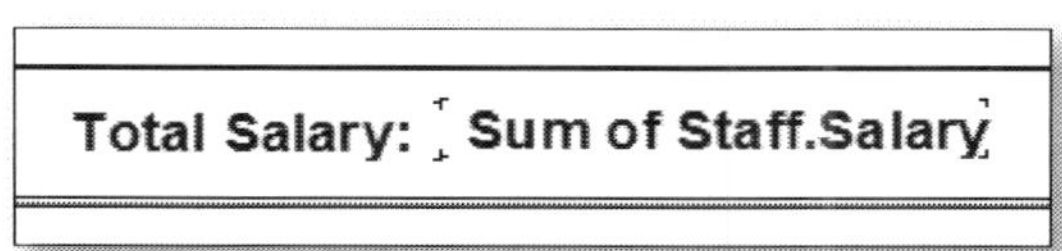

Figure 8.3 The report with the summary field in place

CHANGING SUMMARY FIELD OPERATIONS

After a summary field has been inserted into your report, you can change the summary operation by locating the summary field that you want to change on your report, clicking to select the field, and clicking Edit > Edit Summary. In the Edit Summary dialog box, use the drop-down list to change the summary operation, then click OK to accept your changes. The operator change should be reflected immediately in your report, as shown in Figure 8.4.

Figure 8.4 Summary field options

An important point to remember is that selecting a different summary operator may mean specifying additional parameters, depending on the summary operator you choose.

INSERTING SUMMARY FIELDS SHOWN AS PERCENTAGES

The information in a summary field can be shown as a percentage of a grand total. For example, you may want to see what percentage a certain customer contributes to your total business. Or, using our earlier example, what percentage of sales a certain salesperson has contributed. If you insert a summary field shown as a percentage of the grand total of your report, that information is readily available.

To insert a summary field to be shown as a percentage in the Sales Summary report we have been working with, use the following steps:

1. Open Crystal Reports, and open the report we have been working with (STAFFPAYROLL.RPT), or open a copy from the book download files.
2. Click Insert > Summary.
3. In the dialog box, select the summary type you want to insert from the Calculate this Summary box. In this example, select the Sum operator.

4. Choose the field you want to summarize, and select a summary location. For this report, select the Staff.Salary field, and for the location, select all group levels.
5. At the bottom of the dialog box, check Show as Percentage Of. Using the drop-down list under Show as Percentage Of, select the grand total field that you want to use to calculate the percentage (Grand Total: Sum of Salary), as shown in Figure 8.5.
6. Click OK to accept your changes and to add the summary field to your report.

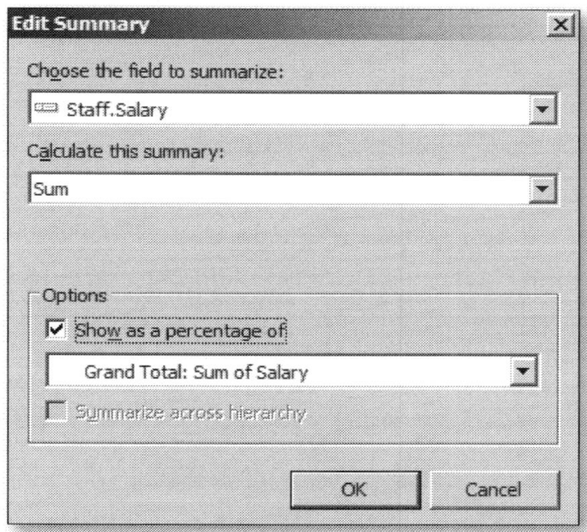

Figure 8.5 Options for showing a summary field as a percentage

The report will now show how much each salesperson has contributed to the overall sales of the company. You can now format this field as a percentage using some of the techniques you learned in Chapter 5.

The show-as-percentage feature is especially handy when used alongside sums and averages. For instance, if you were creating a report on international sales, you could show the dollar amount (sum) for each country and the average sales as well as a percentage representing that country in the total sales.

INSERTING GRAND TOTAL FIELDS

Grand total fields appear in the report footer at the end of your report and are used to summarize the contents of your report. To insert a grand total field, use the following steps:

1. Open Crystal Reports, and open the report we have been working with (SALESSUMMARY.RPT), or open a copy from the book download files.
2. Locate and click the field you want to summarize with a grand total, in this case, the Order Amount field.
3. Click Insert > Summary. Using the same Insert Summary dialog box we have been working with, select a summary type and a summary location of Grand Total. Click OK to accept your changes. A summary field representing the grand total is inserted into your report footer.

The term *grand total* is a little misleading. A grand total in Crystal Reports can be a value calculated with any of the summary operators, including Sum, Average, and so on.

WORKING WITH STATISTICAL SUMMARIES

For the statistical functions available for use within summary fields, you may want to pull on a pair of boots, because we will be wading into deeper water. If you work with statistics, this information will make more sense, but if you are just starting out, check out the About.com site available at http://bit.ly/40ksDO which has links to detailed explanations of all of the statistical terms.

Correlation, covariance, and weighted averages are all related because they require a field to serve as the basis of the summary as well as a second field that is related.

To insert a correlation, covariance, or weighted average summary field, locate and select the field that will serve as the basis for your correlation, covariance, or weighted average. Right-click the field and click Insert > Summary. In the dialog box that opens, select the summary function you want to use.

When you select correlation, covariance, or weighted average, a With drop-down list appears. Select a field to be used when calculating your summary field from this list.

For a correlation or covariance, this will be the field against which you want to compare. For a weighted average, choose the field that contains the values that will weight the average denominator. (In a normal average, this defaults to 1 for each value, but for a weighted average it can be any number you specify.)

Select the summary location and click OK to accept your changes and return to your report. Your summary field is inserted into the group footer you specified.

Another handy statistical summary is the Pth percentile. The Pth percentile summary function can be used to determine the value of P in a numeric or currency field. For example, suppose you want to see where an employee's age falls within your company's distribution. If you enter 50 for the P value in your summary, Crystal Reports will return a value from the 50th percentile (for example, 42, meaning that 50 percent of your employees are younger than 42).

In addition, you may want to look at statistical functions that center around size and frequency. When creating a report, you may want to know what is the largest, smallest, or most frequent data item. Although you can use Group Sorting (which we will talk about later) to obtain similar information from your report, it is much easier to use a summary field.

To insert one of these types of summary fields onto your report, use the same method to insert a regular summary field and then select the Nth Largest, Nth Smallest, or Nth Most Frequent Summary function. When you select any of these summary operators, an N Is text box appears. Enter a value for N in the box. It's that simple.

ANALYZING REPORT DATA

In addition to calculated fields like summaries and formulas, Crystal Reports also includes a number of other options for analyzing information that appears in your report. Analysis methods range from simply reordering the data to presenting running totals alongside the data to highlighting areas of your report based on pre-set criteria. Regardless of which analysis options you choose to use in your report, you can quickly see the value they add.

USING TOPN/BOTTOMN ANALYSIS ON GROUPS

Group Sorting is a function of Crystal Reports used to sort groups according to a summary field that has been created based on that group. Most often, this function is used to determine the top 20 customers (that is, Top N where N is 20) or the top (or bottom) five products. Before you can use Group Sorting analysis in your report, you need to make sure that you have two things inserted onto your report: a group and a summary field. Without both of these present, you cannot use Group Sorting analysis.

To add Group Sorting analysis to your report, use the following steps:

1. Open Crystal Reports, and open the ANALYSISREPORT.RPT report from the book download files.
2. With any report, the next step should be to verify that your report has at least one group and summary inserted (otherwise, we wouldn't have any way to perform our analysis).
3. Click Report > Group Sort Expert. From the dialog box shown in Figure 8.6, you can select either Top N, Bottom N, Top N Percentage, or Bottom N Percentage. In this example, select Top N.

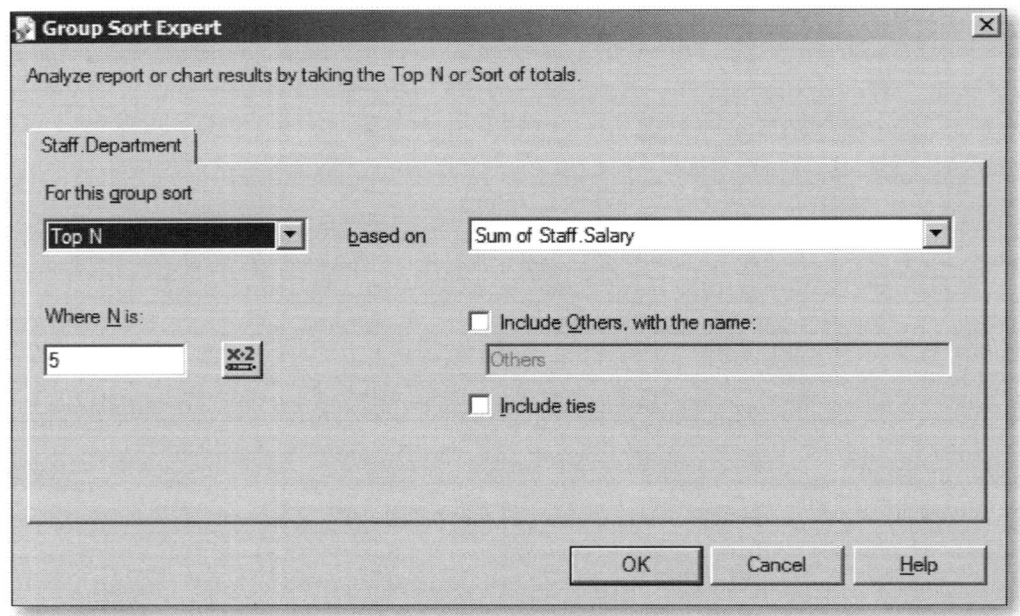

Figure 8.6 Options for Group Sorting analysis

4. Using the drop-down list of available summary fields, select the field on which your analysis will be based. (You'll notice that this list is limited to the summary fields you have inserted in your report.) For this report, select Sum of Orders.Order Amount.
5. Next, enter a value for N. Because we want to create a Top 10 customer report, enter 10 for n.
6. There may be more than 10 groups in the report, so you can use the checkbox provided to include the other groups in your report; enter your own name for this group. If a

group is not included in the Top (or Bottom) N you specify, it will get lumped into this group. If you would like to display the others, check the box provided and enter a name for the group.

7. Click OK to accept your changes and to apply Group Sorting analysis to your report.

Your report will now only display the Top 10 customers as well as any other records (if you selected that option). You could use this same method to create a Bottom N report or top or bottom percentage, that is, a Customers in the Top 50 Percent report.

SORTING GROUPS BY SUBTOTAL OR SUMMARY FIELDS

When working with groups, you may sometimes want to sort the groups according to some summary field that you have inserted in your report. This functionality is similar to Group Sorting analysis, except that it isn't limited to just a certain number of groups. Using the Crystal Reports Sort All functionality, you can order all of the groups by a summary value.

To sort the groups in your report by the value of a summary field, verify that your report has at least one group and summary inserted. Click Report > Group Sorting Expert. Then, in the dialog box shown in Figure 8.7, Select All from the For This Group Sort drop-down list.

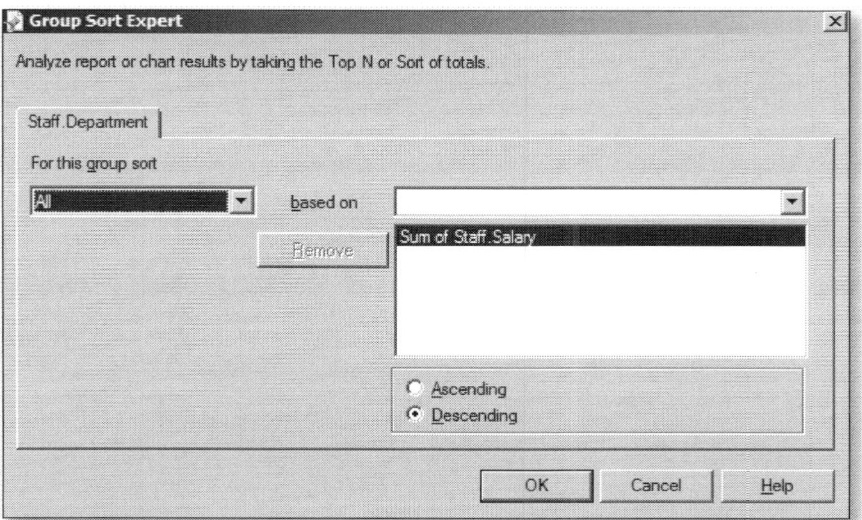

Figure 8.7 You can sort groups by a summary field that you have inserted into your report

Using the drop-down list of available summary fields, select the field on which your analysis will be based. Select a sort order (Ascending or Descending), and click OK to accept your changes and to apply Group Sorting analysis and Sorting to your report. Your groups will now be ordered by the summary field you selected.

RUNNING TOTALS

Running totals provide an at-a-glance look at cumulative values in your report and display a running summary beside each record. With each release of Crystal Reports, running totals have grown in functionality and can now be used for a wide range of summary and analysis tasks. Running totals also feature a flexible evaluation and reset function that makes complex analysis easier.

By using running totals in your report, you can quickly give users the information they need without having to wade through the report to get to a summary field or the end of a section. For example, suppose you want to create a running total that runs alongside a list of your customers and their last year's sales. With each record, you want this last year's sales figure added to the running total, as shown in Figure 8.8.

Customer Sales Report

Customer Name	3Yr Avg Sales	Running Total
abraKIDabra Toys Inc	$12,928.00	$12,928.00
Arkasi Toys and Games	$9,289.52	$22,217.52
Black Forest Books & Toys	$4,013.44	$26,230.96
Boystowne Toys	$8,246.77	$34,477.73
Bugaboo Babies	$7,057.72	$41,535.45
Chimney Rock Toys	$1,385.98	$42,921.43
Dancing Bear Toys, Ltd	$7,244.07	$50,165.50
Destination ImagiNation	$4,510.49	$54,675.99
Dig-It!	$5,951.49	$60,627.48
Gateway Toy Store	$6,539.39	$67,166.87

Figure 8.8 An example of a report with a running total

You can also insert running totals from the Field Object menu (Insert > Field Object). Locate the section for running totals in the Field Explorer, right-click, and select New.

The first step in creating a running total is to locate and select the field that will serve as the base field. Right-click the field, and click Insert > Running Total. The dialog box shown in Figure 8.9 opens.

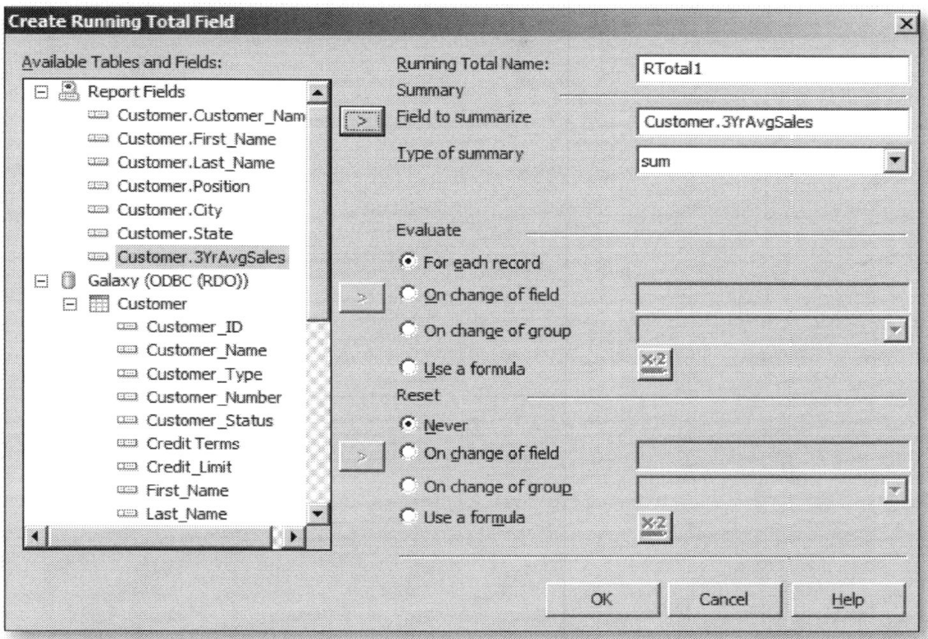

Figure 8.9 Running total options

First you must type a name for your running total in the Running Total Name box. It can be any name you like, as long as it makes sense to you.

Crystal Reports will put the hash symbol (#) in front of your running total name so you can easily identify this field as a running total when it is inserted into your report.

The next step in creating a running total is selecting a field to summarize and then choosing a summary option, as shown in Figure 8.10. To select a field, locate it in the list on the left and then click the top-right arrow to move the field to the text box on the right.

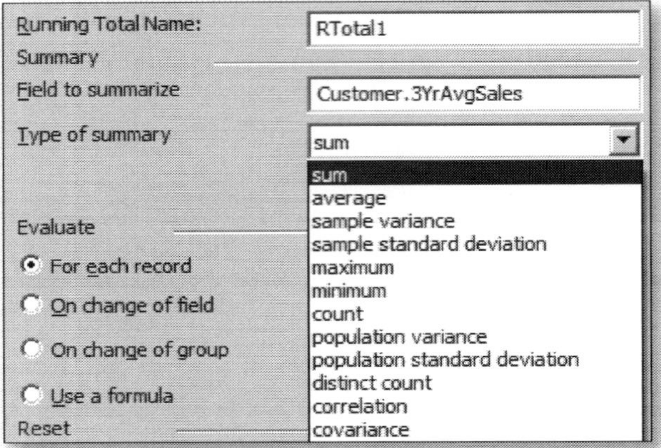

Figure 8.10 You need to select a field to summarize and a summary operator for your running total

From the drop-down list immediately below the summary field, select a summary operator. All of your old favorites are here: Sum, Average, and so on.

In this example you are inserting a running total that will run down the page, so you don't need to worry about the Evaluate and Reset options for the running total. Click OK to accept your changes and return to your report's Design or Preview. You'll notice that your running total field has been inserted into your report in the Details section.

As you created your first running total field, you may have noticed two sections in the Create Running Total Field dialog box, marked Evaluate and Reset. These sections are for setting the options related to when your running total will be evaluated and when the total will be reset.

For evaluation times, you can select a calculation time for your running total from the following options:

- For Each Record
- On Change of Field
- On Change of Group
- Use a Formula

For example, you would want to use these options if you were creating a running total to sum all of the international sales in a report. You could select the Use a Formula option and enter the following criterion:

```
{Customer.Country} <> "USA"
```

The resulting running total field would be evaluated only for those customers who are not in the United States.

Likewise, you can reset your running total field using the following options:

- Never
- On Change of Field
- On Change of Group
- Use a Formula

For example, you could reset the running total for each change of the Country field in your list. By using the Evaluation and Reset options, you can create running total fields for just about any use you can imagine.

To use these options, in most cases you will need to select the option and the corresponding field or group. For the Use a Formula option, you will need to select the option and then click the X+2 button to open the Crystal Reports Formula Editor and enter your criteria.

Just as with record selection, the formula you create here needs to return a Boolean value: either true or false. If the value is true, the record will be evaluated or the running total reset (depending on the option you are working with); likewise, if the condition evaluates to false, the action will not take place.

INSERTING RUNNING TOTALS FOR A GROUP

In addition to setting running totals for a list of fields, you can set running totals for a group, allowing you to quickly summarize the contents of multiple groups in your report. To insert a running total for a group, select the field you want to summarize, right-click the field, and click Insert > Running Total.

Because you are inserting a running total for a group, select the For Each Record Evaluate option, and under Reset, select On Change of Group and select the group you want to use, as shown in Figure 8.11.

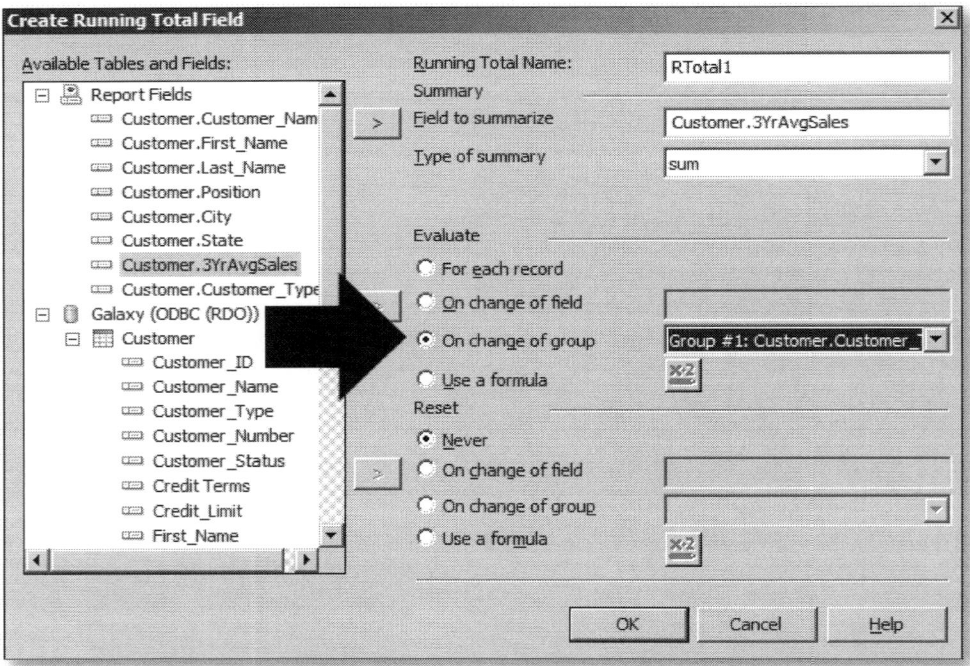

Figure 8.11 An example of creating a running total on a group

Click OK to accept your changes. Your new running total should appear in your report in the group footer of the group you specified earlier.

HIGHLIGHTING YOUR REPORT

Just like a highlighter can be used to mark up a report physically, indicating values to be scrutinized, the Crystal Reports Highlighting Expert can do the same thing by changing the font, background colors, and borders when criteria are met. Created through a simple interface, multiple highlighting criteria can be established to mark problem areas that need to be addressed or data to be reviewed further. You may want to highlight customers whose sales have been under expectations or customers who have had exceptionally large sales in the past year. The criteria you specify is completely up to you.

To use the Highlighting Expert in your report, locate and select the numeric field that you want to highlight. Right-click the field, and from the menu that appears, select Highlighting Expert. The Highlighting Expert opens, shown in Figure 8.12.

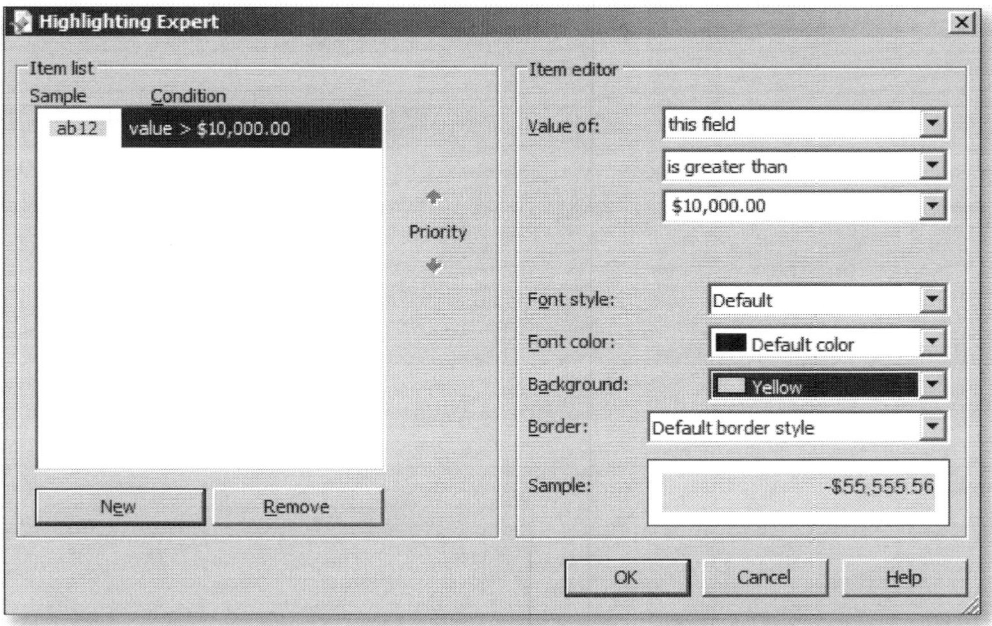

Figure 8.12 The Highlighting Expert

The Highlighting Expert is easy to use. You simply enter a condition on the right side of the dialog box and then specify the formatting options that you want used when that condition is true. If you would like to enter multiple criteria, in the item list click New Item.

To change the order of precedence for highlighting criteria (items), use the up and down arrows. After you have finished entering all of your criteria, click OK to accept your changes. The field that you selected originally should now reflect the options set in the Highlighting Expert.

SUMMARY

With the techniques you have learned in this chapter, you should now be able to add real value to the data that you are reporting from, providing users with summaries that can condense thousands of rows of data into only a few lines. These can then be highlighted based on exceptions that you create.

While summaries are great for adding value to your report data, chances are you don't want to see every bit of data in your database. That is where the next chapter comes in — with the ability to filter your report with record selection, you can get down to just the data you want.

Chapter 9
Record Selection

In This Chapter

- Working with the Select Expert
- Record Selection on Single Values
- Record Selection on Multiple Values
- Record Selection on Date Fields
- Writing Record Selection Formulas
- Complex Record Selection Formulas

INTRODUCTION

When creating a report, chances are good that you don't want to use all of the records that are stored in your database. You may want to use a subset of records for a particular state, region, date range, and so on. The process used to cut down the number of records returned is called record selection. Record selection uses the Crystal Reports formula language to create a logical statement against which records are evaluated. A record selection formula might look something like the following code:

```
{Customer.Country} = "USA"
```

As records are read from the database, this formula is evaluated, and where it is true, those records are returned to Crystal Reports from the database. When the report is printed, you will see only records of customers in the United States.

WORKING WITH THE SELECT EXPERT

If formulas make you squeamish, you are in luck—you don't have to write complex formulas to use record selection (although you can if you want to). Crystal Reports features a Select Expert that will do most of the work for you. The Select Expert, shown in Figure 9.1, is actually a set of specialized dialog boxes designed to help you quickly create record selection criteria without writing your own formula.

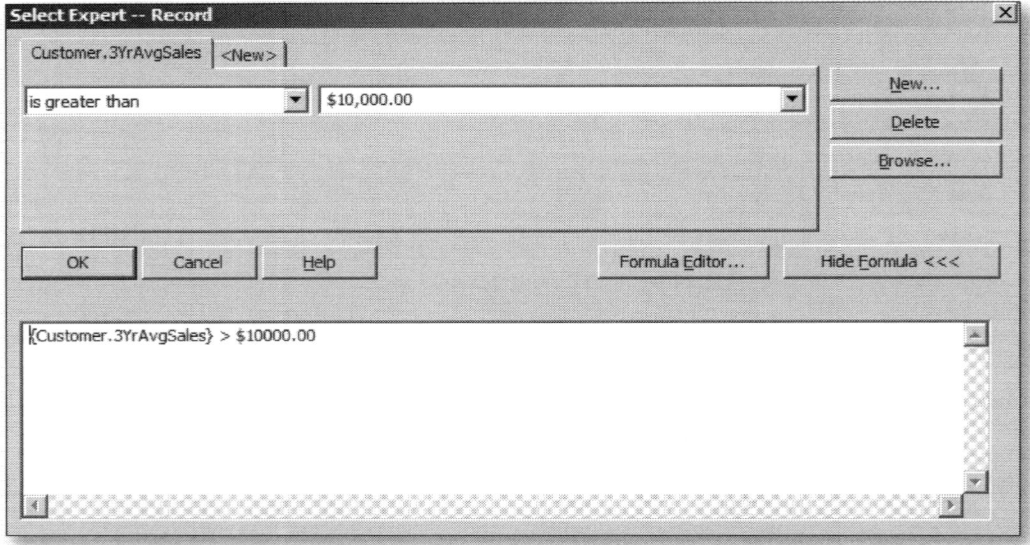

Figure 9.1 The Crystal Reports Select Expert

The Select Expert can be used to apply record selection to a single field or multiple fields with a number of different record-selection operators. Regardless of how many fields or how complex the criteria, the Select Expert can take the choices you make and actually write the selection formula for you. The sections that follow describe some of the most common uses of record selection.

APPLYING RECORD SELECTION TO A SINGLE FIELD

Basing your record selection on a single field is the easiest way to see how the Select Expert works. Our earlier example showed a record-selection formula created on a single field that returned only customers within the United States (`{Customer.Country} = "USA"`). We could enter this formula directly, but we are going to let the Select Expert do the work for us.

To get started, click Report > Select Expert. A list of fields that are available for you to use for record selection opens, as shown in Figure 9.2.

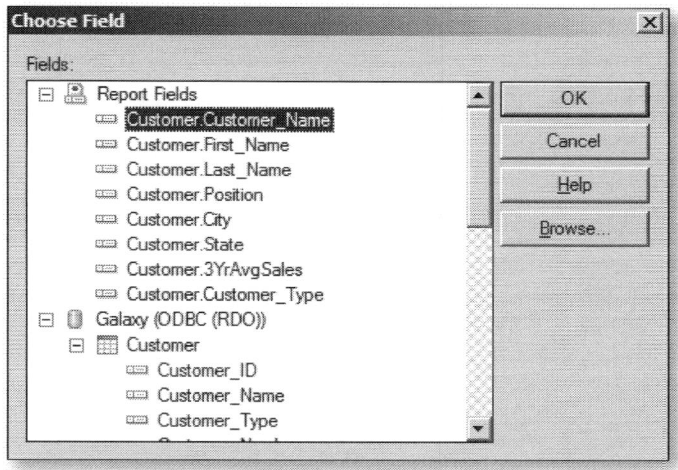

Figure 9.2 You need to choose the field that your record selection will be based on

If you don't see this list of fields, you may have had a field selected before you clicked Report > Select Expert. In this case, Crystal Reports will assume you want to perform record selection using this field.

Select the field that will be used for your record selection criteria, and click OK. Using the dialog box shown in Figure 9.3, select a record selection operator and enter the required criteria.

From our earlier example, we could select the operator Equal To and then type in USA as our criteria. Click OK to accept your changes. Your report will now be filtered for only those customers that are located in the United States.

You can also use the drop-down list to select values directly from the database, but this feature only returns around 200 records from the underlying table, so the value you are looking for may not be there.

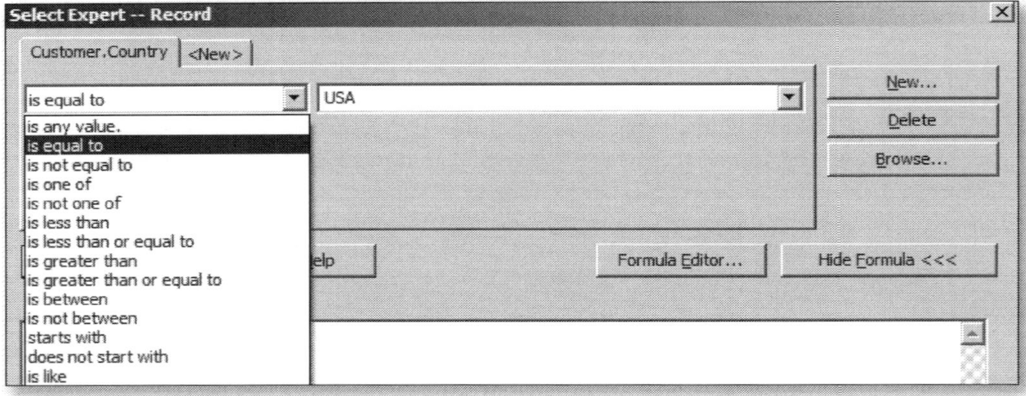

Figure 9.3 The Select Expert

In addition to Equal To there are a number of other record-selection operators that can be used to narrow your report records to display only the records you need. Table 9.1 on the adjacent page describes these operators.

Table 9.1 Record Selection Operators

Operator	Description
Is any value	Default option for record selection, allowing all records to be returned, regardless of value
Is equal to	Looks for an exact match to the criteria entered
Is not equal to	Looks for all records except those matching the criteria specified
Is one of	Is used to build a list of criteria, allowing you to select multiple values from one field (for example, is one of "USA", "Canada", and "Mexico")
Is not one of	Is used to build a list of criteria you don't want (for example, is not one of "Australia", "New Zealand", and "Japan")
Is less than	Brings back any records less than the criteria entered
Is less than or equal to	Brings back any records less than or equal to the criteria entered
Is greater than	Brings back any records greater than the criteria entered
Is greater than or equal to	Brings back any records greater than or equal to the criteria entered
Is between	Used to specify inclusive values as criteria; any records between the two inclusive criteria are returned
Is not between	The opposite of Is between; any records outside the two inclusive criteria are returned
Formula	Used to enter a record selection formula directly without using the Select Expert

APPLYING RECORD SELECTION TO MULTIPLE FIELDS

The Select Expert can also be used to apply selection criteria to multiple fields. To establish criteria for multiple fields, you perform the same process as for a single field: Click Report > Select Expert. Then click the New button, shown in Figure 9.4.

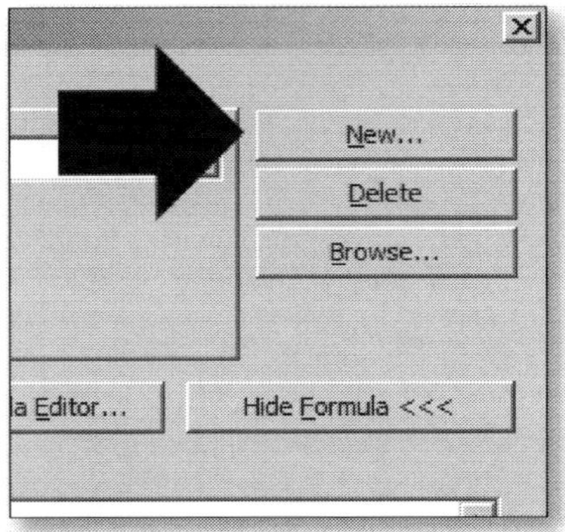

Figure 9.4 You can click the New button to add criteria for multiple fields

After you have clicked the New button, a field list will appear, from which you can choose the second field you want to use in your record selection; then a second tab will appear, and you can specify the operator, values, and so on.

Whenever you use multiple fields for record selection, Crystal Reports makes the relationship between these two criteria an AND relationship (that is, Condition1 and Condition2 must be true for Crystal Reports to return a record).

To delete criteria, click the tab for the field you want to delete and use the Delete button to remove that tab.

USING AN OR STATEMENT IN YOUR RECORD SELECTION FORMULA

When using the Select Expert and using record selection on multiple fields, Crystal Reports assumes that the relationship between these two fields is AND, as shown in Figure 9.5.

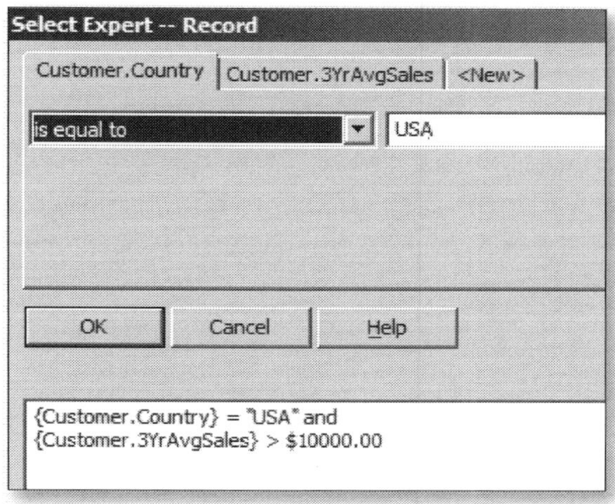

Figure 9.5 Crystal Reports places an AND between multiple fields by default

In this example, the records would be returned to your report only where the Country was equal to the USA AND the company's last year's sales was over 10,000.

If you do need to use the OR operator, you can click the Show Formula button in the bottom-right corner and manually edit the record-selection formula in the box shown.

Once you manually edit the record-selection formula, you may not be able to use the Select Expert to edit the formula. A message may appear stating "Composite Expression: Please use formula editor to do editing." This message indicates that the Select Expert doesn't know how to represent your record-selection formula in the tabbed interface provided.

APPLYING RECORD SELECTION BASED ON DATE FIELDS

For date fields, an additional record selection operator is available, called In the Period. This operator addresses some of the common types of record selection based on dates.

To make your life easier, Crystal Reports has a number of predefined periods that can be used for record selection. These periods each generate their own internal list of dates based on each period's definition and the current date. For example, if you select the period MonthToDate, Crystal Reports builds a list of all of the dates that have passed since the first of the month and subsequently uses that list to select records from the database.

One of the most frequently asked questions about periods is: "Can we add periods to this list or change the definitions?" The answer is no. The periods reflected in the list are hard-coded in Crystal Reports. You can, however, create your own user-defined functions and include similar functionality.

The following periods are available for use when performing record selection on dates:

- WeekToDateFromSun
- MonthToDate
- YearToDate
- Last7Days
- Last4WeeksToSun
- LastFullWeek
- LastFullMonth
- AllDatesToToday
- AllDatesToYesterday
- AllDatesFromToday
- AllDatesFromTomorrow
- Aged0to30days
- Aged31to60days
- Aged61to90days
- Over90Days
- Next30days
- Next31to60days
- Next61to90days
- Next91to365days
- Calender1stQtr
- Calendar2ndQtr
- Calendar3rdQtr
- Calendar4thQtr
- Calendar1stHalf
- Calendar2ndHalf
- LastYearMtd
- LastYearYtd

To use these periods with a date field in your report, click Report > Select Expert. From the Choose Field dialog box, select the date field that you want to use in your record selection, and click OK. The Select Expert opens. Using the first drop-down list, shown in Figure 9.6, select either Is in the Period or Is Not in the Period. A second drop-down list appears with the periods listed.

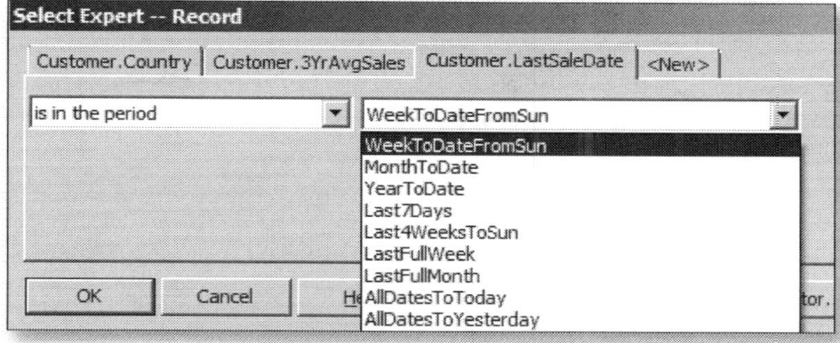

Figure 9.6 When you use "Is in the Period" for record selection, a second drop-down list appears showing all of the available periods

Select the period you want to use for record selection. Click OK to accept your changes to the record selection criteria and to return to your report, which should now reflect the data only from the period you specified.

WRITING RECORD SELECTION FORMULAS

In addition to the Select Expert, Crystal Reports provides a second method of record selection by allowing you to edit the record-selection formula directly. When looking at the Select Expert, you may have noticed a button in the bottom-right corner marked Show Formula, shown in Figure 9.7.

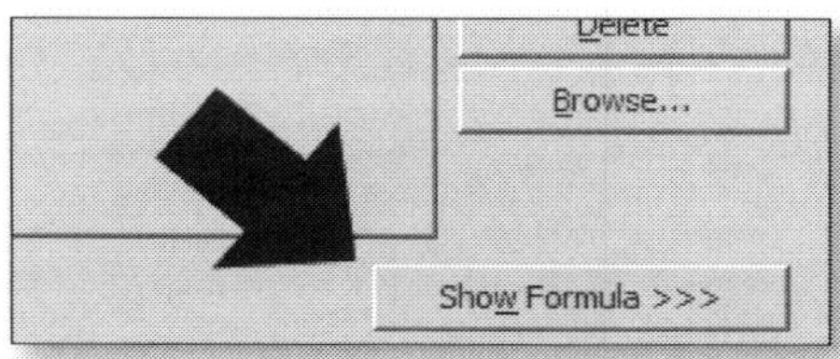

Figure 9.7 You can view the formula the Select Expert has created by clicking the Show Formula button

By looking at the formula that Crystal Reports has created, you can pick up clues about how record-selection formulas work. To write your own selection formula, click Report > Selection Formulas > Record. The Record Selection Formula Editor opens.

You may be warned that you will not be able to keep any drill-down tabs that are open. Click OK to acknowledge this message and to continue writing your record-selection formula.

Using the Formula Editor, shown in Figure 9.8, you can create a record-selection formula that results in a Boolean value (meaning it can be evaluated as either true or false). Use the X+2 button on the toolbar to check the syntax of your record-selection formula. When you have finished working with the selection formula, click the Save button and the close button in the top-left corner to save your formula and close the Formula Editor. Your report should now reflect the record selection criteria you created.

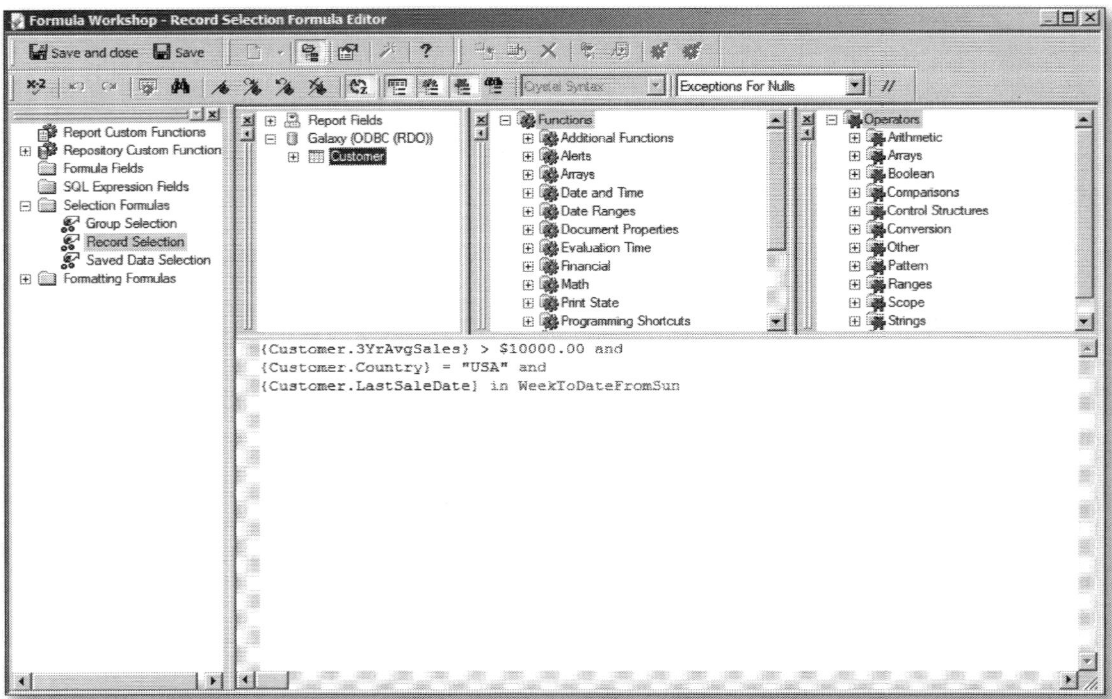

Figure 9.8 The Crystal Reports Formula Editor

In the following sections, we will walk through creating some of the most common types of record-selection formulas.

If all this is just a bit too much, check out Chapter 11 for more information on working with the Crystal Reports formula language and then revisit this section with those new-found formula skills.

SELECTING RECORDS BASED ON DISCRETE VALUES

For creating record-selection formulas based on a single, discrete value, you have a number of operators that are available for your use, and they are pretty easy to use because they match the mathematical operators you are probably already used to working with.

From our earlier example, if you wanted to create a report that returned records only from the United States, you could create a record-selection formula that looks like this:

```
{customer.country} = "USA"
```

In addition to the equals operator, you could also use other operators to retrieve records. For example, if you wanted to retrieve all of the records where the customer's last year's sales were greater than $10,000, the formula would look like this:

```
{customer.last year's sales} > 10000
```

As you have probably already noticed, for comparison against string fields, we place the value in quotes; for numeric fields, no quotes or demarcation is required.

When working with date or date-time fields, we can also use these same operators. For example, to return all database records for purchases made before August 1, the formula would look like this:

```
{Purchases.PurchaseDate} < Date (2003, 08, 01)
```

And if the Purchase Date field were a date-time field, it would look like this:

```
{Orders.Order Date} = DateTime (2003, 08, 01, 00, 00, 00)
```

As you have probably already noticed, using a date field in a record selection formula requires that we convert the values to a true date format, either using the Date or DateTime functions provided by Crystal's formula language.

SELECTING RECORDS BASED ON MULTIPLE VALUES

If you have multiple values that you want to use to select records, you can use a couple of different methods, depending on how the values are arranged.

If the values you want to use are naturally arranged in a range of values (that is, from 1 to 30, A to K, and so on), you can use the `In` operator and compare the database field against a range of values, separated by the keyword `to`.

```
{Customer.Last Year's Sales} in 10000 to 30000
```

And a similar formula using strings:

```
{Customer.Initial} in "A" to "K"
```

For values that don't fall into a range, you can also use the `In` keyword to compare a database field against an array of objects, separated by commands and enclosed in square brackets. For example, if you wanted to show records only for customers in the United States, Canada, and Mexico, the formula would look like this:

```
{Customer.Country} in ["Canada", "Mexico", "USA"]
```

Or for numeric values:

```
{Customer.Last Year's Sales} in [10000, 20000, 30000]
```

To make things really confusing, you can also combine the two methods, using both a range and an array in the same statement:

```
{Customer.Last Year's Sales} in [10000 to 20000,
30000, 40000]
```

As mentioned earlier in the chapter, Crystal Reports includes a number of built-in periods (MonthToDate, YearToDate, etc.) and treats these as arrays, so you can use these in your record-selection formula the same way (that is, `{Order.Order Date} in MonthToDate`**).**

WRITING ADVANCED RECORD SELECTION FORMULAS

As you start to work more with Crystal Reports and record selection, your record selection formulas will get more and more complex. Since record selection relies heavily on the Crystal Reports formula syntax, you can apply what you learn in Chapter 10 (Working with Parameters) and Chapter 11 (Formulas and Functions) to create more complex record selection formulas, like the one shown below:

```
If (not isnull({?Company}) and {?Company} <> "")
and (not isnull({?Month}) and {?Month} <> "")
then (Month({tbl_sales_fact.invoice_date}) = {@
Month}) or (Month({tbl_sales_fact.invoice_date}) <=
{@Month}))
    and Year({tbl_sales_fact.invoice_
date})=Year(CurrentDate)
    and {tbl_company.company_id} = {?Company}
Else If (not isnull({?Company}) and {?Company} <> "")
and (isnull({?Month}) or {?Month} = "") then
   {tbl_company.company_id} = {?Company}
```

While this may look like it is complicated, the basic tenets for record selection are the same — your record selection formula should always return a True/False result, indicating the records you would like to select.

In the example above, the report developer is checking to see if values were passed to various parameter fields, then selecting records based on what the user has selected.

SUMMARY

So with your new found skills in record selection, you should now be able to filter your report to see just the data you want. This is one of the critical skills required to turn your data into decision making information. In the next chapter, we'll look at how to extend this even further with the use of parameter fields. Using parameter fields, you can create one report that can be used for many different uses, allowing the user to select what data they want to see in the report, as well as control report features and choose the level of detail they want to see.

Chapter 10
Working with Parameter Fields

In This Chapter

- Using Parameter Fields in Your Report
- Working with Static Parameter Fields
- Working with Dynamic Parameter Fields
- Introducing Interactive Parameters
- Parameter Display Options
- Parameter Fields and Record Selection

INTRODUCTION

One of the goals of a good report design is that a single report should be able to deliver information to a number of people, eliminating the need for multiple reports that basically show the same information. One method we have for creating reports that fit many different types of situations is through the use of Parameter Fields.

In the following sections, you will learn about Parameter Fields, how they are created, and how they can be used to enhance your report's design and usefulness.

Parameter fields are used in reports to prompt a user for information using either the standard parameter dialog or through the new parameter panel that appears on the left-hand side of the report viewer.

Parameter fields are just like any other fields in use in your report and can be displayed in your report, used in record selection, and more. With the release of Crystal Reports 2008, parameter fields have been enhanced to be even more powerful and can add some very useful features and functionality to your reports (even existing reports you may have created in a previous version.

UNDERSTANDING PARAMETER FIELDS

Parameters are used to prompt users for information required for a report — below in Figure 10.1 is a typical parameter dialog, prompting the end-user to enter a Country.

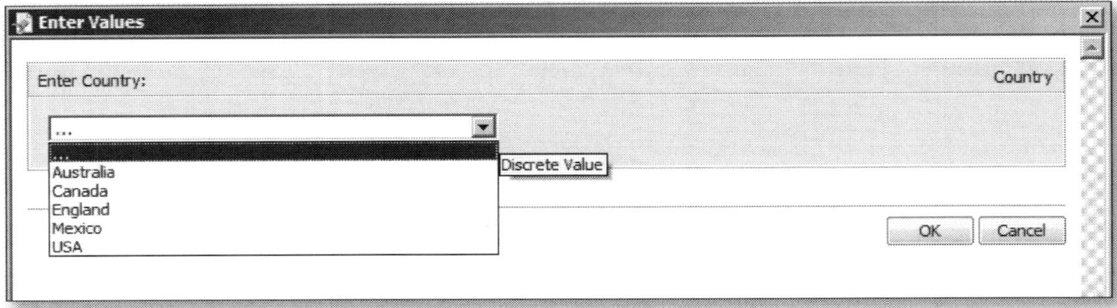

Figure 10.1 Typical parameter dialog

The parameter value which is entered can then be used anywhere in the report — for example, they may want to use this value to filter the report or display the value on the report itself.

When creating a parameter field, you will need to choose a type for the field, based on how you plan to use the parameter field later in your report. The following types are available for parameter fields:

- Boolean
- Currency
- Date
- DateTime
- Number
- String
- Time

Each of these types have specific attributes — for example, a boolean parameter field will only accept a "True/False" or "Yes/No" value, while a date or datetime parameter field will expect a date or date-time respectively. Another key decision on how to create your parameter field is based on the type of list of values you want to create. Crystal Reports actually supports two different types of parameter fields:

Static Parameters — Static parameters can display a list of values that are either entered when you create or edit the parameter or appended from your data source. These values are static and are stored within the report itself.

Dynamic Parameters — These parameters can be associated with your data source, so when you add a new value in the database, it will be added to the parameter pick list. Dynamic parameters can also be cascading, where one value determines what appears in the next list of parameter values.

In the following sections, we will cover parameters with both static and dynamic list of values, as well as how they can be used with record selection to filter your report's results. This is the most common usage of parameter fields, but they can also be used to control report features, which we will also look at a little later in the chapter.

WORKING WITH STATIC PARAMETER FIELDS

Static parameters are good for values that are not set to change in the near future — for example, a "State" field is a good choice for a static parameter value, as the list of states won't change any time soon. To create a static parameter field to be used in your report, click View > Field Explorer. The Crystal Reports Field Explorer opens, as shown in Figure 10.2.

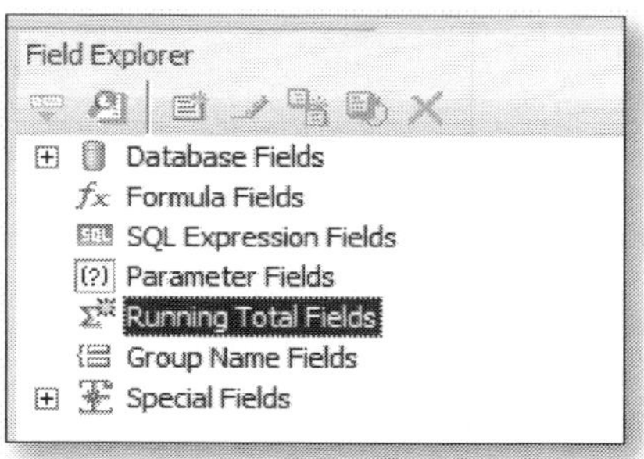

Figure 10.2 The Crystal Reports Field Explorer

Right-click the section of the Field Explorer labeled Parameter Fields, and select New from the shortcut menu. Using the dialog box shown in Figure 10.3 on the following page, type a name for your parameter field. In our example, we have named the parameter field Employee.

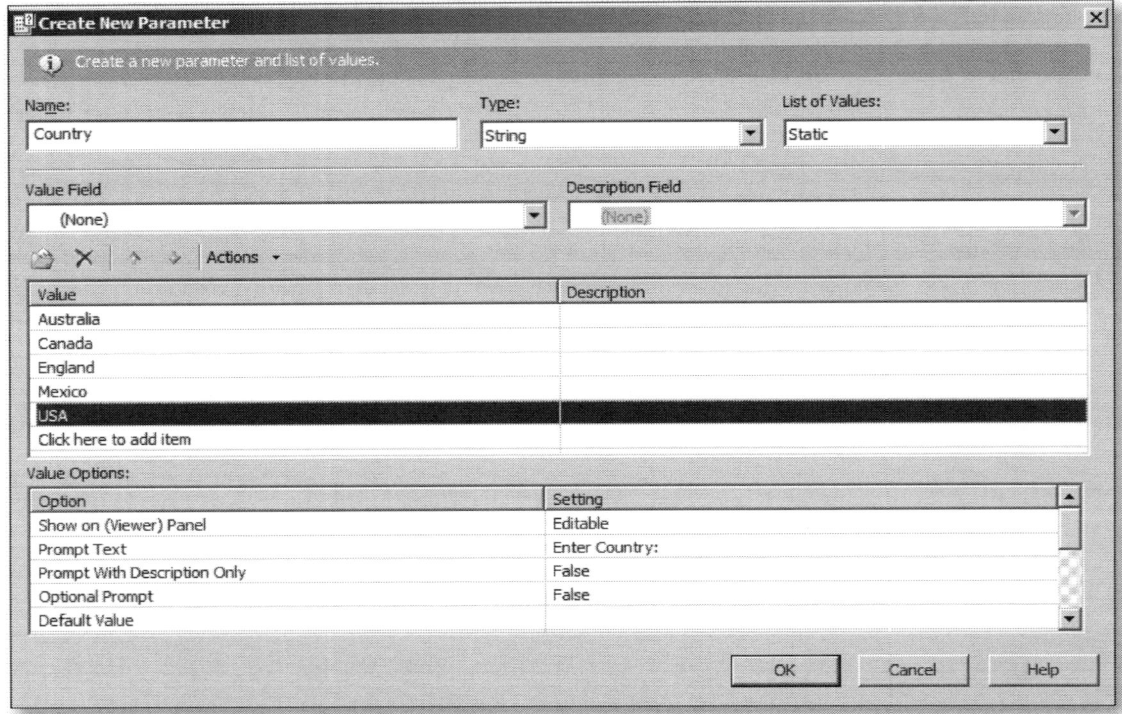

Figure 10.3 You can create a parameter field using this simple dialog box

Next, using the combo-box labeled Value Type, select a data type for your parameter. To create a static parameter field, the two required items are the name and type, so you can click OK to accept your changes and return to your report. Your parameter field should now appear in the Field Explorer, ready to be used on your report.

Inserting a Parameter Field on Your Report

To insert a parameter field that you have created, you can simply drag it from the Field Explorer onto your report's Design or Preview.

It is usually easier to place fields on your report if you use the Design view of the report.

Once you have dragged a parameter field into your report, you will be prompted to enter a value for the parameter the next time the report is previewed, as shown in Figure 10.4.

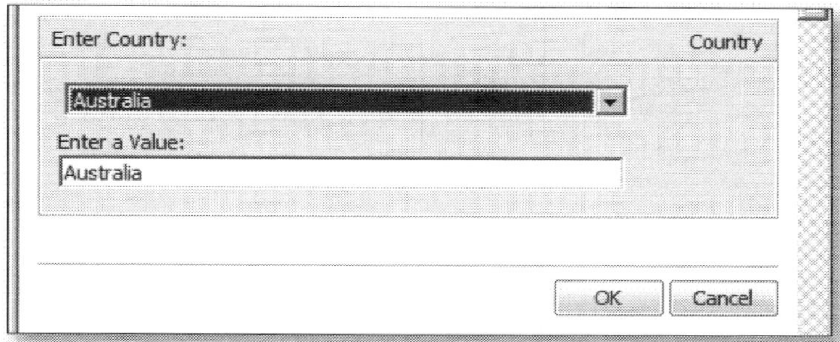

Figure 10.4 A typical Parameter prompt dialog box

After you have entered a value for your parameter field, that value will be displayed in the report preview until you refresh your report and specify that you want to prompt for a new parameter value.

SETTING DEFAULT PARAMETER VALUES

An easy way to help users complete parameter field prompts for static parameters is to give them some default values from which to choose. A list of default parameter values can be read from your database or entered manually, giving the users a list of values. An important concept when working with default parameter values is that this is a manual process with static parameters and can occur only when designing the report.

If you want to enter a list of values, you can click the Insert button in the Parameter dialog box and then enter a value and/or description in the grid area below.

If you would like to follow along with the step-by-step instructions in this chapter, this book includes downloadable files that are available from www.kuiperpublishing.com. Also, check out "Setting up the Samples" section in the front of this book for instructions on how to configure the data sources, reports, etc.

If you would prefer to read these values from your data source, use the following steps:

1. Open Crystal Reports, and open the EMPLOYEEBYDEPARTMENT.RPT report from the download files.
2. Click View > Field Explorer to open a list of fields in your report.
3. Right-click the Parameter Fields heading, and select New from the shortcut menu. The Create New Parameter dialog box opens, shown in Figure 10.5.

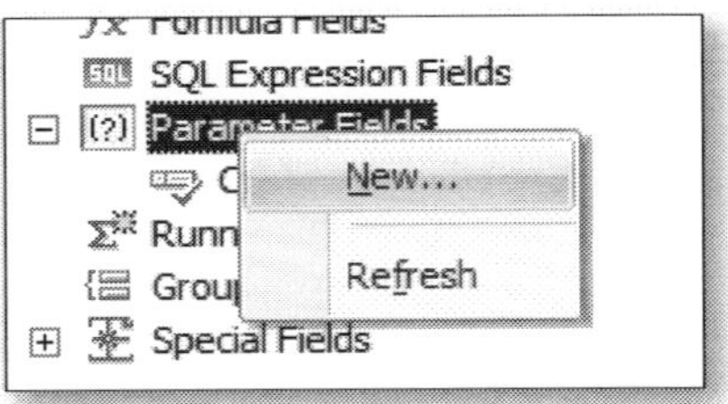

Figure 10.5 The Create Parameter dialog box

4. In the Name field, enter Department, and select a type of Number from the drop-down list on the right side — for the "List of Values" option, select Static.
5. Next, use the Value drop-down list to select the Department field.
6. Click the Actions button, and select Append All Database values.
7. Click OK to return to your report.
8. Drag the Department parameter field into the Page Header of your report, and then preview the report.

The parameter you just created should now prompt you to use the list you just created, as shown on the next page in Figure 10.6.

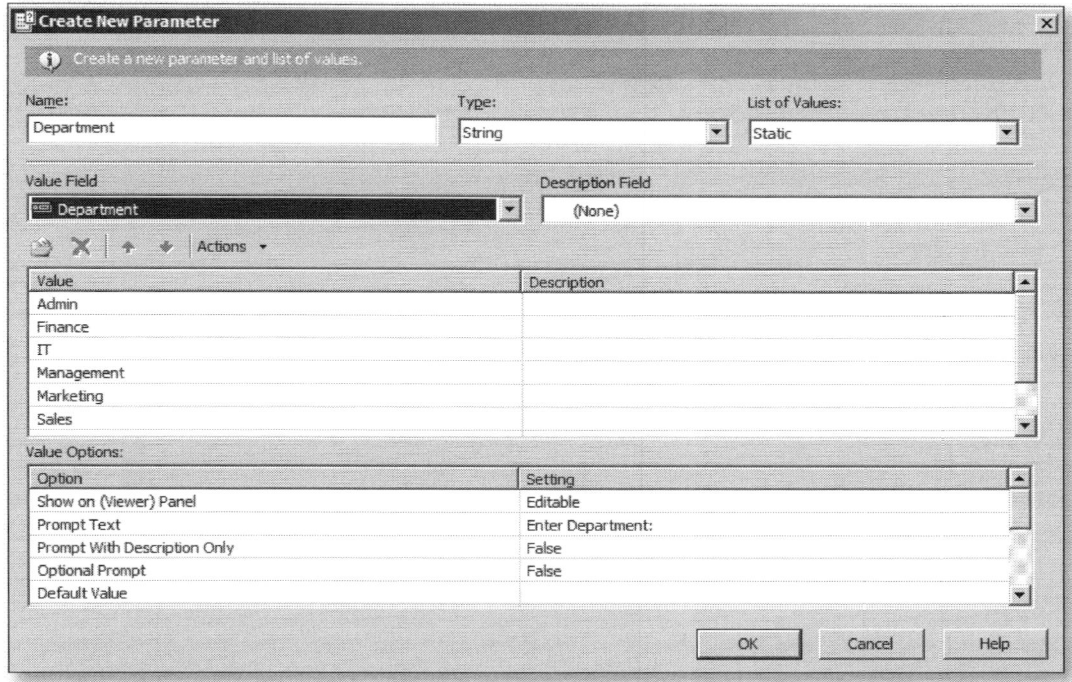

Figure 10.6 The parameter dialog should now show the default values you added

Now, the question is "Why didn't the parameter filter my report?!" The answer is actually quite simple — all we just did was create a parameter field. But we didn't link it to your record selection. So while the parameter field appears on the report, it is not actually filtering the report data, so you can still see all of the departments.

So to use this parameter to filter your report, there is one additional task. Using the same report you have been working on, follow these steps:

1. Select Report > Select Expert > Record.
2. Select the Department field and click OK.
3. In the Record Selection dialog, select Department, the operator "Is Equal To" and then use the drop-down list to select the field "{?Department}" and click OK.

You will be prompted for a parameter value and your report should now be filtered to show only the department you selected.

There are actually some pretty neat tricks we can perform with parameter fields and record selection — that is why this topic gets its own section towards the end of the chapter.

Sorting Parameter Field Default Values

To make looking through these default values a little easier, we can sort the contents of the drop-down lists of values that the user sees. When working with the default values, click the header of either the Value or Description columns in the grid to sort the items to be in either ascending or descending order. This sort order will be reflected in the drop-down list when the user selects a value. With a sort order in place, users will have an easier time selecting the values they want.

Importing/Exporting a Parameter Field Pick List

When working with static parameter fields and default values, you may have multiple reports that use the same default values and descriptions. If you entered these values manually, retyping these values and descriptions in each time you want to use them is time consuming, so Crystal Reports has an easy way to work around this problem. If you frequently use the same parameter fields, you can export and import field pick lists, eliminating the need to establish default values and descriptions each time. The pick list files themselves are simply text files and can be used with any report you create.

To import/export a parameter field pick list, use the following steps:

1. Open the report we have been working on in this section.
2. Click View > Field Explorer.
3. Right-click the Department parameter field, and select Edit to open the Edit Parameter dialog box. Under the Actions button, a drop-down list contains the following three options for working with pick lists:

Clear — Removes all default values from the list.

Import — Allows you to import a text file with default values.

Export — Allows you to export a text file with all the values currently listed in this dialog box.

4. Select Export to export the default values for use another time.
5. Click OK to return to your report design.

You can then use Notepad to open the file to verify that the process ran correctly and check the format of the file, shown here in Figure 10.7.

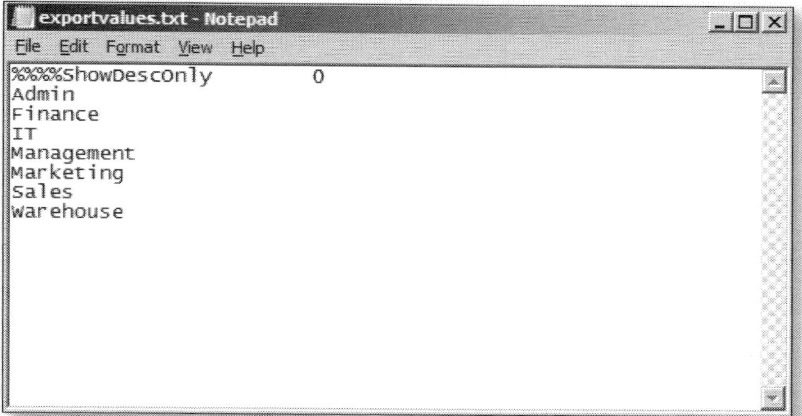

Figure 10.7 You can export pick list values for use in other reports

---

Alternatively, you could also use the Import Pick List button to import an existing pick list as long as the text file you were importing was in the same format.

CUSTOMIZING PARAMETER PROMPTS

Although users are prompted for parameters using a standard dialog box, you do have some control over how the parameter dialog box appears. On the bottom of the New Parameter or Edit Parameter dialog box shown in Figure 10.8, you will find the options you can use to customize the prompt that users see.

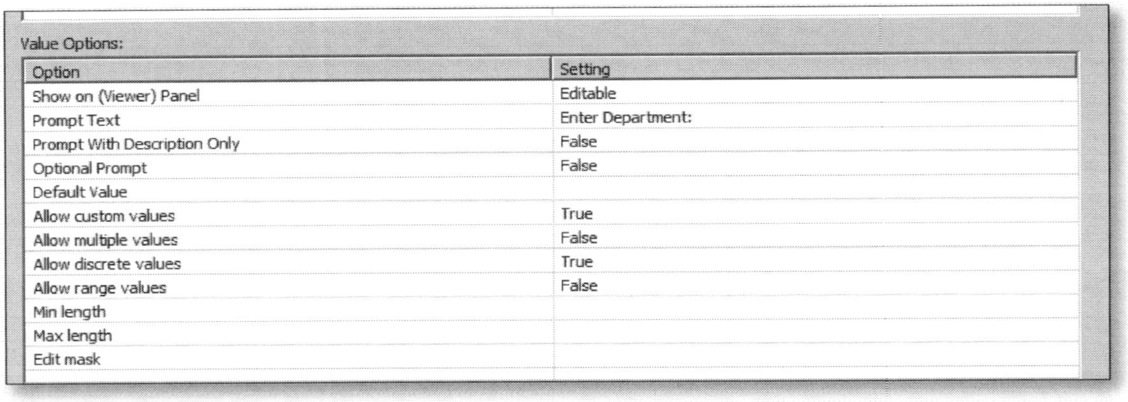

Value Options:

Option	Setting
Show on (Viewer) Panel	Editable
Prompt Text	Enter Department:
Prompt With Description Only	False
Optional Prompt	False
Default Value	
Allow custom values	True
Allow multiple values	False
Allow discrete values	True
Allow range values	False
Min length	
Max length	
Edit mask	

Figure 10.8 Options for customizing parameter prompts

These options include:

Show on (Viewer) Panel — For determining whether or not the parameter field can be viewed or edited through the parameter panel.

Prompt Text — For entering text that will appear to instruct the user on what values to enter.

Prompt With Description Only — For prompting the user with the description only (rather than show the description and value, for example, CA — California).

Default Value — For setting the default value for a field.

Allow Custom Value — For allowing the user to enter his own custom value in the parameter.

There are also a number of other options that you can use with static parameters, but these are a bit more involved, so in the following sections we will look at these options in-depth.

CREATING A MULTIPLE-VALUE PARAMETER FIELD

In addition to single and range values, parameter fields can be created that enable users to enter multiple values. What this means for you is that you can create one parameter field that can accept from one value to as many as you would like. You could create a parameter field for Country, for example and then let users pick for which countries they want to run the report. From one country to three to 30 — it is their choice.

The only downside of using a multiple-value parameter field is that the values you enter can't be displayed on your report, which makes sense because there isn't enough room in the field to display the extra values.

To create a multiple-value parameter field, create a parameter field as you normally would, but this time, change the option Allow multiple values to True. Click OK to accept your changes and return to the Design or Preview of your report. When users are prompted for a parameter field, a dialog box similar to the one shown in Figure 10.9 will appear and allow them to build a list of values.

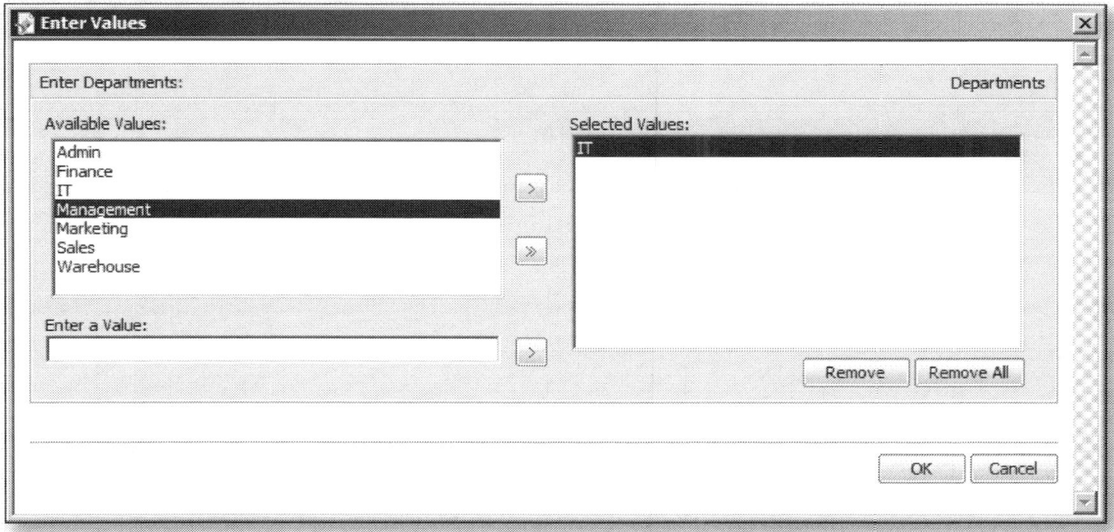

Figure 10.9 Users will be able to enter multiple parameter-field values using this dialog box

Limiting Parameter Input to a Range of Values

When using parameter fields, we can give users the option of entering a start value and end value, allowing them to use this field in record selection. Again, insert a parameter field, as you normally would, but this time use the drop-down list under the Options section of the dialog box to change Allow range values to True. When your parameter is inserted into your report or used with record selection and this setting is in effect, the dialog box shown in Figure 10.10 will prompt you for a range of values.

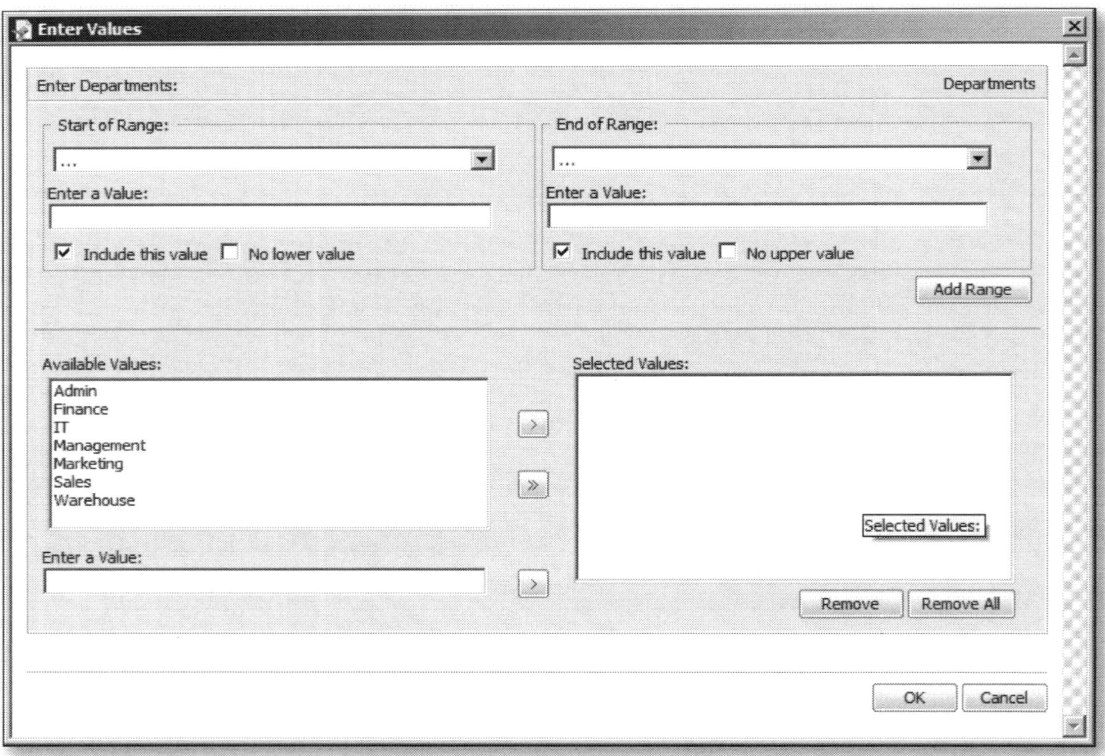

Figure 10.10 You can define a parameter to accept a range of values, a discrete value, or both

A little later in the chapter, we will talk about how to use this type of field with record selection.

USING EDIT MASKS TO CONTROL INPUT

With some parameters, you may want to force the user to enter the values in a specific format. To help with these, Crystal Reports parameters can be used with Edit Masks to force the user to enter the values in a specific format. The table below lists the most frequently used masking characters that can be used to mask parameter input:

Character	Description	Requires Value
A	Alphanumeric character	Yes
a	Alphanumeric character	No
0	Digit (0 to 9)	Yes
9	Digit (0 to 9)	No
#	Digit, Space, Plus/Minus Sign	No
L	Letter only	Yes
?	Letter only	No
&	Any character or space	Yes
C	Any character or space	No
Password	Causes characters to be displayed as a password	No

To use these masking characters, create or edit a parameter field and ensure that the field type is set to "String". When the parameter field type is set to string, an additional property called "Edit Mask" is displayed at the bottom of the add/edit parameter dialog, as shown below in Figure 10.11.

Figure 10.11 Edit Mask options

Enter your masking characters using this dialog, in the format you would like users to respond to your prompts. For example, to require the user to enter a phone number with an area code, the masking characters would be:

```
(999) 999-9999
```

Alternately, if this was not required, you could use the "0" character, as shown below:

```
(000) 000-000
```

When the report was refreshed, the parameter dialog would display this edit mask and the user would be required to fill in the values in the format requested, as shown below in Figure 10.12.

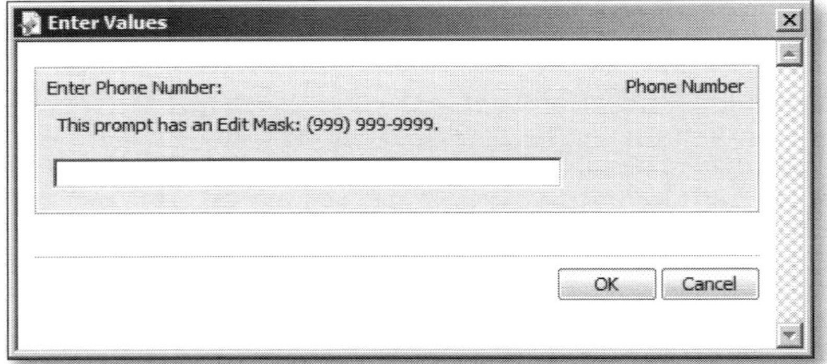

Figure 10.12 A parameter with an edit mask in place

You can use this technique to ensure that whatever your user inputs conforms to the pattern you have set — this should help eliminate any issues when using this parameter value in the rest of your report, as you will know the exact format in which it has been entered.

WORKING WITH DYNAMIC PARAMETERS

Dynamic parameters are a handy tool in every report developers tool set. Using dynamic parameters, you can tie a parameter field to a database table, view, or store procedure and use the data to drive the pick list that is presented to users. For example, if you were to set up a parameter to prompt a user for a customer number, using a dynamic parameter would ensure that every new customer that is added to the database would appear in the parameter pick list.

The downside to dynamic parameters is that this can add to the processing overhead of the report. In this section we will look at how to use dynamic parameters in your reports as well as how to use SQL commands to minimize the amount of processing required to generate parameter pick lists.

But before we can get into that, we must look at how to create a dynamic parameter field. To get started, use the following steps:

1. Open Crystal Reports, and open the SALESBYCOUNTRY.RPT report from the download files.
2. Click View > Field Explorer to open a list of fields in your report.
3. Right-click the Parameter Fields heading, and select New from the shortcut menu.
4. Enter a name for your parameter (in this case, Country), and select a type of String.
5. Use the radio button to select Dynamic, which will change the dialog box to be like the one shown in Figure 10.13.

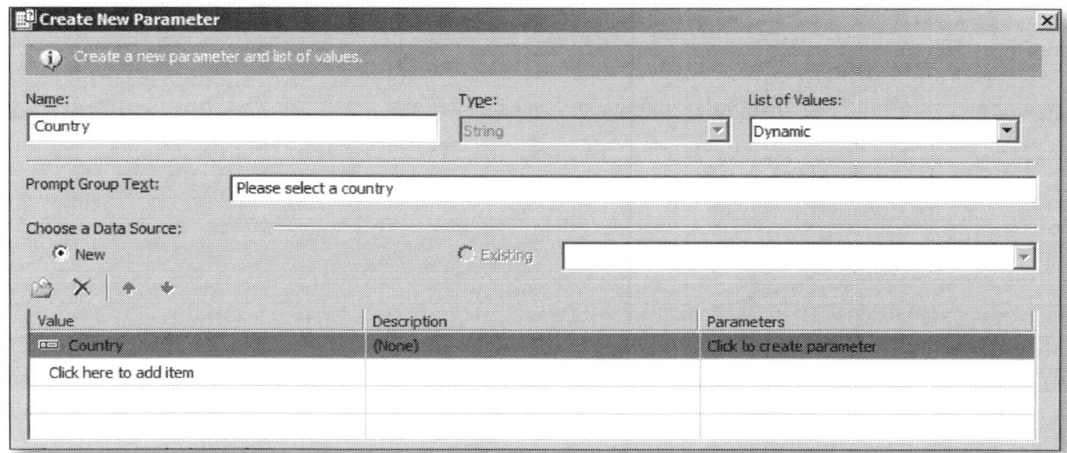

Figure 10.13 Dynamic parameter options

6. Click the link marked Click here to add item, and select the Country field.
7. Next, enter any text that you want to appear when the prompt appears. In this example, enter the text, Please select a country from the list.
8. Click OK to return to your report.
9. Drag the parameter field you just created into your report. A dialog box should appear and prompt you to select a value from a drop-down list.

The list of values that appear in this list is dynamic and will be refreshed from the database each time a user runs the report and is prompted for a parameter.

USING DYNAMIC PARAMETERS WITH CASCADING VALUES

Another way of letting users select values is through the use of cascading parameters. Like water cascades over a waterfall, down one rock to another, so do cascading parameter fields. The selection you make in the first field is used to filter the second field, then the third, and so forth. An example of a cascading parameter field is shown in Figure 10.14.

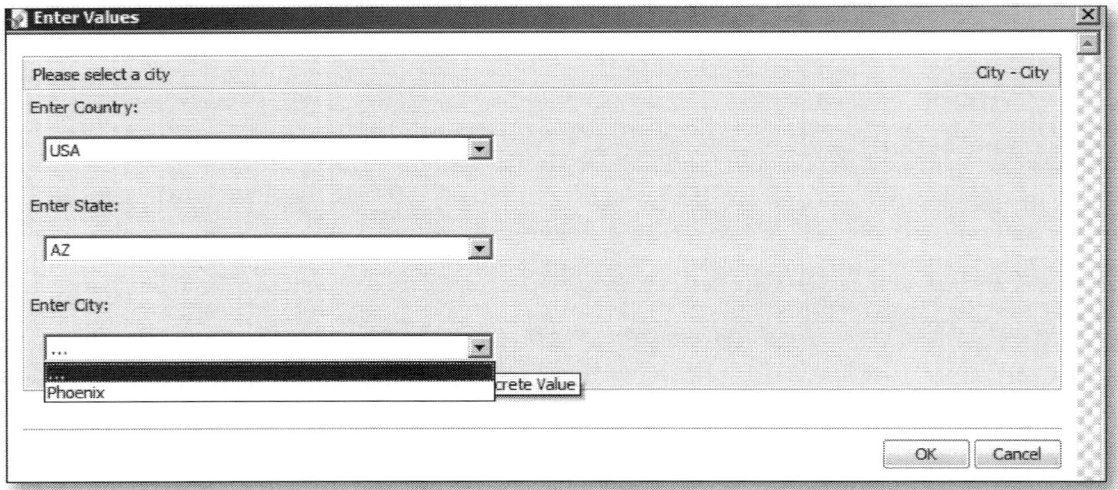

Figure 10.14 An example of a cascading parameter

In this example, the user will select a Country field, which in turn filters the Region field to show only the regions related to that country. Then the user selects a Region field, which then filters the city list to only those cities that exist within that region.

You can easily apply this same technique to your own reports. In the following example we are going to re-create this scenario using a report based on one of our customer reports. To create a cascading parameter, use the following steps:

1. Open Crystal Reports, and open the REGIONALSALES.RPT report from the download files.
2. Click View > Field Explorer to open a list of fields in your report.
3. Right-click the Parameter Fields heading, and select New from the shortcut menu.
4. Enter a name for your parameter (in this case, City), and select a type of String.
5. Use the radio button to select Dynamic.
6. Click the link marked Click here to add item, and use the drop-down list to select the Country field.
7. On the next line, click the same line and use the drop-down list to add the Region field.
8. Finally, on the line below, click the same link and use the drop-down list to select the City field. Your parameter field dialog box should now look like the one shown in Figure 10.15 over the page.

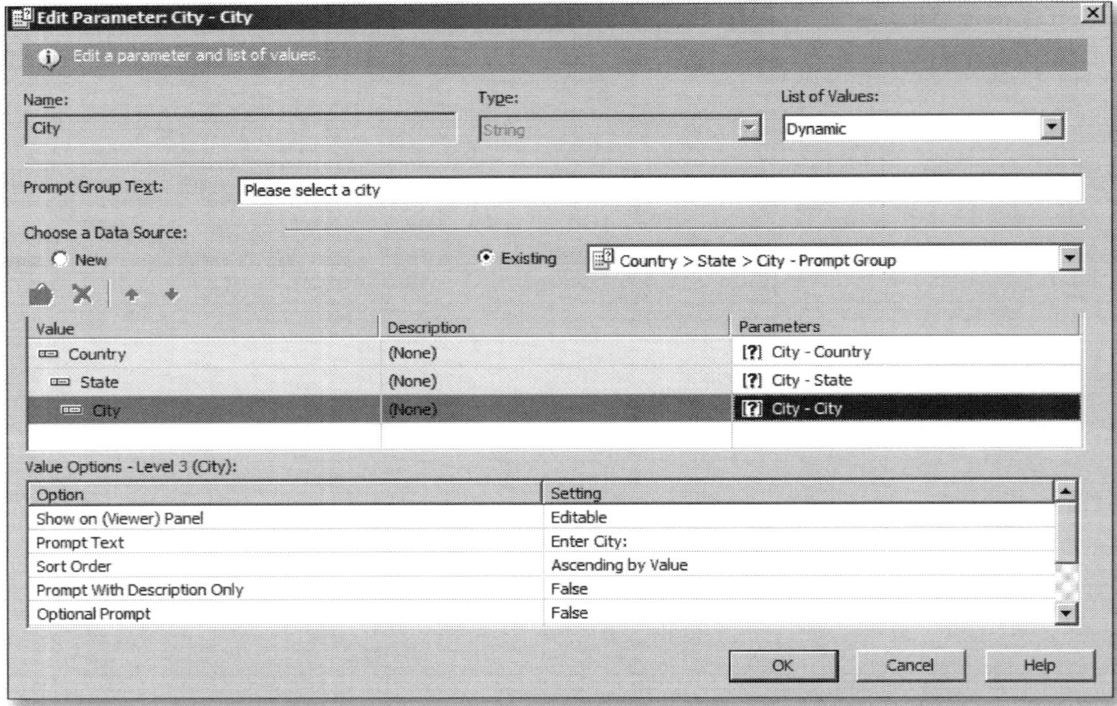

Figure 10.15 Settings for a dynamic, cascading parameter based on the city field

9. Click OK to return to your report.
10. Drag the parameter field you just created into your report design.

A dialog box should now appear and prompt you for the Country, Region, and City fields. As you make a selection in one drop-down list, the selections in the other lists will be filtered based on what you selected.

Splitting Cascading Parameters into Separate Fields

There will be times when you want to separate a cascading parameter into multiple fields. In our previous example, you may want to separate the parameter fields so you can use the values the user selected for the Country, Region, and City parameters.

To split the parameter fields into separate fields, use the following steps:

1. Open Crystal Reports, and open the cascading parameter report you were just working on.
2. Click View > Field Explorer to open a list of fields in your report.
3. Expand the Parameter Fields section, right-click the City parameter, and select Edit from the shortcut menu.
4. From the Edit Parameter dialog box, locate the grid that displays the value, description, and parameters field, and click to create Parameter on the Customer.Country field.
5. Repeat the same operation on the Customer.Region field.
6. Click OK to return to your report design.

Your report will now show the three parameters separately in the Field Explorer, as shown in Figure 10.16.

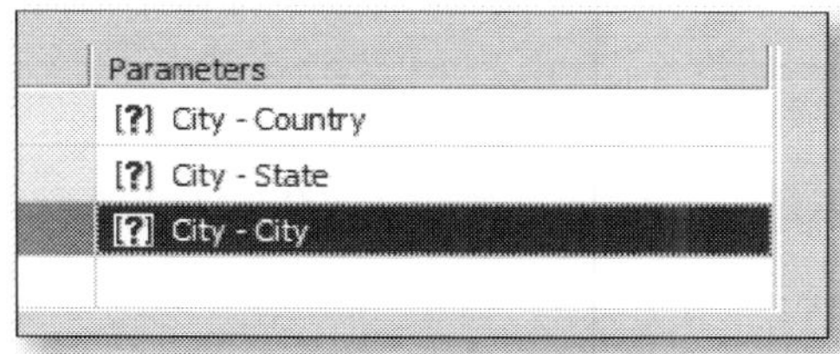

Figure 10.16 The parameters split into separate fields

You can use these parameter fields independently, but they are still part of a Parameter Group, so whenever you edit one of these fields, the entire prompt group will be shown in the Edit Parameter dialog box.

USING COMMAND OBJECTS FOR DYNAMIC PARAMETERS

One way to cut down on the processing overhead required for dynamic parameters is to use an SQL command. We haven't covered SQL commands yet, but they can be used to write an SQL statement that is used as the basis of your report. We will look at this type of data source in-depth in the next chapter.

SQL commands for dynamic parameters are used to ensure that only the required amount of data is returned. For example, if we were to base a dynamic parameter for the user to enter a Country, and we used the customer table in our sample database, it would not be the most efficient way of populating the pick list.

The customer table contains thousands of customer records, and some of these records repeat the same country. A much better technique would be to use a specialized lookup table (that is, a table that listed each country only once) or use an SQL command to bring back only a distinct list of countries from the customer table. To use this technique, follow these steps:

1. Open Crystal Reports, and open the ACTIVECUSTOMERSALES.RPT report from the download files.
2. Click Database > Database Expert.
3. Click Create New Connection > ODBC RDO to open a list of available ODBC data sources.
4. Select the Galaxy option, and click Finish.
5. Below this data source double-click the Add Command option. The dialog box shown in Figure 10.17 opens.

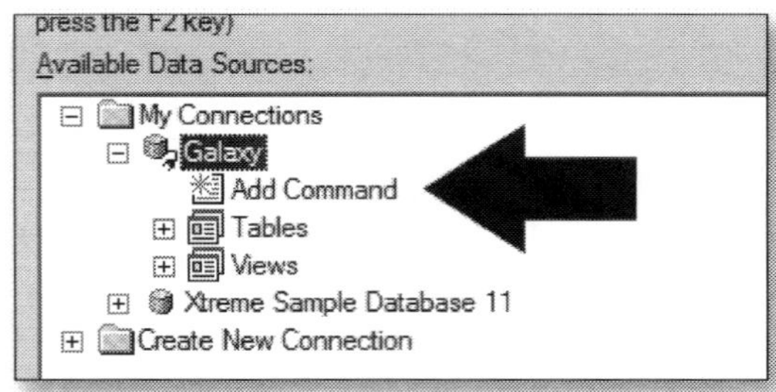

Figure 10.17 The Add Command dialog box

6. Enter the following SQL statement in the text box provided: SELECT DISTINCT COUNTRY FROM CUSTOMER.
7. Click OK to return to the Database Expert, and then click OK twice to return to your report.
8. Next, click View > Field Explorer to open a list of fields in your report.
9. Right-click the Parameter Fields heading, and select New from the shortcut menu.
10. Enter a name for your parameter (in this case, Country), and select a type of String.
11. Use the radio button to select Dynamic.
12. Click the link marked Click here to add item, and use the drop-down list to select the Country field from the Command you entered earlier.
13. Click OK to return to your report design.
14. Drag the parameter field from the list into your report.

Now when the parameter dialog box appears, the SQL command will be used to populate the pick list instead of querying the table underlying the report.

PARAMETER FIELDS & RECORD SELECTION

Using parameter fields for record selection is a popular way to give users more control over the report at runtime. Setting up this functionality is quick and easy — there are only two steps involved.

The first is to actually create the parameter field used to prompt the user for information using the techniques learned earlier in the chapter. You will want to make sure that this parameter field is created with the same type as the field you want to use for record selection. (It will do you no good to create a string-type parameter field when you want the user to enter an invoice number.)

The second step is to set your record-selection formula using this parameter field. In its most simple form, your record-selection formula might look something like this:

```
{Customer.Country} = {?EnterCountry}
```

where `{?EnterCountry}` is the name of the parameter field you have created.

If formulas are not your thing, you can use the Select Expert to do most of the work for you.

In this example, whenever the report is run, a dialog box will open and ask for a country to be entered, which we will assume is entered as USA. Once this data entry has occurred, the record-selection formula will replace the parameter name with the actual value and make a request to the database to retrieve the correct records for companies in the United States.

To make things a little more complicated, parameter fields can also be created with the ability to enter multiple values. In this case, the formula looks exactly the same:

```
{Customer.Country} = {?EnterCountry}
```

When the dialog box opens and prompts for information to be entered for this Parameter Field, as shown in Figure 10.18, you can pick a list of values to be used — in this case we will assume "USA", "Canada", and "Mexico". These values are then used in the record-selection formula, and the appropriate records are returned.

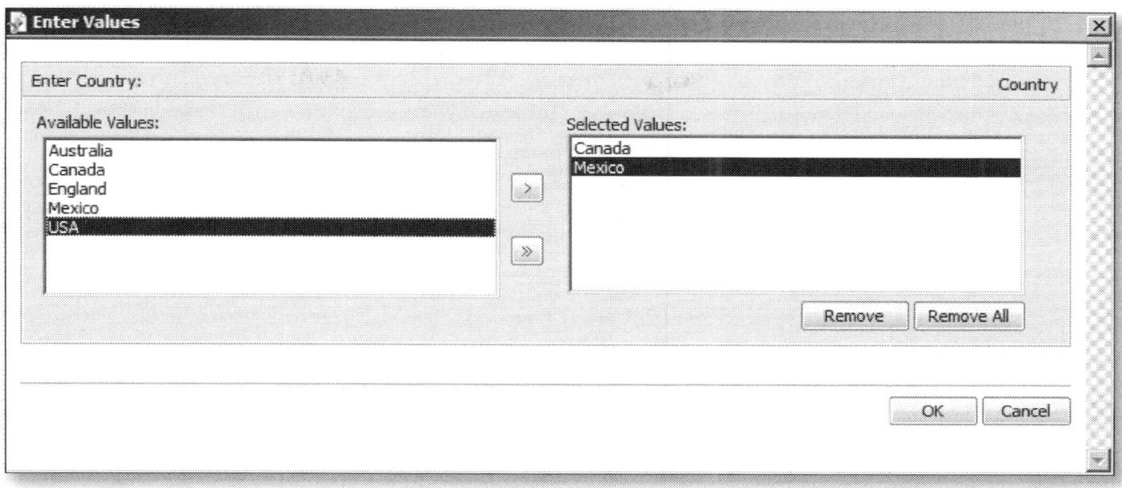

Figure 10.18 You can enter multiple values for parameters using this dialog box

This is often confusing to new report developers, because the proper record-selection operator for multiple values is "One Of". Crystal Reports stores these multiple values in memory and is smart enough to make the translation when processing occurs. Unfortunately, as was mentioned earlier, we are unable to display the contents of this array on our report. It can be used only for record selection.

The same concept also applies when we specify that a parameter field can accept a range of values. You will be prompted for a start value and end value, but Crystal Reports stores these values in its own internal memory and allows you to use the same record-selection formula of:

```
{Customer.Country} = {?EnterCountry}
```

That said, it does not mean that Is equal to is the only record-selection operator you can use with parameter fields. Because parameter fields are treated just like any other field, you can use any record-selection operator or logic to achieve the desired results.

For example, if you wanted to create two parameters in a sales report where you created one parameter field for a StartDate and another one for an EndDate, you could use these in your record-selection formula just like you would any other values, as follows:

```
{Orders.Order Date} in {?StartDate} to {?EndDate}
```

USING A PARAMETER FIELD IN SIMPLE RECORD SELECTION

Parameter fields can be used in a number of different ways, but the most common use of parameter fields is in conjunction with record selection, prompting users for a value that will be used to narrow the results of a particular report.

To use a parameter field with record selection, determine which field from your database you want to use for record selection, and note the type and length of the field. You will then need to create a parameter field with the same field type as your database field. To use this field with record selection, click Report > Select Expert. A list of fields will appear. Choose the field from your database that you want to use for record selection. The Select Expert opens and allow you to choose a record-selection operator (Equal to, Is Not equal to, Is one of, and so on).

Once you have selected an operator, a second dialog box opens, as shown in Figure 10.19, which will allow you to select or enter a value to be used in your record selection. Use the drop-down list to locate and select the parameter field you have created.

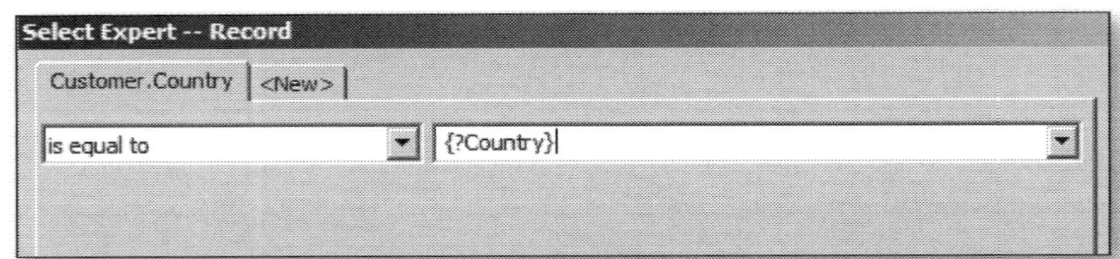

Figure 10.19 A second drop-down list should appear and allow you to select the parameter field you have created

Click OK to accept your changes to your report's record selection and return to the report Design or Preview. When you next preview or refresh your report, a dialog box will open, prompting you for the parameter field you created. Once you enter this value, it will be used in the record-selection formula and subsequently passed to the database.

DISPLAYING PARAMETER FIELDS

To display a parameter field on your report, you can drag and drop the field from the Field Explorer directly on to your report canvas. In addition, you can combine a parameter field with text fields, use them in formulas, etc. to give you more control over how the field is displayed. For example, in this formula, the parameter value is concatenated with some introductory text:

```
"Show Ex-Employees in Report: " & {?ShowExEmps}
```

If you try this technique, you may notice that it works well with single-value parameters, but when you try this technique with a multi-value parameter, you only see the first value that has been selected.

Luckily, we can write a formula to help us display these values. Any values that are entered in a multi-value parameter are actually stored as an array, so we can use the Join function to join these values back together to be displayed. The Join function allows to you select the character that will be used to separate the values — in this case, we have used the forward slash character, but you could use commas or any other character you choose.

```
Join ({?Departments}, " / ")
```

This formula will produce a list of parameter values, separated by the slash character, as shown below in Figure 10.20.

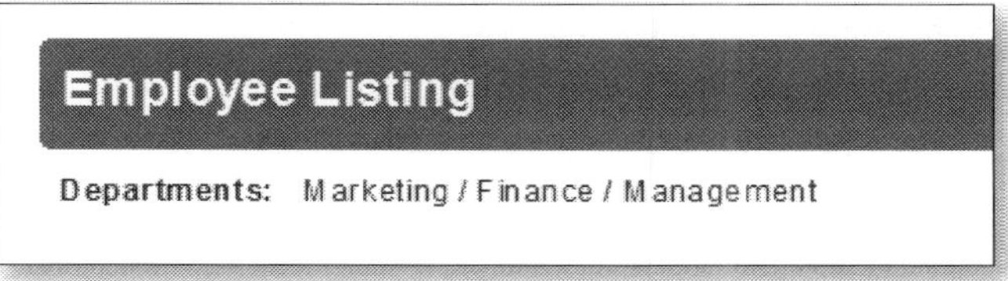

Figure 10.20 Multiple parameter values

Another variation on this technique is to use the CHR() function to throw a line feed, so each value is shown on its own line. For this technique, we would need to pass CHR(13), which is the line feed code, so the formula would look something like this:

```
Join ({?Departments}, Chr(10))
```

When the report is run, the parameter values will each be shown on their own line, as shown in Figure 10.21. For this formula field to display properly, you will need to tick on the "Can Grow" option for the field by clicking to select the field, then right-clicking to select "Format Field" and then checking the "Can Grow" option.

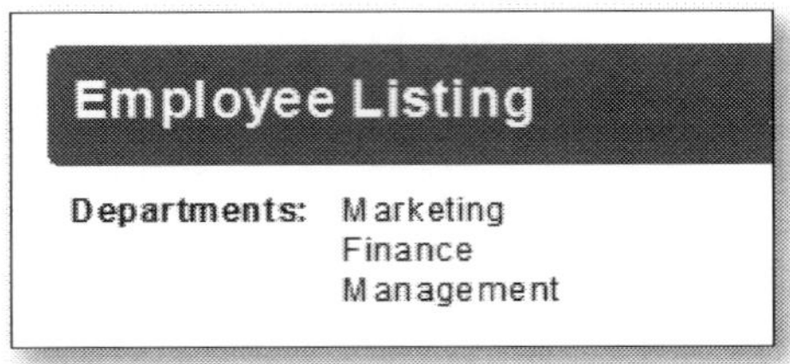

Figure 10.21 Multiple parameter values with a line break

WORKING WITH THE PARAMETER PANEL

The parameter panel, which was introduced with Crystal Reports 2008 gives end-users an easy way to interact with report parameters, including the ability to specify optional parameters when the report is run, as well as parameters that allow them to control report features. The parameter panel (shown in Figure 10.22 on the adjacent page) appears within the Crystal Reports Designer, as well as in most of the Crystal Reports viewers including the .NET and Java viewers, the DHTML viewer used with Crystal Reports Server, BusinessObjects Edge, BusinessObjects Enterprise, etc.

When creating or editing parameter fields, you have the ability to control whether or not the parameter will appear in the parameter panel. You can also mark the parameter as optional, so user input is not required. The setting "Show on (Viewer) panel" can either be set to:

- Do not show
- Editable
- Read Only

Likewise, if you want to make the prompt optional, there is a property for that as well where you can mark "True" or "False". If this optional parameter setting is "True", end-users will not be required to enter a parameter when the report is refreshed.

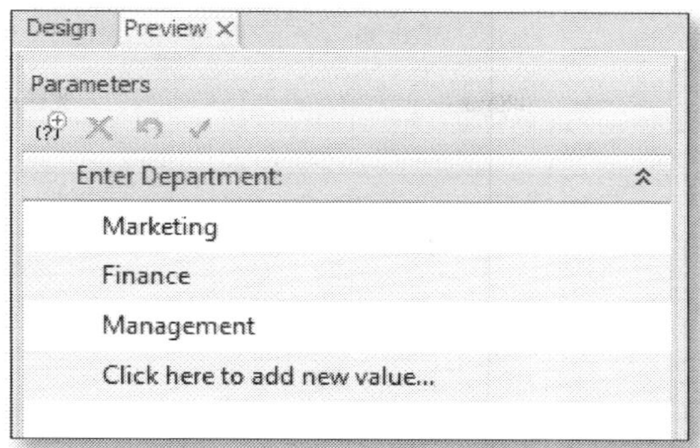

Figure 10.22 Parameter panel in action

If you plan to use optional parameters, just make sure you modify your record selection and other formulas to cater for the situation where no value is actually entered for the parameter field.

USING PARAMETERS TO CONTROL REPORT FEATURES

Another popular use of static parameters is to control actual report features. In this scenario, a user will respond to a parameter prompt and then some report element will be controlled by what they have entered. In the following section, we are going to look at some of the most popular uses of this technique, but it can be applied to a number of different scenarios, so keep this in your bag of tricks for when you need it.

PARAMETERS TO TURN ON/OFF REPORT FEATURES

Boolean parameter fields can be used to prompt users for a true or false response and can be used to control report formatting, summary levels, and more. For example, you may want to create a report that prompts users to answer the question "Show Negative Numbers?" and then use the value they entered in a record-selection formula to filter out any negative numbers that may normally appear in the report.

Creating a Boolean parameter field is just like creating any other parameter field. To start, click View > Field Explorer. The Crystal Reports Field Explorer opens. Right-click the section of the Field Explorer labeled Parameter Fields, and select New from the shortcut menu. Using the Create New Parameter dialog box, type a name for your parameter field and change the type to Boolean. Next, using the Prompt Text option, enter any prompting text you wish to appear when users are prompted for information (for example, "Would you like to print all states?"). See Figure 10.23 over the page.

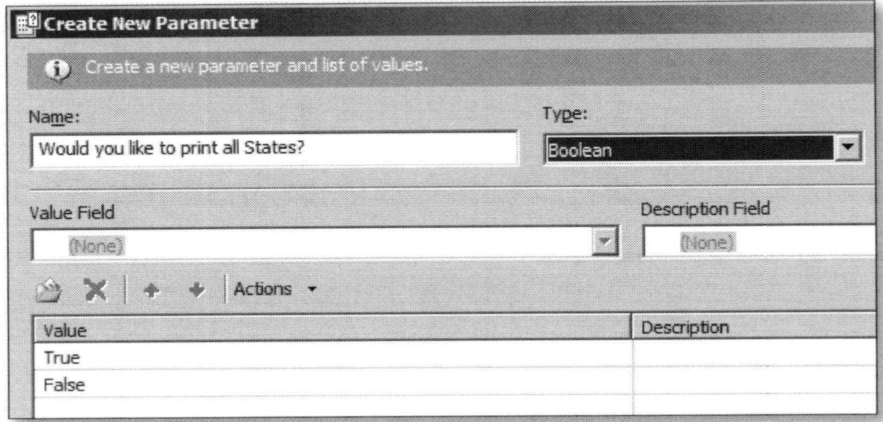

Figure 10.23 Options for Boolean parameter fields

When users are prompted for a Boolean parameter field, a dialog box similar to the one shown in Figure 10.24 will appear, and they can enter a selection of True or False.

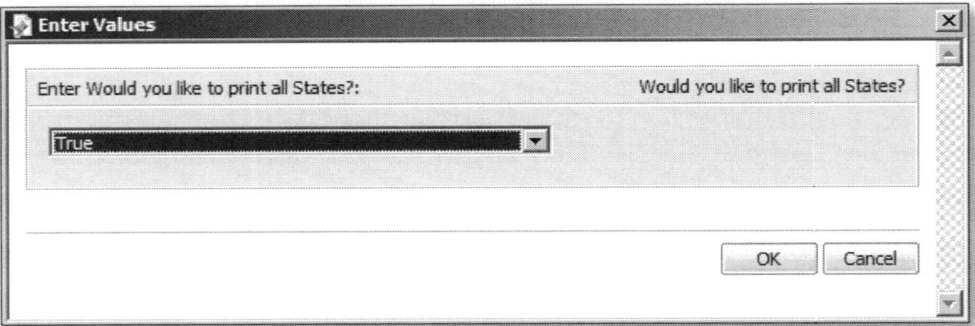

Figure 10.24 A typical Boolean parameter field

If you are not really fond of just True and False, you can select how the Boolean value is entered (True/False, Yes/No, On/Off) by using the options at the bottom of the page. As a final step, you could use this formula field with conditional formatting to determine if a section or area of the report was printed.

PARAMETERS TO CONTROL GROUPING

Another handy feature that you can use parameters for is to control the grouping that is being done in your report. This technique will give users control over how the report is grouped by allowing them to select the fields that will be used for the groups you have inserted in the report. The technique relies on a combination of formulas and parameter fields to make this technique work. In the following example, we are going to use a single parameter field and group, but it works just as well with multiple groups.

To create a parameter that allows the user to control the report grouping, follow these steps:

1. From the book download files, open the PARAMETERGROUPING.RPT report.
2. Select View > Field Explorer to view the available fields for the report.
3. Right-click on the Parameter Fields section and select New > Parameter Field. Name the parameter field "GroupOption" and make it a string-type, static parameter.
4. Under the Values section, enter the following three values: Country, State, Customer Type.
5. Right-click on the Formula Fields section and select New > Formula field and name the formula field "Group1".
6. For the formula, enter the formula text shown below:

```
If {?GroupOption} = "Country" then {Customer.Country}
else
If {?GroupOption} = "State" then {Customer.State}
else
If {?GroupOption} = "Customer Type" then {Customer.
CustomerType}
```

7. Save your formula and close the formula editor.
8. From the report designer, select Insert > Group and select the @Group1 field for your grouping field, then click OK.

Whenever the report is refreshed, this combination of parameter field, formula field and grouping should cause the parameter dialog to open and prompt you for your grouping value — select a value and your report will now be grouped by that particular field. You can extend this technique to multiple parameter fields and then create additional grouping formulas (i.e. Group2, Group3, etc.) and actual groups in the report.

PARAMETERS TO HIGHLIGHT VALUES

Another area where you can use parameter fields is for highlighting values in your report. This will save users getting out their highlighter and manually going through the report, line by line. With this technique, it is a combination of a parameter field with section formatting — when a user enters a threshold amount for a particular value in the report, every detail line which exceeds that amount will be highlighted using the background color of the section, as shown in Figure 10.25.

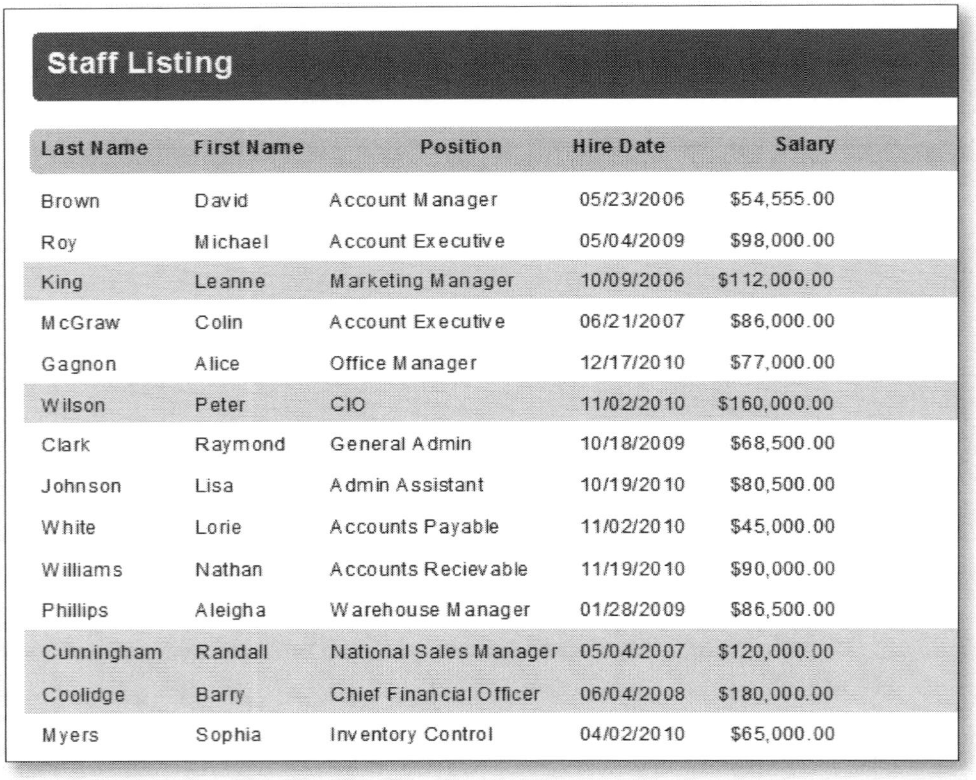

Staff Listing

Last Name	First Name	Position	Hire Date	Salary
Brown	David	Account Manager	05/23/2006	$54,555.00
Roy	Michael	Account Executive	05/04/2009	$98,000.00
King	Leanne	Marketing Manager	10/09/2006	$112,000.00
McGraw	Colin	Account Executive	06/21/2007	$86,000.00
Gagnon	Alice	Office Manager	12/17/2010	$77,000.00
Wilson	Peter	CIO	11/02/2010	$160,000.00
Clark	Raymond	General Admin	10/18/2009	$68,500.00
Johnson	Lisa	Admin Assistant	10/19/2010	$80,500.00
White	Lorie	Accounts Payable	11/02/2010	$45,000.00
Williams	Nathan	Accounts Recievable	11/19/2010	$90,000.00
Phillips	Aleigha	Warehouse Manager	01/28/2009	$86,500.00
Cunningham	Randall	National Sales Manager	05/04/2007	$120,000.00
Coolidge	Barry	Chief Financial Officer	06/04/2008	$180,000.00
Myers	Sophia	Inventory Control	04/02/2010	$65,000.00

Figure 10.25 Report with highlighting

This parameter can be used in conjunction with the parameter panel, so after the report is run and has saved data, you can then change the parameter and highlight all of the values in your report. To create a report that uses a parameter to highlight values, follow the steps as described on the following page.

1. From the book download files, open PARAMETERHIGHLIGHT.RPT
2. Right-click on the Parameter Fields section and select New > Parameter Field. Name the parameter field "Highlight Threshold" and make it a numeric-type, static parameter.
3. From within the report, right-click on the Details section and select the Section Expert.
4. Click on the "Color" tab and then click the conditional formatting icon (X+2). Enter the following conditional formatting formula:

```
If {Customer.3YrAverage} > {?Highlight Threshold}
then crYellow else crWhite
```

5. Click OK to return to your report.

You now can refresh your report and enter a threshold value — your report should change to show a yellow background in the details section for every value over the threshold, as shown below in Figure 10.26.

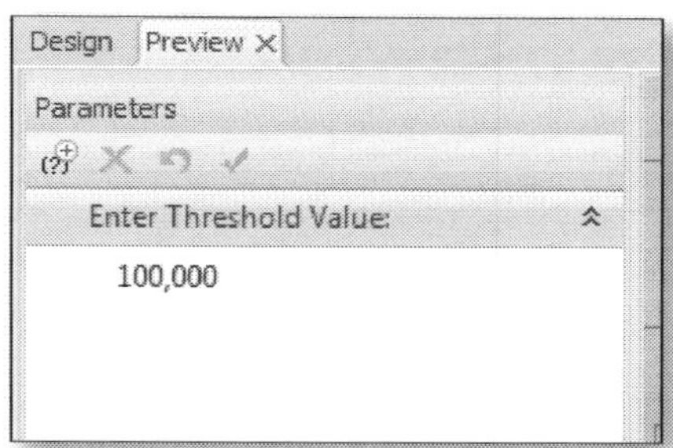

Figure 10.26 Report with values greater than the threshold highlighted

Another variation of this technique would be to create two parameters — the first would be a Boolean parameter that controls whether or not to show the highlighting, while the second would be the threshold value itself. In this instance, the conditional formatting formula would look like this:

```
If {?Show Highlighting} = "True" then
If {Customer.3YrAverage} > {?Highlight Threshold}
then crYellow else crWhite
Else crWhite
```

A sample report demonstrating this technique is available from the book download files and is named 2PARAMETERHIGHLIGHT.RPT

SUMMARY

Analyzing report data can be an effective way to bring real meaning and value to the information presented. In this chapter we looked at two different techniques to make that happen — analyzing report data using summaries, running totals, and so forth; and filtering reports using parameter fields. In the next chapter we will continue on with this theme, looking at how you can add calculations to your report with functions, formulas and SQL expressions.

Chapter 11
Formulas, Functions and SQL Expressions

In This Chapter

- Formula Overview
- Working with Formulas
- Creating Simple Arithmetic Formulas
- Using Crystal Reports Functions
- Adding Logic to Formulas
- Understanding Formula Evaluation Times
- Debugging Formulas
- Working with Custom Functions
- SQL Commands and Expressions

INTRODUCTION

It isn't very often that you will just want to present the data from your data source — often you will want to add your own calculations and formulas to the report, to add some value to the data that is being presented.

Luckily, Crystal Reports includes not one but two different ways you can add these calculations, either through a Crystal Reports formula or an SQL Command. The difference between the two is that a formula is created within Crystal Reports using one of two formula languages and is evaluated by the Crystal Reports print engine when the report is run. An SQL Command on the other hand is written using SQL and is evaluated by the database server itself.

In this chapter, you'll learn how to use both calculation methods. We're going to start off with learning about the Crystal Reports formula language, and how to create and debug formulas to add calculations and logic to your reports. You'll also learn how to you can reuse your formula code to create custom functions that can be used over and over again.

Later in the chapter, we'll look at how to create SQL commands and expressions and use them in your reports. SQL commands offer a way to reuse SQL statements as the data source for your report and push more processing back to the database server.

If you don't want to write full-blown SQL, you can still use SQL expressions to add calculations to your report. SQL expressions offer only a subset of the functions and operators found in Crystal's own formula language, but they do have the advantage of being evaluated by the database server (instead of locally by the print engine). At the end of the chapter, we'll look at how to create and use SQL commands and expressions, including some common usage scenarios and examples.

FORMULA OVERVIEW

By now, you are already somewhat familiar with the Crystal Reports formula language— it is used throughout the product in record selection, conditional formatting, and more. But the Crystal Reports formula language goes beyond those uses, giving you the ability to add complex calculations to your reports and manipulate report fields and elements.

If you have ever worked with a programming language or development tool, the Crystal Reports formula language will seem familiar. Likewise, if you are new to software development but have had some experience creating formulas in Excel, you should be able to transfer your skills to Crystal Reports easily (in fact, many of the functions work just like the ones you find in Excel).

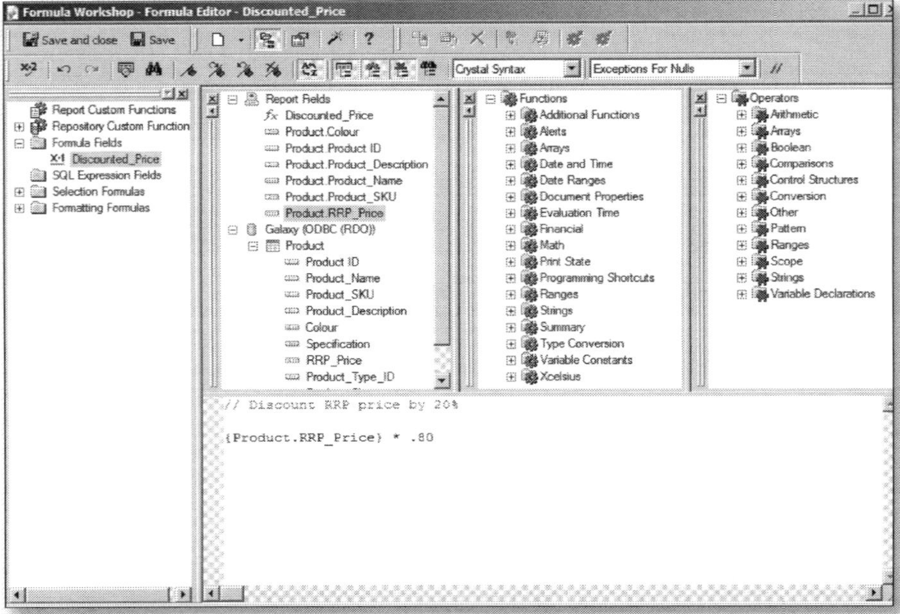

Figure 11.1 The Crystal Reports Formula Editor

Formulas are written using the Crystal Reports Formula Editor, shown in Figure 11.1. A formula can consist of any number of database or other fields, operators, functions, text, numbers, and control structures such as If...Then statements. Before we can start our discussion of how formulas are put together, we need to look at the Formula Editor and see how it works.

USING THE FORMULA EDITOR

The Formula Editor is the tool you will use to add or edit formulas that appear in your report. To open the Formula Editor, use the following steps:

1. Open Crystal Reports and any existing report on which you have been working.
2. Click View > Field Explorer.
3. Locate the Formula Fields section, right-click the section header, and choose New.
4. You will be prompted for a name for your formula. Enter a name and click OK. The Crystal Reports Formula Editor opens.

The Formula Editor consists of five main areas:

The Toolbar — Contains icons for creating a new formula, switching between formulas, finding and replacing, and more (see Figure 11.2 over the page).

The Workshop Tree — Located on the left side of the page, this allows you to navigate through and access formulas and custom functions wherever they may appear in your report.

The Report Fields section — Lists all fields present in your report, followed by your data source and all of the tables and fields contained within.

The Functions section — Lists all of the available functions. They range from simple summaries (sum, average, and so on) to type conversion and field manipulation functions to functions for complex statistical and financial analysis.

The Operators list — Contains a hierarchical view of all of the operators available in Crystal Reports (all of the arithmetic operators, variable declarations, comparison operators, and so on).

Some of the operators, such as +, –, /, and *, may be easier to just type, but you can double-click any operator in this list to add it to your report.

The largest section in the Formula Editor is used for the formula text you enter. This area behaves similar to other text editors (such as Notepad) or word-processing applications you may have used in the past.

Figure 11.2 The Formula Editor toolbar

When working with formula text, you may notice that Crystal Reports uses different colors for words or phrases in your formula text. This color-coding is designed to identify reserved words, functions, and comments. You can control this and many other aspects of the Formula Editor's appearance by clicking File > Option > Editors.

Over the past few versions of Crystal Reports, the Formula Editor itself has come a long way in terms of functionality and features for the developer. For example, the editor has an auto-complete function that you can use to complete code, similar to the Intellisense feature in Visual Studio. To use the auto-complete function, press CTRL+spacebar to open a drop-down list with the most likely text to complete the text you are entering.

CRYSTAL VS. BASIC SYNTAX

Just as English has its own syntax that dictates how words and sentences are put together, so does Crystal Reports — in fact, it has two types of syntax: Crystal syntax and Basic syntax. Crystal syntax has been around the longest, and up until version 8.0, it was the only choice for report developers. Crystal syntax has no direct relationship to any programming language (although it does resemble Pascal or dBase at times), and for report and application developers, it was difficult to learn.

With the release of Crystal Reports 8.0 came a new formula syntax, Basic, with structures and functions that closely resemble those used in Visual Basic. For application developers, the Crystal Reports Formula Editor then became familiar territory, because they could apply the concepts and functions they knew from Visual Basic. For new report developers, Basic syntax provides a better frame of reference and some additional functionality.

In Crystal Reports 2008, both types of syntax can be used, side by side, in different formulas in your report, according to your needs. To select a syntax for your formulas, locate the drop-down list in the top-right corner of the Formula Editor and select the appropriate syntax. When it comes to actually choosing a particular syntax, there is no clear winner — you will find most of the functions you need in both.

If you are just starting out with Crystal Reports, it would probably be a good idea to learn Crystal syntax first, because the record-selection formulas in Crystal Reports are written using only this syntax. However, Basic is easy to pick up if you have some programming experience.

To illustrate the difference between the two syntaxes, an example works best. The same formula is shown in the next examples, first written in Crystal Syntax:

```
{Invoice.Amount} * 1.06
```

and then again in Basic Syntax;

```
Formula = {Invoice.Amount} * 1.06
```

You can see that the syntax is almost identical, except the Basic syntax requires a `Formula =` to indicate what the output of the formula should be. To make things even easier, most of the functions in Crystal Reports are the same for both Crystal and Basic syntax, as shown first in the following sample formula with Crystal Syntax:

```
Sum({Invoice.Amount})
```

and then again in Basic syntax;

```
Formula = Sum({Invoice.Amount})
```

Some cracks start to show, however, as formulas become more complex. In the next two examples you can see the difference between how Crystal and Basic syntax handle returning information to the report. First, in Crystal syntax we see a simple If... then that returns a string if the condition is met:

```
If Sum({Invoice.Amount})>10000 then "Free Shipping"
```

In this example, the string to be returned is simply enclosed in quotes, and it is assumed that this will be the value returned to the formula field. In the second example, you see the same formula in Basic syntax, only this time `Formula =` is placed at the start of the formula text, and is used to assign what will be returned to the report:

```
If Sum({Invoice.Amount})>10000 then Formula = "Free
Shipping"
```

Finally, the last major difference we are going to point out between Crystal and Basic syntax is the way variables are declared. To declare, use, and display a variable in Crystal syntax, you may write something that looks like the following:

```
CurrencyVar OrderAmountWithTax;
OrderAmountWithTax = {Invoice.Amount} * 1.06;
OrderAmountWithTax
```

The first line dimensions a variable called `OrderAmountWithTax` and then performs a calculation and returns it to the report. Note that each line has a semi-colon as the continuation character, and to display the results on the report, you just need to repeat the variable name. The Basic syntax version of the same formula would look something like the following:

```
Dim OrderAmountWithTax As Currency
OrderAmountWithTax = {Invoice.Amount} * 1.06
Formula = OrderAmountWithTax
```

You can see that the Basic syntax example follows conventions you may have used in Visual Basic, including the Dim statement to dimension a variable with a particular type.

So now that you have seen a little bit of the difference between the two, it is time for you to groan. For the examples and projects in this chapter, we will use Crystal syntax: The main reason is that key Crystal Reports features that are formula based (such as record selection) rely on this syntax. Therefore, if you want to use these features, you need to know Crystal syntax.

And we haven't left out Basic Syntax entirely—in downloads from our website (www.kuiperpublishing.com) you will find all of the examples from this chapter shown in Basic syntax as well. A number of reports are also included that demonstrate formula concepts covered in the chapter as well as some we didn't have room for, including both Crystal and Basic syntax examples.

Another key point is that this chapter is not an exhaustive reference on every function and operator available within Crystal Reports; rather it covers the most popular functions and operators. If we were to cover every function and operator in depth, that could be a book in itself.

WORKING WITH FORMULAS

Before we can get into the nitty-gritty and look at the specific operators, functions, and text that make up formulas, we need to look at some of the procedures for working with formulas.

If you would like to follow along with the step-by-step instructions in this chapter, this book includes downloadable files that are available from www.kuiperpublishing.com. Also, check out "Setting up the Samples" section in the front of this book for instructions on how to configure the data sources, reports, etc.

CREATING A NEW FORMULA

Like most fields in Crystal Reports, formula fields can be created and inserted using the Field Explorer. To create a new formula in a report, use the following steps:

1. Open Crystal Reports and from the book download files, open the STAFFLISTING.RPT report.
2. Click View > Field Explorer; you should see a section named Formula Fields.
3. To insert a new formula, right-click the section marked Formula Fields, and select New from the shortcut menu, shown in Figure 11.3.

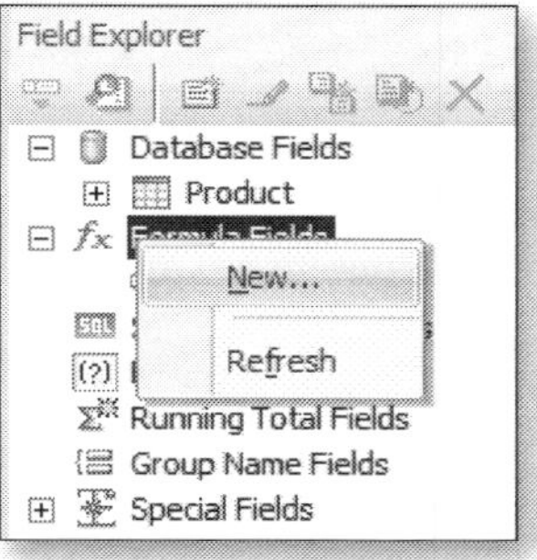

Figure 11.3 You create new formulas from the Field Explorer dialog box

4. In the next dialog box that appears, enter a name for your formula, and click OK. The name you select can be anything that makes sense to you and can include spaces or special characters. In this example, we have a report that lists all of their employees and

salaries. We are going to create a formula to apply a "cost of living" increase 3% to their salary as a simple first example.

5. The Formula Editor appears, and you can enter your formula text. Expand the Report Fields section, and under the Product table, double-click the Staff.Salary field to insert it into your formula.
6. Next, type an asterisk and then a space and 1.03. Your formula should now look like the following:

```
{Staff.Salary} * 1.03
```

At any time, you can check the syntax of your formula by clicking the X+2 button on the toolbar, but remember that Crystal Reports also performs a syntax check whenever you exit the Formula Editor.

7. When you have finished editing your new formula, click the Save and Close icon located in the top-left corner of the Formula Editor. You will then be returned to your report's Design or Preview view, where you will be able to insert your newly created formula onto your report.

INSERTING A FORMULA INTO YOUR REPORT

Inserting your formula into your report may be easier from the Design tab of your report. From there you can see all of the sections clearly and understand where you are placing your formula field.

To insert the formula field you just created, use the following steps:

1. In the Field Explorer (which you open by clicking View > Field Explorer), your formula should be listed under the section named Formula Fields.
2. From the Field Explorer, drag and drop your formula field onto your report into the details section.
3. Alternatively, you can click to select the field and then press ENTER, which will attach the field to the tip of your mouse. When you have the field positioned on your report, click once to release the field and place it in your report.

EDITING AN EXISTING FORMULA

Crystal Reports identifies formula fields you insert into your report by placing the @ symbol in front of the name and { } (curly braces) around the name. For instance, the field name for the Discounted_Price formula would appear as {@Discounted_Price}.

You can edit any existing formula that has been inserted into your report by locating the formula field you want to edit and right-clicking that formula field. Then select Edit Formula from the shortcut menu, shown in Figure 11.4. This opens the Crystal Reports Formula Editor, where you can edit the formula field.

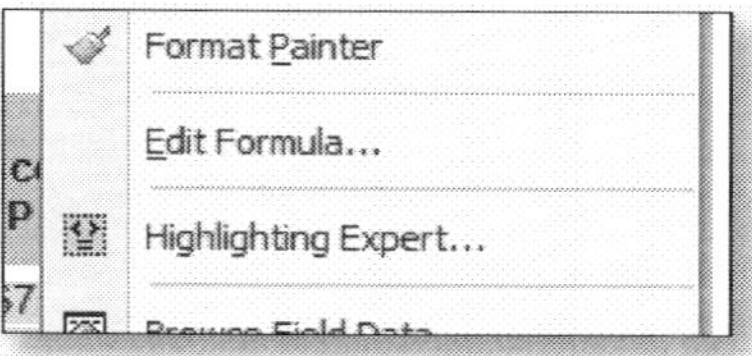

Figure 11.4 To edit an existing formula, right-click the formula shown in your report

When you have finished editing your formula, click the Save, and then the Close icon in the top-left corner to close the Formula Editor and save your changes. You will be returned to your report's Design or Preview view, and your changes should be reflected in the formula results.

RENAMING A FORMULA

Often you will want to revisit your report design and clean up the names of formulas, parameter fields, and running totals that appear in your report, to make it easier for other report designers or users to understand the logic behind your report design. To rename a formula, click View > Field Explorer to open the Field Explorer, locate the Formulas section, and find the formula you want to rename.

If you right-click the formula name, you should see the Rename option in the shortcut menu, shown in Figure 11.5, over the page. Select Rename to edit the formula name.

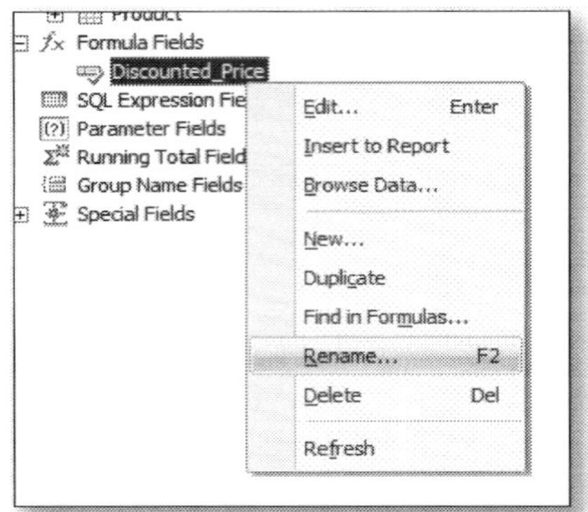

Figure 11.5 You can edit a formula name from the Field Explorer

When you have finished editing the formula name, click anywhere outside of the formula name to accept your changes.

Even if the formula is used on your report multiple times or referenced in multiple other formulas, the name change will be propagated everywhere it used.

CREATING SIMPLE ARITHMETIC FORMULAS

The most basic formulas use one of the simple arithmetic operators (+, – , *, /) and perform a calculation. To see how arithmetic formulas are written, we are going to write a Crystal Reports formula that will add together all of the different parts of an employee's compensation package, including Insurance, Retirement and Other Benefits and then add this together with a Salary increase of 3% to get each employee's "Total Compensation Package."

To create a simple arithmetic formula, use the following steps:

1. Open Crystal Reports, and open the STAFFLISTING.RPT report that we have been working with, or open a copy of the report from the book downloads folder.
2. Click View > Field Explorer and, from the Field Explorer, right-click the section named Formula Fields and select New from the shortcut menu.
3. In the next dialog box, you enter a name for your formula field—in this case, Sales Tax—and click OK. The Formula Editor will appear and allow you to enter your formula text.

4. In this example, we need to locate the fields for Salary, Insurance, Other Benefits and Retirement. Look in the Fields pane of the Formula Editor, shown in Figure 11.6; locate the fields listed above. All of the fields that appear in your report are located in the top section, named Report Fields. All of the fields in your data source, but NOT showing in the report will be listed below.

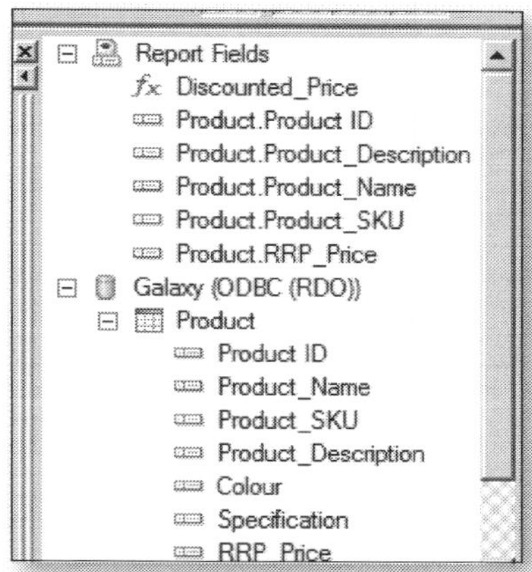

Figure 11.6 The Fields pane of the Formula Editor contains all of the fields that appear in your report as well as all of the fields available from your report's data source

If you want to use a field that does not appear in your report, you can do so. All of the tables and fields in your data source are available in the list.

5. To place a reference to a particular field in your formula text, double-click the field name. The reference should immediately appear in the formula text below. In this example, double-click the Staff.Salary. You'll notice that database fields are represented in the format of TableName.FieldName, with a set of curly braces around the entire lot. This format indicates that this is a database field. All of the Formula Editor panes — Fields, Functions, and Operators — behave the same way; double-click any of the items listed, and it will be inserted into your formula text.

6. For this formula, not only do we need a reference to the Staff.Salary field, we also need to do some simple arithmetic to multiply this field by 1.03 (3 percent increase) and then add the three other fields for Insurance, Retirement an Other Benefits on to this total. To do this, move to the Operators pane.

The Operators pane, shown in Figure 11.7, contains all of the operators that are available in Crystal Syntax. These operators are separated into categories: Arithmetic operators, Conversion operators, and so on.

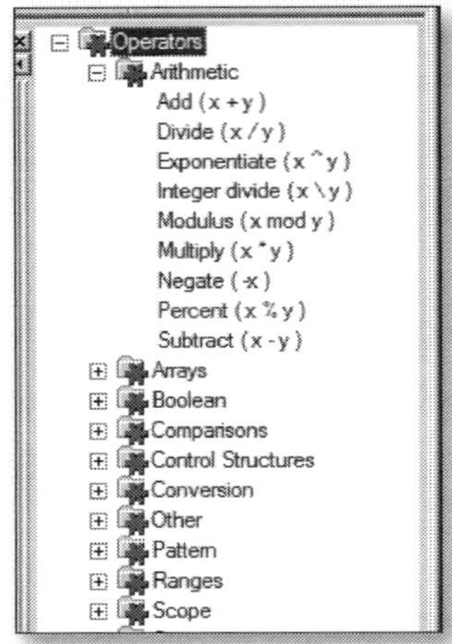

Figure 11.7 The Operators pane lists all of the operators available in Crystal Reports

7. In this instance, we know that multiplication is an arithmetic operator, so we can expand the Arithmetic section, locate the operator for multiplication (*), and then double-click it to insert it into our formula text. It is also sometimes just as easy to type the operator yourself, once you are familiar with them.
8. Now all we need to do is enter the 1.03 value, to show the salary amount plus 3%. The formula should then look something like the following:

```
{Staff.Salary} * 1.03
```

9. Next, we need to add the three other fields to our formula. Because Crystal Reports follows a logical order of operation for all calculations, it is always a good idea to use parentheses to indicate which part of the calculation should be performed first. So, surround your formula text with parentheses and then add + (the plus operator) and the 20 value. Your formula should now look like the following:

```
( {Staff.Salary} * 1.03 ) + {Staff.Insurance} +
{Staff.Retirement} + {Staff.OtherBenefits}
```

Congratulations — that's all you need to do. If you were to place this formula in the Details section of your report, it would calculate the cost of living increase on the Salary field and add this up with all of the other compensation for each employee, like in the report shown in Figure 11.8.

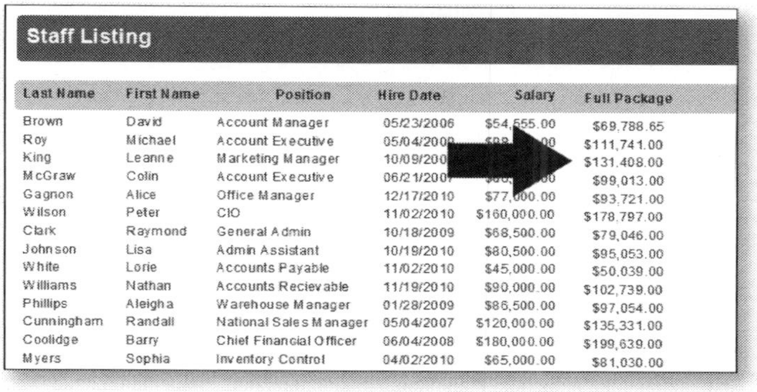

Staff Listing

Last Name	First Name	Position	Hire Date	Salary	Full Package
Brown	David	Account Manager	05/23/2006	$54,555.00	$69,788.65
Roy	Michael	Account Executive	05/04/200[illegible]	[illegible]	$111,741.00
King	Leanne	Marketing Manager	10/09/200[illegible]	[illegible]	$131,408.00
McGraw	Colin	Account Executive	06/21/200[illegible]	[illegible]	$99,013.00
Gagnon	Alice	Office Manager	12/17/2010	$77,000.00	$93,721.00
Wilson	Peter	CIO	11/02/2010	$160,000.00	$178,797.00
Clark	Raymond	General Admin	10/18/2009	$68,500.00	$79,046.00
Johnson	Lisa	Admin Assistant	10/19/2010	$80,500.00	$95,053.00
White	Lorie	Accounts Payable	11/02/2010	$45,000.00	$50,039.00
Williams	Nathan	Accounts Recievable	11/19/2010	$90,000.00	$102,739.00
Phillips	Aleigha	Warehouse Manager	01/28/2009	$86,500.00	$97,054.00
Cunningham	Randall	National Sales Manager	05/04/2007	$120,000.00	$135,331.00
Coolidge	Barry	Chief Financial Officer	06/04/2008	$180,000.00	$199,639.00
Myers	Sophia	Inventory Control	04/02/2010	$65,000.00	$81,030.00

Figure 11.8 An example of a formula in action on a report

ORDER OF OPERATIONS

Crystal Reports follows the standard order of operations, reading formulas from left to right and in the following order:

- Parentheses (any formula text enclosed in parentheses)
- Exponents (such as in X^2)
- Multiplication
- Division
- Addition
- Subtraction

When working with the order of operations, make sure that you use parentheses to force calculations that may not fall under the scope of normal algebraic equations.

For example, if you are attempting to calculate someone's age from a database field that holds the person's birthday, you could use a formula that looks like the following:

```
Today - {Staff.BirthDate}
```

The only problem with this formula is that when you insert it onto your report, it shows the number of days, instead of years. An easy solution would be to divide by 365.25 (the 0.25 accounts for leap years), making your formula read as follows:

```
Today - {Staff.BirthDate} / 365.25
```

When you attempt to save this formula or perform a syntax check, an error will occur, due to the order of operations. When Crystal Reports attempts to calculate the division part of the formula first, it doesn't understand how to divide a date field by 365.25, so an error results.

If you add parentheses to your formula, as shown in the following code, the formula will work correctly:

```
(Today - {Staff.BirthDate}) / 365.25
```

In these special cases, where a function or operator cannot be immediately used with the field you need, you will need to use parentheses to force a type conversion or other manipulation and then use the result in your formula.

As long as you keep the order of operations in mind and plan what calculations need to occur first, everything should work fine.

USING CRYSTAL REPORTS FUNCTIONS

Functions extend the Crystal Reports formula language and can be used to simplify complex calculations. If you expand the Functions pane of the Crystal Reports Formula Editor, you will see all of the available functions, arranged by function type. To insert a function into your formula text, double-click the function name.

Functions generally require one or more arguments, enclosed in parentheses and separated by commas. When you insert a function into your formula text, Crystal Reports automatically adds the parentheses and commas to indicate the arguments required, as shown in the following code for the `Round` function:

```
Round ( , )
```

In this example, you would need to specify a number to be rounded and the number of decimal places to be used, as follows:

```
Round ({Invoice.Amount}, 2)
```

Crystal Reports includes more than 200 functions, and keeping track of all of their names, parameters, and syntax can be tough. To find an explanation of a Crystal Reports function, go to Crystal Reports online Help. First press the F1 key from within Crystal Reports (or click Help > Crystal Reports Help) to display the main Crystal Reports Help screen, shown in Figure 11.9.

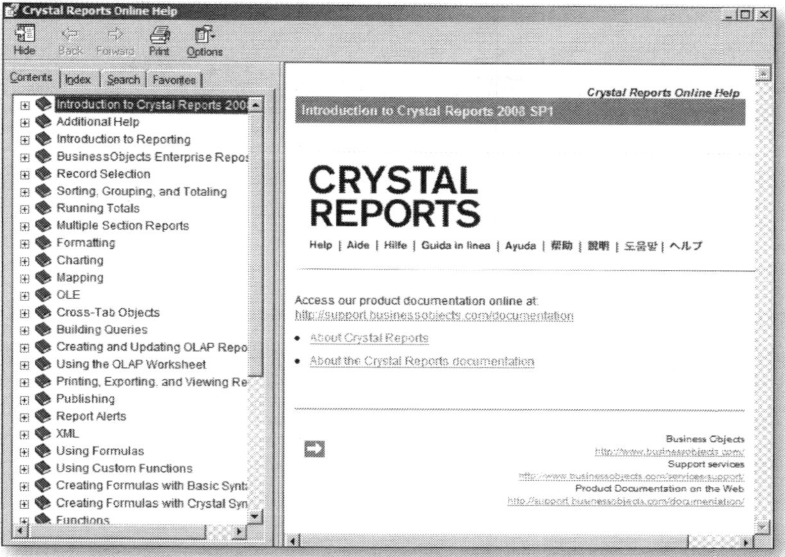

Figure 11.9 The Crystal Reports Help screen

Select the Index tab, type Functions in the text entry box, and press ENTER. This action will take you to the function listing by category. Click a category to see all of the functions of that type; click the link to go to the specific property page for a function. The Help page lists the function's required and optional arguments, the information that is returned, and some examples of the function. Click the close button in the top-right corner to close the Crystal Reports online Help when you are finished.

In addition to Crystal Reports own built-in functions, you can also create your own custom functions, which we will look at a little later in this chapter. These functions can be reused in a report or shared in the repository to be used in multiple reports.

SUMMARY FUNCTIONS

You can use summary functions to summarize fields that appear in your report. The most common summaries are performed on numeric or currency fields, but you can also apply summary fields to other field types, for example, to create a count of countries represented in a report. Numerous summary functions are available in Crystal Reports, but the ones listed in Table 11.1 over the page are the most popular.

Table 11.1 Common Summary Functions

Summary	Syntax	Usage
Sum	`Sum()`	Calculates the sum of a particular field; for example, `Sum({Invoice.Amount})`
Average	`Avg()`	Calculates the average for a particular field; for example, `Avg({Invoice.Amount})`
Min	`Min()`	Finds the minimum value; for example, `Min({Invoice.Amount})` would return the lowest Order Amount value
Max	`Max()`	Finds the maximum value; for example, `Max({Invoice.Amount})` would return the greatest Order Amount value
Count	`Count()`	Returns a count of all values; for example, `Count({Orders.OrderId})` would return a count of all orders
Distinct Count	`DistinctCount()`	Returns a distinct count of all values, meaning that each item is counted once; for example, if you create a formula using `DiscountCount({Orders.OrderId})`, and if OrderId includes duplicates, each distinct record will nevertheless only be counted once

STRING FUNCTIONS

The Crystal Reports formula language also includes a number of functions for manipulating string fields. String fields contain alphanumeric characters and are the most common field type; you can put just about anything in a string field.

One of the handiest tricks for working with strings is concatenation. Using special concatenation operators, you can combine two or more string fields for use in your report. For example, you may have separate First Name and Last Name fields in your data, but when this information is displayed on your report, you may want the information from the First Name field to appear followed by a space and then the information from the Last Name field. Crystal Report's special concatenation operators make this possible.

Two concatenation operators are available within Crystal Reports: + and &. The plus sign (+) operator works just like the operator for adding together two numbers, but it applies to string fields. Using this operator, the formula for the example mentioned earlier would look as follows:

```
{Staff.FirstName} + " " + {Staff.LastName}
```

The formula text in the middle—the space surrounded by double quotation marks—adds a space between the first and last names. Likewise, if you want to include any other specific text in the concatenation, you would enclose it in quotation marks, for example, ("Mr" + {Staff.LastName}).

The ampersand (&) concatenation operator can be used just like the plus sign:

```
{Staff.FirstName} & " " & {Staff.LastName}
```

The difference between the two operators is that the ampersand is a bit more flexible and can be used to concatenate numeric and other types of fields as well (without having to convert them to strings). Extending the earlier example, for instance, we can show the staff first name, last name, and phone extension using the ampersand.

```
{Staff.FirstName} & " " & {Staff.LastName} &
{Staff.PhoneExtension}
```

If you tried to create this same formula using the plus sign operator, Crystal Reports would return an error, because the `Customer.ID` field is numeric, and the + operator works only with strings.

Now that you know how to put strings together, what about ripping them apart? Crystal Reports includes a function called Subscript that numbers each position within a string.

Using the Subscript function, you can rip off a particular position of a string. For example, you can display the first initial of a customer's first name by using the following formula:

```
{Staff.FirstName}[1]
```

Likewise, if you want to display both the first initial of a customer's first name followed by the customer's last name, you can combine the two types of string functions we have worked with so far, as follows:

```
{Staff.FirstName}[1] + " " + {Staff.LastName}
```

In addition to ripping strings apart and putting them back together, Crystal Reports also includes functions for converting strings to all uppercase or lowercase characters, determining the length of a string, and trimming blanks from the start and end of a string. Table 11.2 lists some of the most commonly used string functions, along with an example of how each is used.

Table 11.2 Common String Functions

Function	Purpose	Usage
Uppercase( )	Converts strings to all uppercase	`Uppercase({Table.FieldName})`
Lowercase( )	Converts strings to all lowercase	`Lowercase({Table.FieldName})`
Length( )	Calculates the length of a string	`Length({Table.Fieldname})`
Trim()	Deletes extra spaces at the start and end of a string	`Trim({Table.Fieldname})`
Left(String, Length)	Returns the number of characters from the left side of the string	`Left({Table.Fieldname}, 4)`
Right(String, Length)	Returns the number of characters from the right side of the string	`Right({Table.Fieldname}, 6)`

TYPE-CONVERSION FUNCTIONS

In using Crystal Reports, you may run into problems related to the types of fields contained in a particular database. The field types may be set by the database or application developer, and you can't change them without changing the database or application itself. For example, you can find numeric information such as an order amount, stored in a field that has been defined as a string field. With the information held as a string, you can't apply all of the handy summary functions within Crystal Reports.

If your organization has developed the database or application from which you are reporting, you may be able to submit a change request to get the information stored in a more appropriate field type. But even if you are using a commercial application, or if your own database or application can't be changed, don't give up. Instead of changing the type in the database, you can let Crystal Reports do the type conversion.

The Conversion section of the function list includes a number of functions that can convert field types. To find the appropriate function, first determine the target type (that is, what type do you want the field to be when you are done). Then select a function from Table 11.3 and create a formula to perform the conversion.

Table 11.3 Type-Conversion Functions

Target Type	Function	Input
Text	`ToText( )`	Number, Currency, DateTime, Date, Time
Number	`ToNumber( )`	String, DateTime, Date, Time
Boolean	`Cbool( )`	Number, Currency
Currency	`Ccur( )`	Number, Currency, or String
Date Time	`CdateTime( )`	Number, String, DateTime, Date, Time
Integer	`CDbl( )`	Number, Currency, String, or Boolean
String	`Cstr( )`	Number, Currency, String, DateTime, Date, or Boolean
Date	`Cdate( )`	Number, String, DateTime, Time, Ctime() Number, or String

With all of these functions, the formula text will look something like the following:

```
ToText({Invoice.Amount})
```

In this example, the values in the Order Amount field would be converted to text and displayed on your report.

Some of the functions listed in Table 11.3 may have additional, optional parameters that can be passed to control the output. For example, the `ToText()` function can be passed a number of decimal places to convert, as shown in the next example:

```
ToText({Invoice.Amount},0)
```

In this example, no decimal places will be displayed.

Remember that you can find a complete list of functions and their parameters by clicking Help > Crystal Reports Help > Functions and searching on the function name.

PERIOD FUNCTIONS & DATE FIELDS

Crystal Reports has a number of predefined periods for use with dates. Until now, we have applied date periods only to record selection. However, you can use these same periods in the Formula Editor.

When you work with periods, Crystal Reports does all of the hard work for you. When you use the MonthToDate period, for example, Crystal Reports goes behind the scenes to check today's date and then builds a list of all of the dates that should be in MonthToDate — you don't need to lift a finger.

Periods in the Formula Editor are used most often in conjunction with an operator called `In` that determines whether a specific date is within that period. An example of the `In` operator is shown here. This snippet of formula text looks at the Order Date field, and if a date is in the period Over90Days, then it displays the words PAST DUE ACCOUNT! on the report.

```
If {Invoice.InvoiceDate} in Over90Days then "PAST DUE
ACCOUNT!!"
```

You can also use this technique with the other date periods, listed here and found in the Date Ranges section of the Function list:

- WeekToDateFromSunday
- MonthToDate
- YearToDate
- Last7Days
- Last4WeeksToSun
- LastFullWeek
- LastFullMonth
- AllDatesToToday
- AllDatesToYesterday
- AllDatesFromToday
- AllDatesFromTommorow
- Aged0to30Days
- Aged31to60Days
- Aged61to90Days
- Over90Days
- Next30Days
- Next31to60Days
- Next61to90Days
- Next91to365Days
- Calendar1stQtr
- Calendar2ndQtr
- Calendar3rdQtr
- Calendar4thQtr
- Calendar1stHalf
- Calendar2ndHalf

In addition to using these period functions with dates, Crystal Reports allows you to perform some simple arithmetic on date fields. For instance, Crystal Reports allows you to calculate the difference between two dates as well as add a number of days to a particular date, where the result is also a date field. Date arithmetic is especially handy when calculating aging or an invoice due date.

For example, suppose you want to look at the difference between when an order was placed and when it was actually shipped. Using the subtraction operator (–), you can find out how many days have passed.

```
{Invoice.ShipDate} - {Invoice.InvoiceDate}
```

The value returned will be in days. If you want to determine how many years this represents (as when calculating age, for example), enclose the existing calculation in parentheses and divide by 365.25.

Likewise, if you want to calculate a due date, say, in 30 days, you can add 30 days to the ship date, as follows:

```
{Invoice.ShipDate} + 30
```

You can add and subtract dates, but you cannot multiply or divide them.

Keep in mind that when performing calculations between dates, you may need to use parentheses to force the order of operation. Crystal Reports will display an error message when you try to combine calculations involving dates with calculations involving numbers.

ADDING LOGIC TO FORMULAS

At this point, we have talked about simple arithmetic formulas, strings, and date fields and periods, but we really haven't got into adding any logic to your formulas. Crystal Reports has a number of different ways you can add logic, and we're going to start our coverage with the most common, the If...then...else statement.

WRITING FORMULAS

In a few examples earlier in this chapter, you might have noticed the use of If...then statements. These statements work on the simple premise that if some condition is true, then something will happen (if A is true, then B will happen).

To see how If...then statements can be used, we are going to look at a common example. In the previous examples, we worked with an Order Amount field. If we want to flag all Invoice Amount values that are over $1,000 for discounted shipping, we can use an If...then formula that looks something like the following:

```
If {Invoice.Amount} > 1000 then "Free Shipping"
```

If we place this formula beside the Invoice Amount field on the detail line, this formula will be evaluated for every record in the table. Where the condition is true (for orders greater than $1,000), then the message "Free Shipping" will appear on the report, as shown in Figure 11.10.

With If...then statements, we also have the option of adding an Else statement to the end. An If...then formula states some condition and what will happen if the condition is true; an Else statement goes into effect when the If condition is not true.

Using the previous example, we can add an Else statement that prints "Discount" for all of the Invoice Amount values less than $1,000. That formula would look like the following:

```
If {Invoice.Amount} > 1000 then "Free Shipping" else
"Discount Shipping"
```

In this case, if the condition is true (the Order Amount value is greater than $1,000), then the first condition will fire, printing "Discount Shipping" on the report; otherwise, if the condition is false (the Order Amount value is less than $1,000), the "Free Shipping" message will be printed.

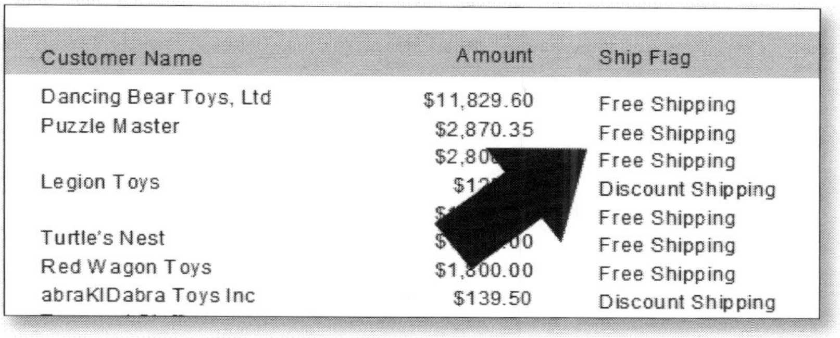

Customer Name	Amount	Ship Flag
Dancing Bear Toys, Ltd	$11,829.60	Free Shipping
Puzzle Master	$2,870.35	Free Shipping
	$2,80[illegible]	Free Shipping
Legion Toys	$1[illegible]	Discount Shipping
	[illegible]	Free Shipping
Turtle's Nest	$[illegible]00	Free Shipping
Red Wagon Toys	$1,800.00	Free Shipping
abraKIDabra Toys Inc	$139.50	Discount Shipping

Figure 11.10 An example of an If... then formula in action

Regardless of whether you use the Else statement on the end, If... then formulas can be combined with other functions and operators you have learned about in this chapter to create complex formulas to calculate the values you need.

USING SELECT... CASE STATEMENTS

Another handy control structure is the Select... Case statement, which is an alternate method to using a standard If...then statement. Using the Select...Case statement, you can specify a value for the Select, which is then evaluated by the corresponding Case statements.

In the following example, the WorkStatus field contains a flag for the employee's work status which needs to be abbreviated, as shown below:

```
SELECT ({Staff.WorkStatus})
Case "Full Time":
        "FT"
Case "Part Time":
        "PT"
Case "Seperated":
        "SEP"
Default:
        "***"
```

If the value in the Select does not match one of the Cases, you can also use the Default option to return a default value.

And when using the Select statement, you can also specify a range of values, as well as greater than/less than, etc. In the example below, there is a Salary field and the Select…Case statement will return a different string depending on the salary range.

```
Select {Staff.Salary}
   Case 0 to 50000:
        "Low Salary"
   Case 50000 to 100000:
        "Mid Salary"
Case Is > 100000:
        "High Salary"
```

Select...Case statements are often used to get you out of some tricky situations, where an If…Then statement may have proved too complex.

In addition to the two control structures we have looked at, there are a few more we didn't have room for in this chapter. For examples of these other types of control structures and loops, including more advanced Select statements and looping, see the reports located in the CONTROLSTRUCTURE.ZIP file, located on the downloads from our website (www.kuiperpublishing.com) in the Chapter 11 folder.

WORKING WITH VARIABLES

Another feature of the Crystal Reports formula languages is the use of Variables. A variable is a placeholder for a value that you may want to use in your report. A variable value can be assigned and re-assigned as many times as you like in a report.

When you declare variables, you can select a scope for where the variable can be used. The following variable scopes are available:

Local — Limited to a single formula, can't be used elsewhere.

Global — Can be used in multiple formulas, throughout the report.

Shared — Can be used in multiple formulas through the report and any subreports.

When you declare a variable, it is assumed to be a Global variable, so you don't need to include the scope in the declaration. In the example below, a variable named "TotalOrderAmount" is being declared as a Currency type:

```
CurrencyVar TotalOrderAmount
```

You can then make an assignment to the variable as follows:

```
TotalOrderAmount:= {Invoice.UnitPrice} * {Invoice.
Qty}
```

When declaring variables, you can declare the following types:

- Boolean
- Number
- Currency
- Date
- Time
- DateTime
- String

When declaring a variable, the scope is listed first, followed by the type and variable name — the examples below show the different declarations for different types (where Varname is the name of the variable you are declaring):

- Global BooleanVar Varname
- Global NumberVar Varname
- Global CurrencyVar Varname
- Global DateVar Varname
- Global TimeVar Varname
- Global DateTimeVar Varname
- Global StringVar Varname

There are a couple of rules to keep in mind with variables — first, your scope determines where they can be used and secondly, they must be declared each time, in every formula where you intend to use them. Also remember that you use := to assign a value to a variable.

WORKING WITH ARRAY VARIABLES

In addition to single value variables, you can also declare and use Array-based variables in your report. These array-based variables share the same types as the single value and the example below show the different declarations for the different types (where Varname is the name of the variable you are declaring).

- Global BooleanVar array Varname
- Global NumberVar array Varname
- Global CurrencyVar array Varname
- Global DateVar array Varname
- Global TimeVar array Varname
- Global DateTimeVar array Varname
- Global StringVar array Varname
- Global NumberVar range array Varname
- Global CurrencyVar range array Varname
- Global DateVar range array Varname
- Global TimeVar range array Varname
- Global DateTimeVar range array Varname
- Global StringVar range array Varname

In the example below, we are declaring a range variable and setting it to a value between 0 and 10,000:

```
CurrencyVar range array DiscountRange
DiscountRange:= := [0 to 10000];
```

You could then use this variable in a formula with an If...then statement to calculate a discounted amount, as shown on the next page:

```
CurrencyVar range array DiscountRange
DiscountRange:= := [0 to 10000];
If {Invoice.Amount} in DiscountRange then {Invoice.
Amount} * .90
```

Whether you need to create a simple single-value variable or something a bit more complex, variables are a great little tool that every report developer should have in their arsenal.

UNDERSTANDING FORMULA EVALUATION TIMES

Crystal Reports is a multipass reporting tool. Traditionally, it has been agreed that when a report is processed, Crystal Reports will perform two passes to accommodate advanced features like grouping, sorting, summaries, and grand totals.

If you have been working with Crystal Reports formulas for a while, you may have noticed that when you place a formula on your report, it does not give the results that you expected. One of the major reasons for this (other than the formula being incorrect) is actually the time at which the formula is processed by the report.

You can correct this problem and force your formulas to be processed using Evaluation Time functions. You can specify the following four evaluation times:

- BeforeReadingRecords
- WhileReadingRecords
- WhilePrintingRecords
- EvaluateAfter (formula name)

Each of these evaluation functions can be used in conjunction with your existing formula text and will determine when the formula will be processed during report execution. An example of one of these functions follows:

```
WhilePrintingRecords;
NumberVar SalesCount
SalesCount := SalesCount + 1;
SalesCount;
```

The formula itself is used to perform a manual running total on the number of sales orders placed. In order to understand how these functions work, we need to drill down a little further into how a report is processed.

BeforeReadingRecords

Formulas marked to evaluate BeforeReadingRecords will be evaluated before any database records are returned to Crystal Reports. This evaluation time can be used only to evaluate "flat" formulas that do not include any database records or calculated elements from the report. Because most formulas include some reference to database or other fields that require processing at a later time, this evaluation time is used infrequently.

WhileReadingRecords

Formulas marked WhileReadingRecords will be evaluated as the database records are read from the database. This evaluation time function is most often used when you want to perform a calculation or conversion on a database field (that is, changing a string to a number), as shown in the next example:

```
WhileReadingRecords;
ToText({Invoice.Amount})
```

This evaluation time is often used for formulas behind the scenes that need to be calculated early in the report processing, but not necessarily shown on the report itself to be printed out.

WhilePrintingRecords

The most popular evaluation time function is WhilePrintingRecords. This evaluation time is used to force formulas to run when the report records are printed on the page. This function can be used with manual running totals, counters, and so on and is often required to make formulas on your report return the required results. An example of this evaluation time formula in action follows:

```
WhilePrintingRecords;
NumberVar StatusCounter;
If (Customer.Status} = "Overdue" then StatusCounter
:= StatusCounter + 1;
StatusCounter;
```

For an example of a manual running total field created with WhilePrintingRecords, open the MANUALRUNNINGTOTALS.RPT report from the Chapter 11 directory in the book download files.

EvaluateAfter (formula name)

The EvaluateAfter function is used to force the processing of one formula after another. For example, if you had a formula that calculated an Order Total, you may want to set your formula for Sales Tax to evaluate after, so you could ensure that the Order Total was calculated before the sales tax was calculated and added on.

The book downloads (www.kuiperpublishing.com) include sample reports that use all three of these evaluation time functions. They are located in the EVALUATIONTIME.ZIP file in the Chapter 11 folder.

DEBUGGING FORMULAS

When you use formulas, you need to make sure that the syntax you have entered for your formulas is correct. As luck (or good design) would have it, Crystal Reports includes its own syntax checker, and it can be invoked in two ways.

When you save your formula by clicking the Save icon on the toolbar, shown in Figure 11.11, Crystal Reports automatically performs a syntax check, just to make sure there are no missing parentheses, misspelled words, and so on.

The second method of invoking the syntax checker is to click the X+2 icon, also shown in Figure 11.11. You can click this icon at any time while working in the Formula Editor. If you are building a complex formula, you may want to check the syntax each time you add a major piece, to make sure that what you have entered is correct syntactically.

Regardless of which method you use, the syntax check Crystal Reports performs is very simple: It makes sure you have spelled all of the function and field names correctly, that you have used the correct function values, and so on.

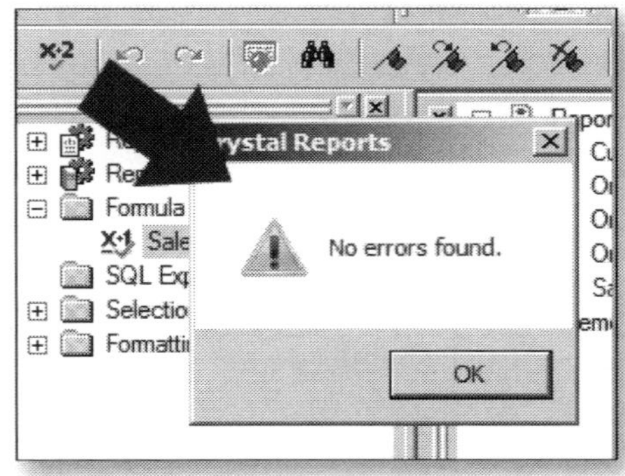

Figure 11.11 Make sure you save your formula before exiting the Formula Editor

If the syntax checker does return an error, your cursor will be moved to the place in the formula that the error occurs to pinpoint which part of the formula you need to modify.

Another method of checking your syntax is the Dependency Checker, shown in Figure 11.12.

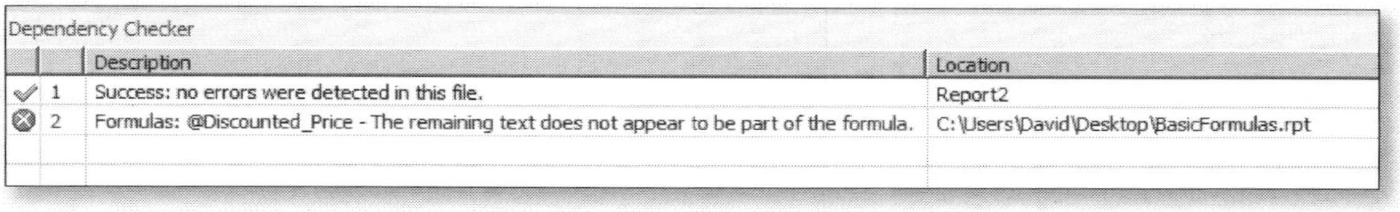

Dependency Checker

		Description	Location
✓	1	Success: no errors were detected in this file.	Report2
⊗	2	Formulas: @Discounted_Price - The remaining text does not appear to be part of the formula.	C:\Users\David\Desktop\BasicFormulas.rpt

Figure 11.12 The Dependency Checker in action

When invoked, the Dependency Checker will run through your report and find any errors or determine any dependencies that are not met. For example, if your report had a formula that references a group summary and you delete the group, the formula would appear in the Dependency Checker as an error, and you would need to fix the formula before it would run. Likewise, if there is a field missing or misnamed, the Dependency Checker will pick this up and provides a quick way of finding all of the errors in a report.

You can set the options for the Dependency Checker by right-clicking the toolbar above the list of dependencies. Select Options from the shortcut menu.

With both of these methods, there is still no guarantee that your formula will return the correct values, but you can be assured that the syntax is correct and that all of the required dependencies have been met — the rest is up to you.

WORKING WITH CUSTOM FUNCTIONS

Crystal Reports has always provided the ability to add your own functions to the Formula Editor. This traditionally has been accomplished through writing a UFL, or User-Function Library, that was compiled to a .dll used with the Report Designer and distributed with your application. This was not always the best method, because UFLs were difficult to create and maintain. A much better alternative was introduced a few versions back with the addition of custom functions.

A custom function is simply a formula field that you have saved and can reuse in other formula fields in your report. In addition, if you are using BusinessObjects Enterprise, there is a piece of server technology called the Repository, which will allow you to save your custom functions and use them in other reports. The advantage of using the Repository to store your custom functions is that when you change the function in the repository, all of the other reports can utilize the updated function, eliminating the need to reopen each report where the function is used.

But before we get too deep into working with custom functions, we need to take a look at how to create them. There are two methods you can use to create custom functions — you can either extract a function from an existing formula field or you can use the Formula Editor to create the custom function from scratch. We'll start our coverage of custom functions looking at the first method, creating functions from existing formula fields.

EXTRACTING FUNCTIONS FROM EXISTING FORMULAS

If you have been working with Crystal Reports for a while, chances are that you have already created a number of formula fields. The easiest method of creating a function is to use these existing formula fields as the basis for custom function and extract all of the logic, variables, and so on that you need from them.

To extract a function from an existing formula, use the following steps:

1. Open the CUSTOMERORDERS.RPT report from the book download files.
2. Click View > Field Explorer to open the list of available fields.
3. Expand the Formula Fields section, right-click any of the formulas, and select Edit. The Formula Editor opens.
4. Using the drop-down menu beside the New icon, select Custom Function, as shown in Figure 11.13 on the next page.

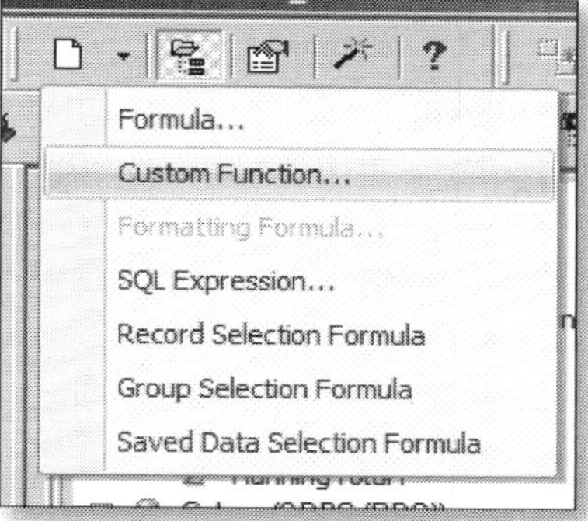

Figure 11.13 Creating a new custom function

5. In the dialog box that opens, enter the name of your custom function — in this example, we'll call it the function ShipMethod.
6. Once you have entered the name of your function, click the Use Extractor button as shown in Figure 11.14. The Extract Custom Function from Formula dialog box opens.

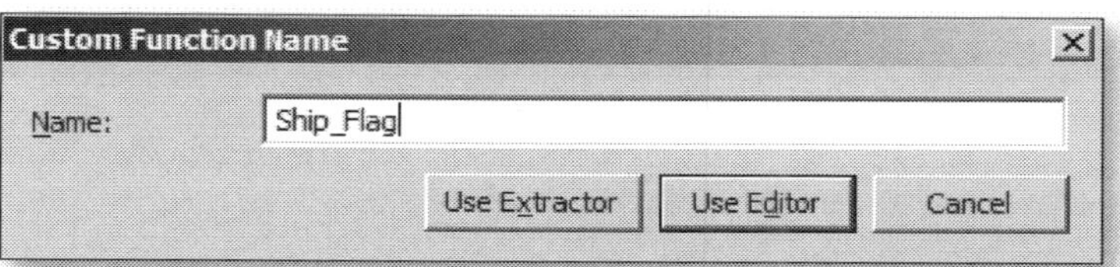

Figure 11.14 The Custom Function dialog box

7. From the left column, click the @Shipping formula field. This will display the formula field in the text box and change the properties dialog box to show any arguments that need to be passed to the function.
8. In the arguments section, the Argument Name is listed by default as V1. Click this box and change the argument name to OrderAmount. This is what needs to be passed to this function to make the logic work.
9. You may notice that there is a button on the top-right corner named Enter More Info. This step is optional; it opens a set of property pages where you can add a description of the function, the author name, and some Help text for the user.
10. Next, click the checkbox in the bottom-left corner to Modify formula to use custom function, and click OK.

11. This will take you to the formula text to review your custom function text, which will look like the following:

```
Function (currencyVar InvoiceAmount)
If InvoiceAmount > 10000 then "Nationwide Couriers"
else
If InvoiceAmount >= 5000 then "Express Shipping" else
If InvoiceAmount >= 2500 then "2-Day Shipping" else
"Standard Postal Shipping"
```

12. Click the Save and Close button to return to your report's Design or Preview view.

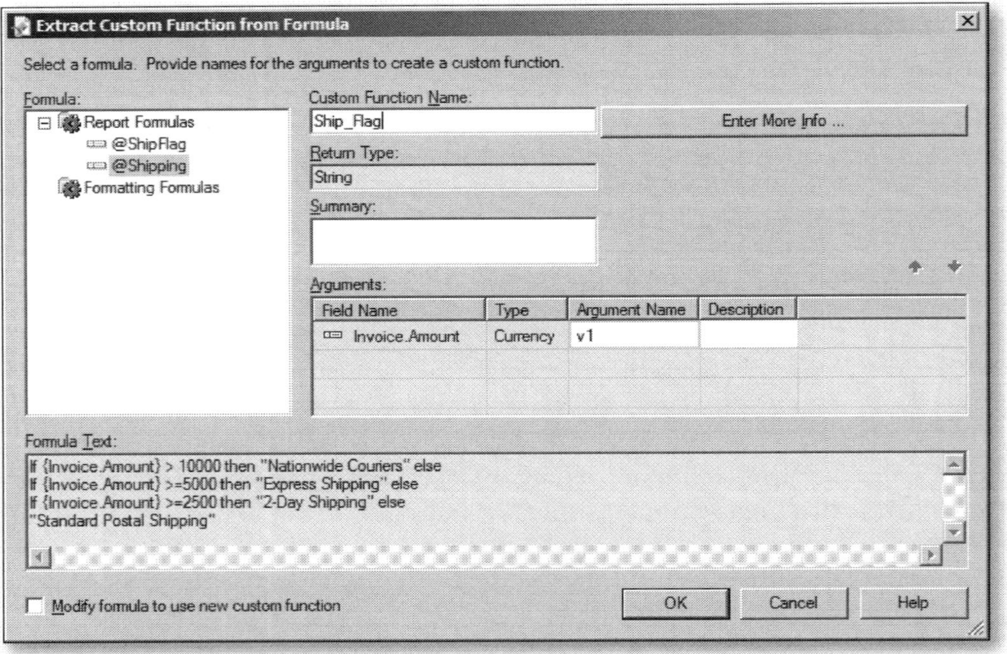

Figure 11.15 Custom functions as they appear in the Formula Editor

The original formula, @shipper, has been modified to use the custom function you created. You can now use this same function in your report without having to redefine or copy and paste the text into other formula fields.

You'll also notice that in the Formula Editor your custom function appears in both the function list as well as in the workshop tree on the left side, as shown in Figure 11.15.

To edit the formula, double-click it from the workshop tree to open the formula in the editor. To add the formula to an existing formula, double-click from the function list in the middle of the page.

CREATING CUSTOM FUNCTIONS FROM SCRATCH

If you don't have any existing formulas to use as the basis of your report, you can create custom functions from scratch, just by entering the formula text yourself.

To create a custom function using this method, use the following steps:

1. Open INTERNATIONALCUSTOMERS.RPT report from the book download files.
2. Click View > Field Explorer to open the list of available fields.
3. Expand the Formula Fields section, and right-click any of the formulas and select Edit. The Formula Editor opens.
4. Using the drop-down menu beside the New icon, select Custom Function. A dialog box appears, prompting you for the name of the function. Enter CountryGrouping, and then click the Use Editor button. The Crystal Reports Formula Editor opens.
5. Enter the formula text as follows:

```
Function (stringVar Country)
If Country= "USA" then "Domestic" else
If (Country = "Mexico" or Country = "Canada") then
"North America" else "Rest of World"
```

6. When finished, click the Save and Close button in the top-left corner.
7. To create a formula to use this function, click View > Field Explorer.
8. Right-click the Formula Field heading, and select New from the shortcut menu.
9. Enter a name for your formula — in this example, name the formula Country Flag.
10. Next, enter the formula text as follows:

```
CountryGrouping({Customer.Country})
```

11. When finished, click the Save and Close button to return to your report's Design view.
12. From the Field Explorer, drag this formula field onto your report into the Detail section. Your report should look something like the one shown in Figure 11.16 on the following page.

Formulas

Customer Name	Order ID	Order Date	Order Amount
Deals on Wheels	1002	12/02/2003	$5,060.28
Warsaw Sports, Inc.	1003	12/02/2003	$186.87
Bikes and Trikes	1004	12/02/2003	$823.05
SAB Mountain	1005	12/03/2003	$29.00
Poser Cycles	1006	12/03/2003	$64.90
Spokes	1007	12/03/2003	$49.50
Clean Air Transportation Co	1008	12/03/2003	$2,214.94
Extreme Cycling	1009	12/03/2003	$29.00
Cyclopath	1010	12/03/2003	$14,872.30
BBS Pty	1011	12/03/2003	$29.00
Piccolo	1012	12/03/2003	$10,259.10
Pedals Inc.	1013	12/03/2003	$1,142.13
Spokes 'N Wheels Ltd.	1014	12/04/2003	$29.00
Cycle City Rome	1015	12/04/2003	$43.50
SAB Mountain	1016	12/04/2003	$563.70
Tyred Out	1017	12/05/2003	$72.00
Has Been Bikes (consignme	1018	12/05/2003	$115.50
Spokes for Folks	1019	12/05/2003	$43.50

Figure 11.16 The finished report with the function-based formula in place

Like functions created using the extract, custom functions written by hand can also be reused in other formulas in your report.

USING CUSTOM FUNCTIONS IN FORMULAS

Because custom functions look like any other function field you would find in Crystal Reports, you can simply add them to your formula text using the same syntax as you just used (that is, function(parameter1, parameter2, etc.)). But if you don't want to go to the trouble of actually typing out a formula field just to use the function, there is another method of using functions.

To see this method in action, use the following steps:

1. Open FUNCTIONWIZARD.RPT report from the book download files.
2. Click View > Field Explorer to open the list of available fields.
3. Expand the Formula Fields section, right-click, and select New. The Formula Editor opens.
4. From the Toolbar, click on the Magic Wand icon to invoke the Custom Function Expert.
5. Expand the list of Report Custom Functions, and click the CountryGrouping function. The bottom of the dialog box displays the values that are expected by this parameter.

6. Click the link Select Field or Enter Value, and enter a value or use the drop-down list to select Choose other field, which will open the Field Selector dialog box shown in Figure 11.17.
7. Select a field from the list, and click OK.
8. Next, click Save and Close to return to your report.
9. You can now drag your formula field onto your report and preview your report to see the results.

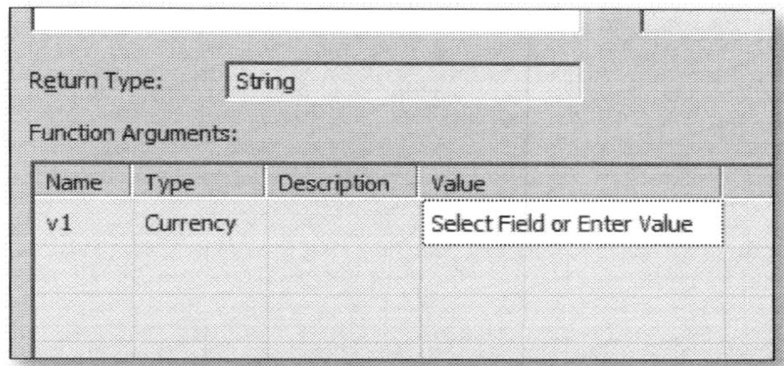

Figure 11.17 The Field Selector

Whether you use the Custom Function Expert or just bang in the formula yourself, the results are the same. If you are planning to create a simple formula where you want to just use the function, the Expert is probably a good choice because it is quick and simple. If you want to write a more complex formula or want more control over the formula text, it is probably better if you write the formula yourself.

SQL COMMANDS & EXPRESSIONS

One of the most powerful features of Crystal Reports is its ability to write the SQL for your report, as you select tables, links, fields, and so on. In early versions of Crystal Reports, this functionality was handy for new users who didn't know how to write SQL, but was frustrating for database administrators and developers who knew exactly what they wanted.

As the product matured, the ability to modify parts of the SQL statement and add expression fields written with SQL was introduced, but this still didn't go far enough. It wasn't until the concept of SQL Commands were introduced that developers were able to fully leverage the power of the database server itself.

WORKING WITH SQL COMMANDS

So what is an SQL Command? In Crystal Reports, it is an SQL statement that is passed to the database server, and the resulting set is treated as a virtual table that can be used as the basis for your report.

For example, you could write the following simple SQL Command to retrieve all of the results from the customer table in your database:

```
SELECT * FROM CUSTOMER
```

When this SQL command is used in your report, all of the resulting fields will appear in the Field Explorer, and the records that are returned can be used as the data set for your report.

Remember as we go through the following sections that the purpose of this part of the chapter is not to teach you how to write SQL — there are entire volumes and references written on the subject, and each SQL syntax may be slightly different. Rather, this section teaches you how to use the SQL you write as the data set for your reports.

Creating SQL Commands

In our first example of creating an SQL Command, we are going to use a similar SQL statement of some fields and a number of records that we can use as the basis for our report. To keep the SQL simple we will select all of the items from a table with a `where` clause to filter the number of records returned.

To create a report based on an SQL Command, use the following steps:

1. From the Start menu, open Crystal Reports.
2. Click File > New > Blank Report. The Database Expert opens.
3. Double-click Create New Connection, and then expand the ODBC (RDO) folder to open a list of available data sources.
4. Select the data source named Galaxy and click Finish. A node is added underneath the data source name in the folder.
5. Click the Add Command option, as shown in Figure 11.18. The SQL Command window opens.

Figure 11.18 The Add Command option appears below your data source name

6. Enter the following SQL code in the large text box on the left side:

```
SELECT * FROM CUSTOMER WHERE COUNTRY='USA'
```

7. The SQL command window should now look like the one shown in Figure 11.19 over the page.
8. Click OK to return to the Database Expert, and then OK to return to your report design.
9. Click View > Field Explorer to open a list of available fields. The available fields will be listed under the heading Command. You can drag and drop them on your report.
10. Save your report as CustomerListing. We will be adding to the report as we go along in this section.

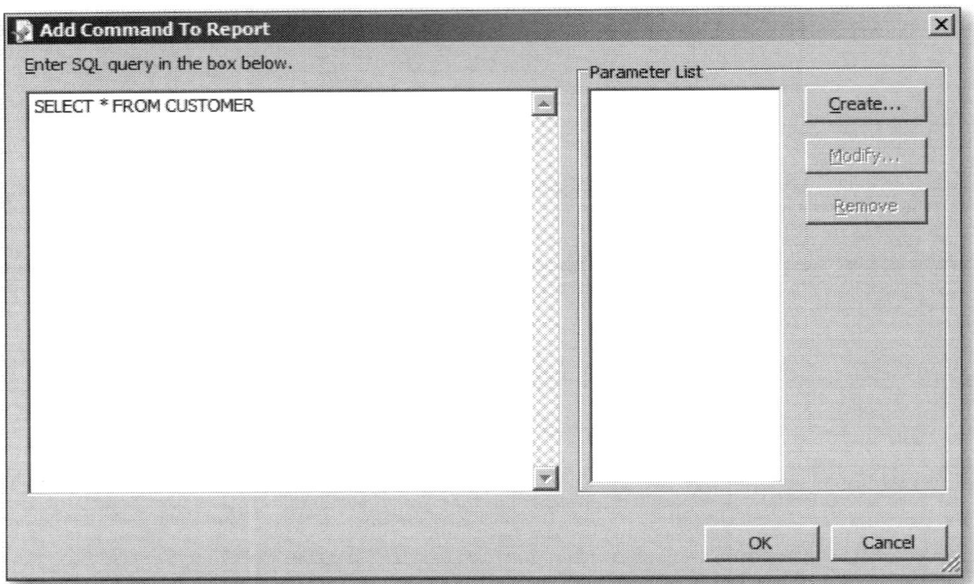

Figure 11.19 The SQL command dialog

This is just a simple example of an SQL Command. The rules of creating an SQL Command are pretty simple. Firstly, the SQL command itself must contain SQL code that can be processed by your database. For example, if you had some SQL syntax that was specific for DB2 and tried to use it to write an SQL Command, it would not work. The SQL syntax that you write for an SQL Command must run exactly as it is entered.

A good practice is to test your SQL statement in a query tool before you actually paste it into an SQL Command. If there is an error in your SQL Command, Crystal Reports will return the error message from the database, but sometimes it is easier to debug SQL in a "native" environment. For example, Query Analyzer for SQL Server has definite advantages when writing SQL, because you can browse databases and tables and confirm field names. Once you have the SQL query running there, paste it into an SQL Command.

Second, the SQL Command should be able to run and return a result set to Crystal Reports. You can use SQL as complex as required, using unions, subselects, and so on, but the query must return a result set. Often developers new to Crystal Reports will get frustrated because they see SQL Commands as a way to run SQL on the database server and will try to update records and use complex SQL normally reserved for stored procedures in their SQL Commands.

Remember, SQL Commands form the data set for your report and need to return columns, records, and so on. If you need to do some complex processing where you need to insert, update, or otherwise manipulate records or advanced multipass SQL usually reserved for processing records, use a stored procedure. You can then use an SQL Command to return a data set from the table or view that you created.

Editing SQL Commands

Because SQL commands can be revised a number of times during the course of writing a report, Crystal Reports makes it easy to edit the SQL Command through the Database Expert. To open the Database Expert, click Database > Database Expert, then right-click the Command object itself. In the shortcut menu you will see options to Edit or View the SQL Command that will open the SQL Command window.

Another handy trick for SQL Commands: To rename an SQL Command, click the command, then press F2 and type a new command name. When finished, press the ENTER key.

From the SQL Command window, you can edit your SQL statement as required and then click OK to return to your report.

ADDING PARAMETERS TO SQL COMMANDS

Parameters are a popular feature in Crystal Reports, because they allow you to enter and select information when the report is run. With SQL Commands, you can create parameters that are included as part of the SQL statement itself, so the processing of these parameters will always occur on the database server itself.

To add parameters to an SQL statement, use the following steps:

1. Open the CUSTOMERLISTING.RPT report that we were working with earlier in this section.
2. Click Database > Database Expert.
3. Right-click the Command object on the right side of the dialog box, and select Edit Command from the shortcut menu to open the SQL Command window.
4. From the right side of the dialog box, click Create to create a new parameter and open the dialog box shown in Figure 11.20 on the following page.

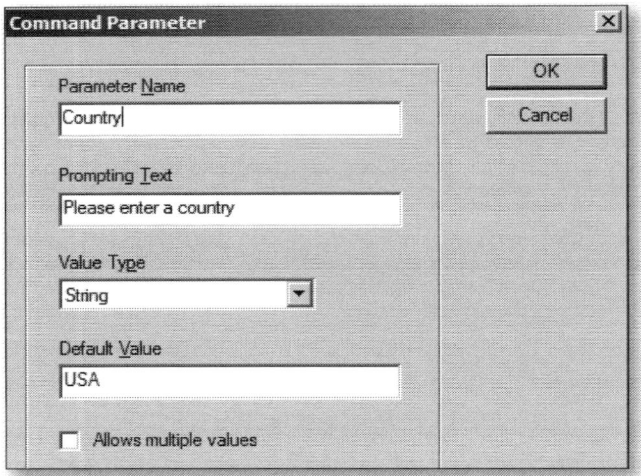

Figure 11.20 The SQL Command parameter dialog box

5. Enter a name for your parameter, as well as some prompting text, a value type, and any default value you want to assign to the parameter field. In this example, name the parameter Country and define the type as "String".
6. When finished, click OK to return to the SQL Command window.
7. Next, modify your SQL Command to include the following parameter in the `where` clause.

```
Select * from customer where Country = '{?Country}'
```

8. Click OK to return to the Database Expert and OK again to return to your report design.
9. Press F5 to refresh your report. You should be prompted for a Country value, which will be used to filter your report.

The key points to remember about parameters in SQL Commands is that they are referenced just like other Crystal Reports parameters (that is, with a question mark and curly braces around the field name (`{?country}`), and the values entered are translated literally. For example, in the SQL Command we just saw, the parameter itself was enclosed in quotes in the `where` clause. The reason for this is that when the SQL Command is executed, it will replace the parameter field with the actual values you enter.

You will also notice that when you are setting up parameters in SQL Commands, the dialog box does not have as many options as when you created a normal parameter in the

Report Designer. Parameters in SQL Commands are passed directly back to the database, so you won't find all of the capabilities you might find with normal parameter fields.

However, you can use a combination of SQL Command parameters and normal parameter fields in your reports if required, and in some instances, you can use normal parameter fields in record selection that can be evaluated on the database server for more efficient processing and flexibility.

Parameters can be used with SQL Commands for all sorts of things, including creating Top N reports that are processed on the database, executing stored procedures, and passing stored procedure parameters. For more examples of using advanced SQL Commands, check out the book download, in particular the Chapter 11 folder.

WORKING WITH SQL EXPRESSIONS

SQL Expressions are SQL statements that are like formulas that get submitted to and evaluated by your database server. The difference between SQL Commands and SQL Expressions is that SQL Expressions can be used with reports created from tables, views, and so on and are an easy way to push more processing back on the database server. It's important to note that you can't have SQL Expressions in a report that is based on an SQL Command (because if you are using an SQL Command, you can just put the required SQL code in the command itself).

For reports based on database tables however, SQL Expressions can definitely reduce report-processing time, because they are evaluated on your database server, as opposed to Crystal Reports formulas, which are evaluated locally.

To create a new SQL Expression in your report, use the following steps:

1. Open the CUSTOMERANALYSIS.RPT report from the book download files.
2. Open the Field Explorer by clicking View > Field Explorer, and then right-click the node marked SQL Expression and select New. A dialog box appears and prompts you for the name of your SQL Expression.
3. Enter a name and click OK to open the SQL Expression Editor, as shown in Figure 11.21 over the page. If this interface seems familiar, it should — the SQL Expression Editor is based on the Formula Editor and features the same panes across the top of the dialog box, showing Fields, Functions, and Operators.

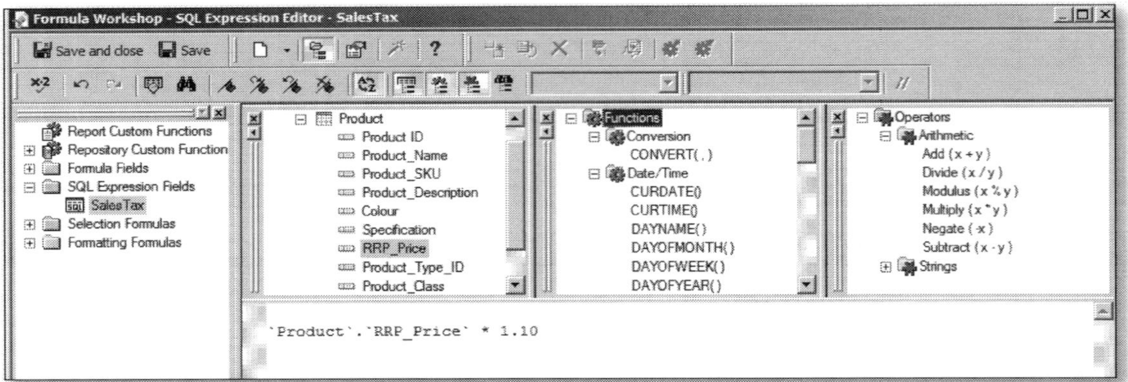

Figure 11.21 The SQL Expression Editor

You'll notice that the functions list displays generic SQL functions that you can use in your SQL Expression instead of the usual Crystal Reports functions.

---

4. Using the text box at the bottom of the editor, enter your SQL expression, following correct SQL syntax and operator usage. If you are unfamiliar with which SQL Expressions are available for your particular database, your database administrator should be able to help you find out.

5. Once you have finished entering your SQL Expressions, click the Save and Close button to return to your report design. You can now drag your SQL Expression into your report as you would any other field.

SUMMARY

Adding calculations and logic to your reports is an easy way to add value and provide insight into the data presented. Using the formula languages available in Crystal Reports, you can create formula fields quickly using the hundreds of functions and operators that are available, or you can create your own custom functions that can be shared between formulas or even reports.

Finally, for complete control over the report data set and to push processing back to the database server, SQL Commands and Expressions return control of the data set squarely back into the developer's hands.

This all provides a good background for the next chapter, where we start looking at producing more complex reports and build on the concepts learned so far.

Chapter 12
Using Subreports

In This Chapter

- Working with Subreports
- Using Unlinked Subreports
- Creating Linked Subreports
- Formatting Subreports
- Subreport Performance Considerations

INTRODUCTION

In the past few years, organizations have seen an explosion in the amount of data that they must analyze and present. Often this data is from multiple data sources, including order entry applications, accounting systems, manufacturing systems, databases, e-mails, and spreadsheets. Often, they are required to show information from these different sources side-by-side in a single report. Because these data sources can be stored in many different ways, it has previously been difficult to display heterogeneous data in a single report. Fortunately, Crystal Reports has a feature called subreports that make this possible.

Subreports are a flexible feature in Crystal Reports that allow you to insert multiple instances of reports into a single report file. Subreports have many different uses, including displaying related or unrelated data in a single report and providing at-a-glance reports with multiple pieces of information displayed on a page to create packages of reports that allow multiple reports to be run and printed at the same time.

In the first half of the chapter, you will learn how to create both linked and unlinked subreports and how to use them in a report. We'll also see how to format a subreport and learn about techniques for managing the use of subreports.

WORKING WITH SUBREPORTS

You may sometimes want to display information from two different data sources in the same report. If these two data sources have a common key and are compatible, you can create a report easily using both of these data sources. For example, if you have an employee database that contains employees' names, employee numbers, and so on and a payroll database that also stores the employee numbers in the same format, you can easily join these two data sources together to create a report that lists information from both sources.

But what if the payroll database doesn't store the employee number in the same way? For example, suppose that the employee table stores the number in a true number field, and the payroll database stores it as a string. This is where a subreport comes in handy. You can create a report from the employee table and then insert a subreport to display the payroll information from the other data source, as shown in Figure 12.1.

Payroll Report

Galaxy Toys

Last Name	First Name	Position	Hire Date
Brown	David	Account Manager	05/23/2006

Pay Period Start	Pay Period End	Gross Pay	Tax	Nett Pay
21-Oct-2005	28-Oct-2005	$383.53	$84.38	$299.16
28-Oct-2005	04-Nov-2005	$211.94	$46.63	$165.32
04-Nov-2005	11-Nov-2005	$782.78	$172.21	$610.57
11-Nov-2005	18-Nov-2005	$283.29	$62.32	$220.97
18-Nov-2005	25-Nov-2005	$987.33	$217.21	$770.12
25-Nov-2005	02-Dec-2005	$749.49	$164.89	$584.60
02-Dec-2005	09-Dec-2005	$141.22	$31.07	$110.15
09-Dec-2005	16-Dec-2005	$979.09	$215.40	$763.69
16-Dec-2005	23-Dec-2005	$276.57	$60.84	$215.72
23-Dec-2005	30-Dec-2005	$948.90	$208.76	$740.14
30-Dec-2005	06-Jan-2006	$871.98	$191.84	$680.14

Last Name	First Name	Position	Hire Date
Roy	Michael	Account Executive	05/04/2009

Pay Period Start	Pay Period End	Gross Pay	Tax	Nett Pay
21-Oct-2005	28-Oct-2005	$471.57	$103.75	$367.83
28-Oct-2005	04-Nov-2005	$793.32	$174.53	$618.79
04-Nov-2005	11-Nov-2005	$867.80	$190.92	$676.89
11-Nov-2005	18-Nov-2005	$715.37	$157.38	$557.99

Figure 12.1 An example of a linked subreport

Subreports come in two flavors: linked and unlinked. In the previous example, you would use a linked subreport, so that each employee is shown with the correct payroll details.

When working with linked subreports, Crystal Reports passes a parameter between the main report and subreport, which the subreport uses for record selection.

Crystal Reports sometimes uses the term main report or container report to indicate that a report contains subreports.

For an unlinked subreport, Crystal Reports does not pass any parameters between the main report and subreports, so the subreport displays all of the records available.

If you would like to follow along with the step-by-step instructions in this chapter, this book includes downloadable files that are available from www.kuiperpublishing.com. Also, check out "Setting up the Samples" section in the front of this book for instructions on how to configure the data sources, reports, etc.

In our example report, you could add an unlinked subreport to your employee report to show national salary averages. Because the data in the main report and the subreport is not linked, you can show totally unrelated data side by side, as shown in Figure 12.2.

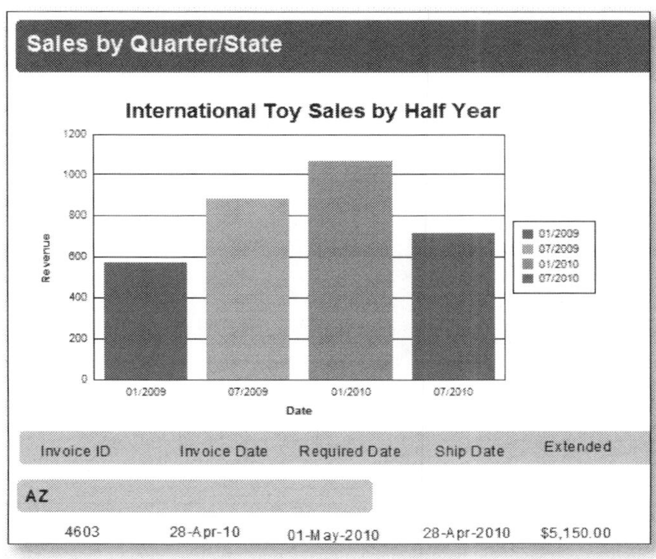

Figure 12.2 An unlinked subreport showing unrelated data

When working with subreports, an important concept to remember is that subreports are both an element of the main report and an individual report in their own right. Each subreport you insert into your main report will be shown in the section where it is inserted, but it will also have its own Design tab, as shown in Figure 12.3.

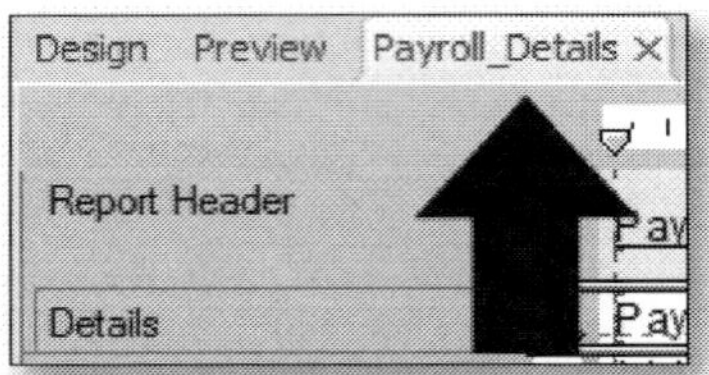

Figure 12.3 A subreport Design tab

If you preview a subreport independently, you will see a separate Preview tab for each subreport as well.

You can edit the attributes of both the subreport object shown in the main report and the subreport itself— but before we get too far down that road, we need to look at how subreports are inserted.

INSERTING AN UNLINKED SUBREPORT

An unlinked subreport doesn't require a parameter to be passed to the main report and can display any information you require. The information can be totally unrelated and can be from different databases with different formats, tables, and so on. When working with unlinked subreports, you can either select an existing report to insert or create one from scratch. To get started, click Insert > Subreport to open the dialog box shown in Figure 12.4.

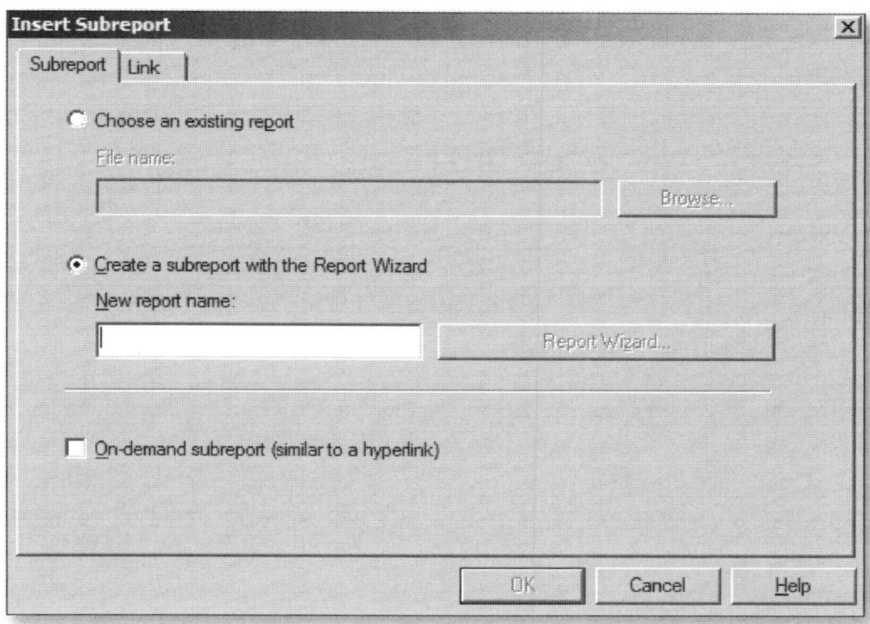

Figure 12.4 You can insert an existing report or create a new one

You may want to switch to the Design tab before inserting a subreport, so you can see the sections and their boundaries easily.

Use the radio buttons to indicate whether you will choose a subreport or create one. If you select Choose an Existing Report, click the Browse button to locate a Crystal Report that you have created earlier. If you select Create a Subreport with the Report Wizard, you will need to enter a name for your subreport and then click the Report Wizard button to invoke an expert to create your report. When you have finished, click OK to return to the Insert Subreport dialog box.

When you have finished with the Insert Subreport dialog box, click OK. Your subreport will be attached to the tip of your mouse pointer. You can then position the subreport on your main report and click to place it.

You should choose the location of your subreport carefully. Where you position it determines how many times it is processed. If you place your subreport in the report header or footer, for example, the subreport will be processed only once for each report. If you place the subreport in the page header or footer, the subreport will be processed for every page, which can cause significant performance problems when your report is run or refreshed.

INSERTING A LINKED SUBREPORT

Working with linked subreports is just as easy as working with their unlinked counterparts. The only difference is that you will need to specify a field in both the main report and subreport that will determine the relationship between the two. In the Insert Subreport dialog box shown in Figure 12.5, there is an additional tab for Link.

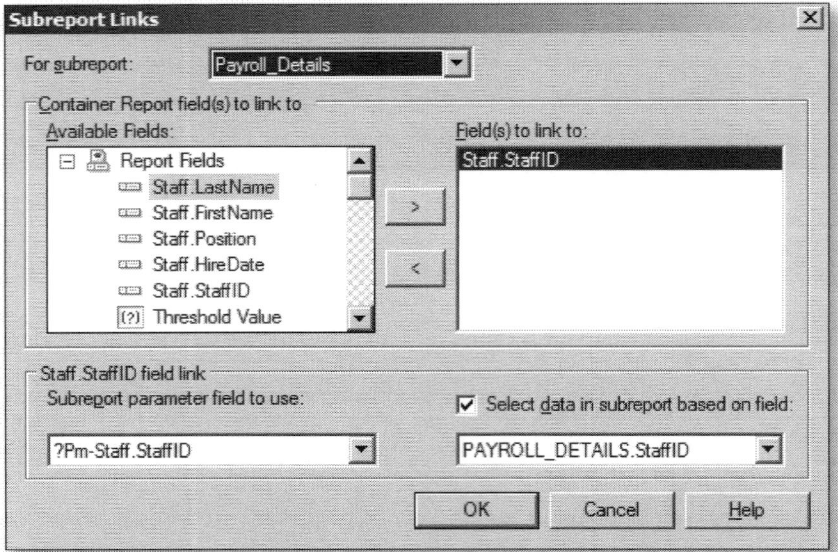

Figure 12.5 You can specify the links between your main report and subreport

The first step in linking the two reports is to decide which field in your main report to use. From the list of available fields on the left side of the dialog box, select a field and use the right arrow to move the field from this list to the one on the right.

Once you have selected a field, additional options at the bottom of the dialog box appear, as shown in Figure 12.6, to allow you to select a field in your subreport. You'll notice that a parameter field is created automatically for you; all you need to do is select a field in use in your subreport.

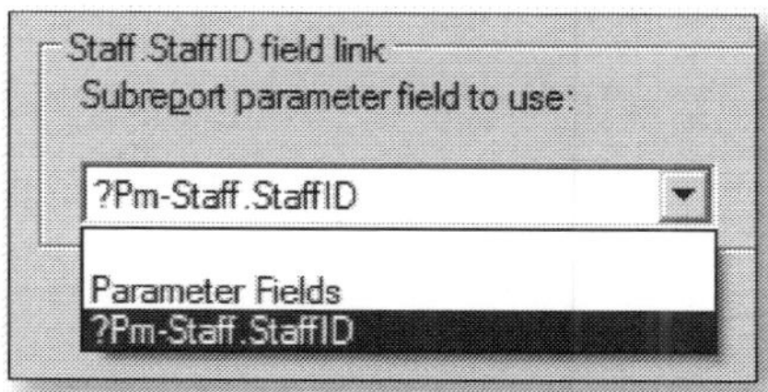

Figure 12.6 You will need to select a parameter field and a field from your subreport

If you don't want the value in the parameter field to be used for record selection, you can remove the check mark from the Select Data in Subreport Based on Field box. The subreport will be linked and the parameter field value will be passed, but the parameter will not be used in record selection.

You can select multiple fields in your main report for linking in the same manner. You will need to specify a field in the subreport for each. When you have finished, click OK to accept your changes and place the subreport in your main, or container, report.

Again, it is important where you place a linked subreport. A subreport will be processed once every time the section appears. If you have a report with 500 detail records and place the subreport in the Detail section, it will be processed 500 times, adding to the total report processing time.

CREATING A LINKED SUBREPORT

To demonstrate some of the techniques you just saw for linked and unlinked subreports, the project that follows recreates the Employee Payroll report we looked at earlier in the chapter. We will start off with a simple employee listing report and then add a subreport to display the payroll details from a separate data source. To recreate this report, use the following steps:

1. Open Crystal Reports, and open the PAYROLL.RPT report from the book download files.
2. Switch to the Design view of your report, and click Insert > Subreport to open the Insert Subreport dialog box.
3. Click the option to Create a subreport with the Report Wizard, and enter a name for your subreport. In this example, call the subreport Payroll Details.
4. Click the Report Wizard button to open the Standard Report Creation Wizard.

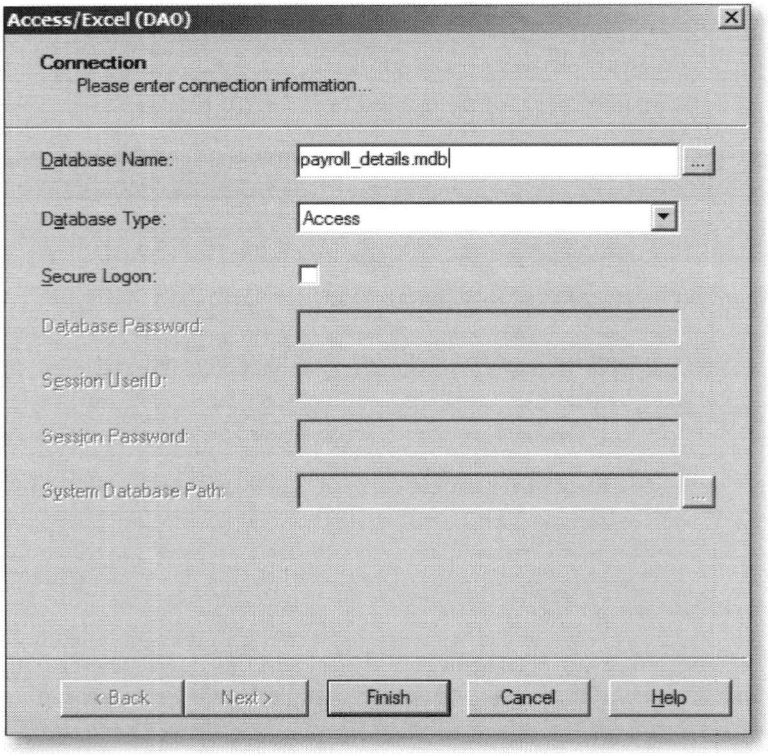

Figure 12.7 Access/Excel Data Connection options

5. In the Data dialog box, expand the node for Create New Connection, then double-click Access/Excel DAO to open the dialog box as shown in Figure 12.7 on the previous page.
6. Click the browse button (…), browse to the book download files, and select the PAYROLL.MDB file. Then click the Finish button.
7. Expand the Tables node of your data source, and double-click the PAYROLL_DETAILS table to add it to your list of selected tables. Then click Next.
8. In the Fields dialog box, click the double-arrow to move all of the fields from the table to the selected area.
9. Click the Finish button to return to the Insert Subreport dialog box.
10. Click OK to finish the expert. The subreport will now be attached to your mouse pointer.
11. Click to place the subreport in the Details section of your report.
12. Preview your report to verify that the subreport is working correctly.
13. Next, we need to set the linking for your subreport, so right-click the subreport, and select Change Subreport Links to open the dialog box shown in Figure 12.8.

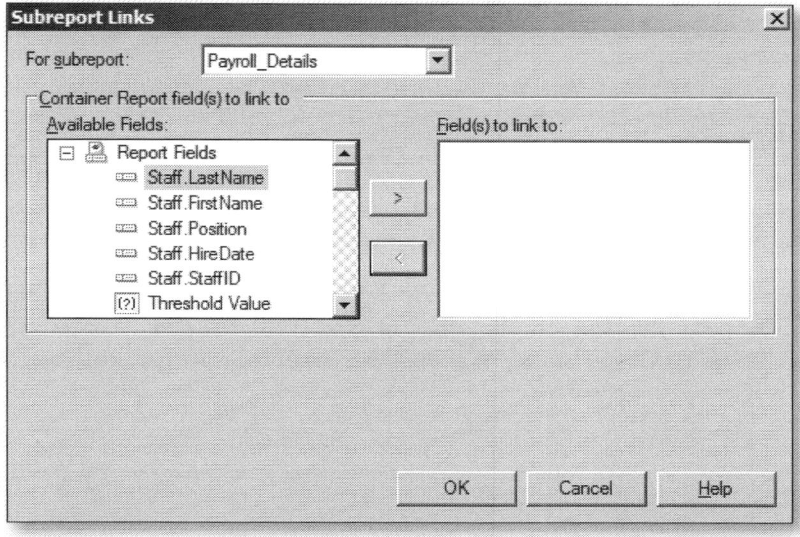

Figure 12.8 The Change Subreport Links dialog box

14. Select the Staff ID field, and use the right arrow to move it to the list of selected fields on the right. A parameter field is created and appears at the bottom of the dialog box.

15. On the right side, select the option for Select data in subreport based on field, and use the drop-down list to select the Employee ID field. Then click OK to return to your report.
16. Your subreport results should now be filtered for each employee, so only the payroll details will be displayed, as shown in Figure 12.9.

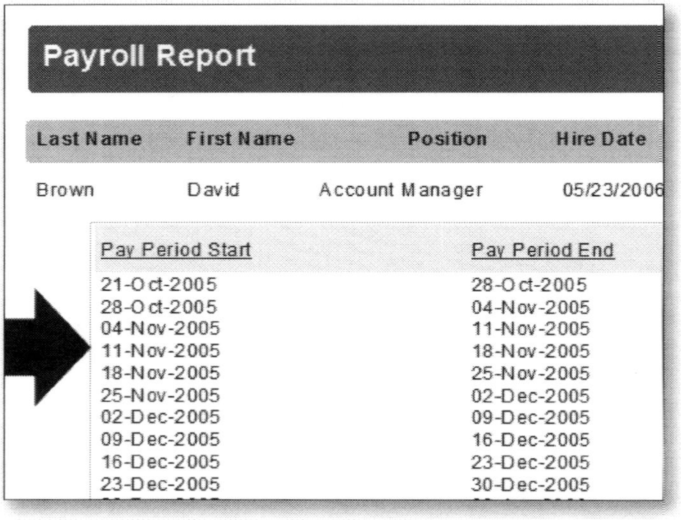

Figure 12.9 The finished report with a linked subreport for the payroll details

Remember, when you create a linked subreport, a parameter field is used in the subreport to pass the value from the main report to the subreport. You can use this parameter field as you normally would and include it in formulas, record selection, etc. In addition, if you were to preview your subreport on its own, you would be prompted to enter this parameter. This can often be a good way to test your subreport independent of the main report.

CREATING AN ON-DEMAND SUBREPORT

On-demand subreports are subreports that are run when the user requests them. Unlike the subreports we have looked at so far in this chapter, on-demand subreports do not add any additional processing overhead to our report, because they are not run until the user requests them.

To create an on-demand subreport, use the following steps:

1. Open Crystal Reports, and open the Payroll report we have been working on in this section.
2. Right-click the subreport, select Format Subreport, then click the Subreport tab to open the dialog box shown below in Figure 12.10.

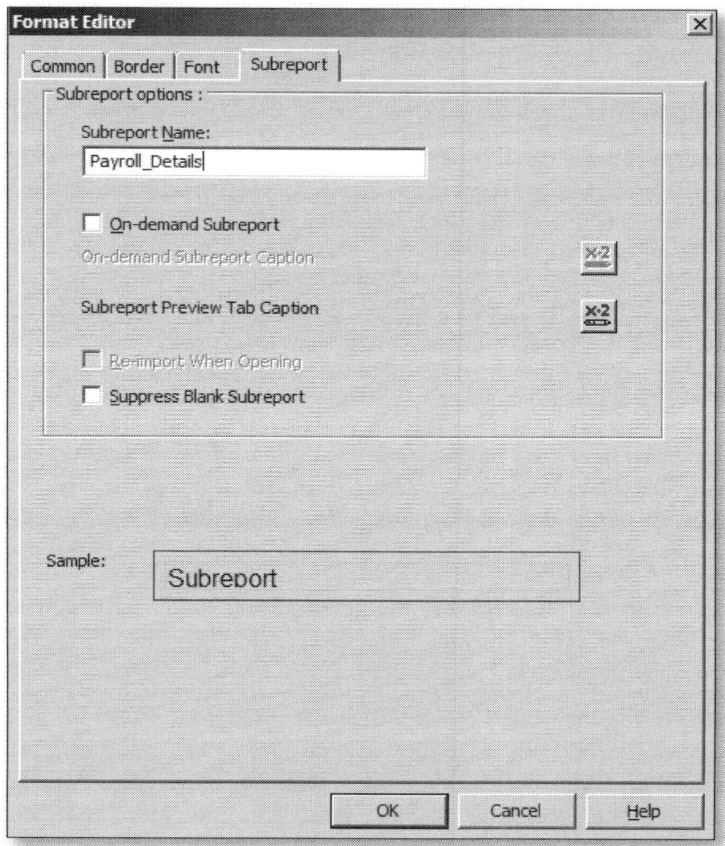

Figure 12.10 Subreport formatting options

3. Select the option for On Demand Subreport. The subreport will then be shown as a link on your report.
4. To control the text that appears either on the on-demand link or Preview tab, click the X+2 option beside either of these options and enter the caption in quotes (for example, "Click here for Payroll Details").
5. Click OK to return to your report, which should now look like the one shown in Figure 12.11.

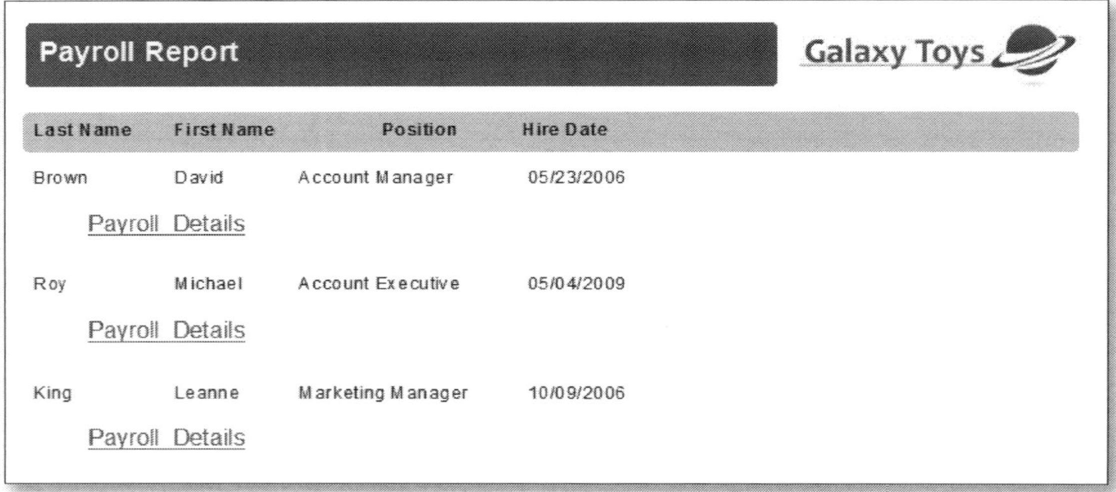

Figure 12.11 A report with an on-demand subreport

FORMATTING SUBREPORTS

Once you have inserted a subreport into your report, you can apply a number of formatting options and techniques to integrate the main report and subreport into one seamless presentation.

CHANGING THE SUBREPORT NAME

When you insert an existing subreport into your main report, by default Crystal Reports names the subreport the same name as the report file you have inserted. For example, if you inserted MYSALES.RPT, the subreport would also be named MYSALES.RPT. To rename the subreport, right-click the subreport and select Format Subreport from the shortcut menu. Using the option shown on the Subreport property page and the Formula Editor, shown in Figure 12.12, you can rename your subreport. Whenever the subreport name is shown (for example, in the main report, tool text, on a design tab for the subreport, and so on), this new name will be used.

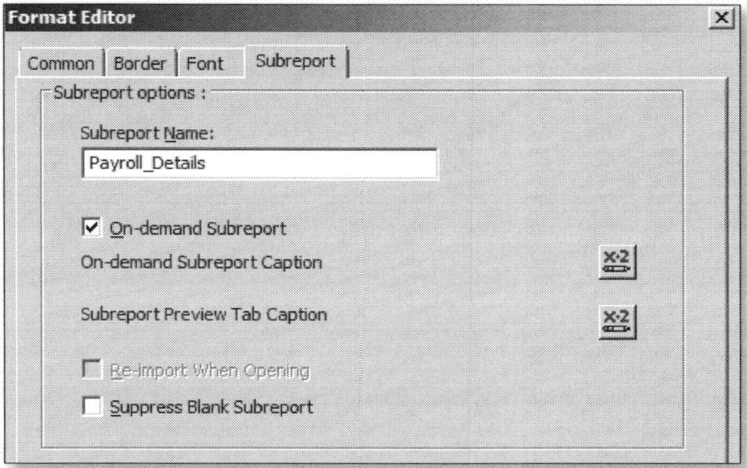

Figure 12.12 You can rename the subreport to something more meaningful using the Formula Editor

CHANGING THE BORDER

By default, Crystal Report places a border around any subreports you have inserted into a main report. This is usually the first default formatting option you will want to turn off. To change the border around a subreport, right-click the subreport, select Format Subreport, then click the Border tab to open the dialog box shown in Figure 12.13.

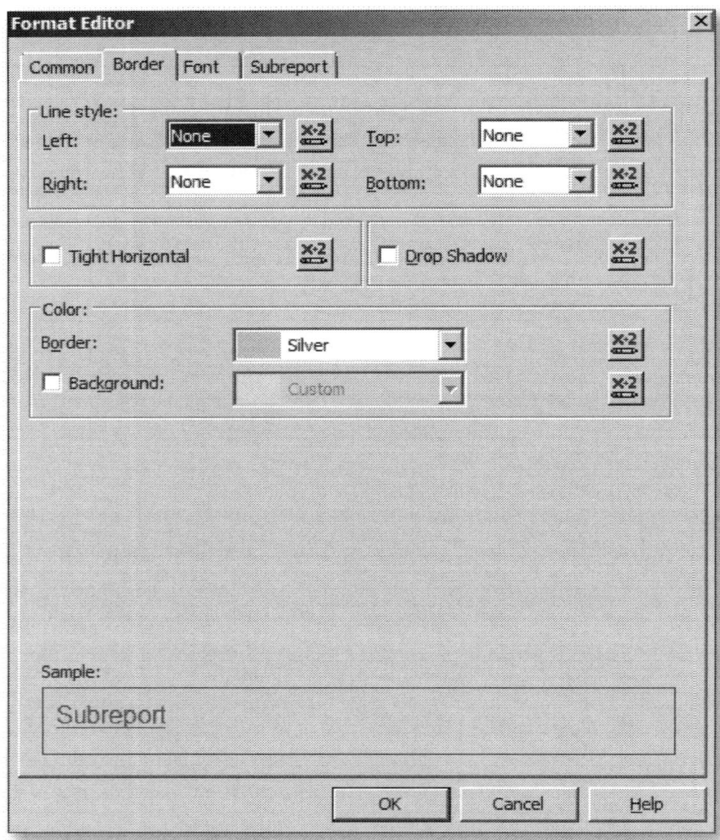

Figure 12.13 You can control subreport formatting from the shortcut menu

From the Border tab in the Format Editor dialog box, you can change all four of the Line Style drop-down boxes from Single to None. Alternatively, if you want a border around your subreport, you can use the drop-down boxes and options to select a line style (single, double, dashed, dotted) and color, as well as a background color and drop shadow. When you have finished editing the borders and colors for your subreport, click OK to return to your report Design or Preview view.

For precise control over where the subreport appears and its size on the page, right click the subreport and select Size and Position from the shortcut menu. Here you can enter precise dimensions and positioning coordinates.

CHANGING SUBREPORT LINKS

Subreport links are usually set up when you first insert a subreport, but you can change subreport linkage as your needs and report structure change. You will need to locate the subreport you want to change, right-click it, and select Change Subreport Links from the shortcut menu. The Subreport Links dialog box you saw earlier opens.

To select a field for linking, locate a field in the list of available fields, click to select it, and use the right arrow to move it to the list on the right. Once you have selected a field, a second set of drop-down lists and options will appear, allowing you to select fields in your subreport.

When you have finished changing the links for your subreport, click OK to accept your changes. The next time your report is refreshed, these changes should be reflected.

You can refresh your report at any time by clicking the Refresh icon or pressing F5.

SUMMARY

While subreports can enable you to combine data from different sources, as well as display data in new and unusual ways, they should be used sparingly in your report, to keep report performance at an optimum. Where you can, it is better to use on-demand subreports, as the subreport will not be run until the user requests it. While they may wait a few seconds for the subreport to run and load, this is a more efficient use of system resources.

With subreports out of the way, it's time to have a look at cross-tabs, which provide the ability to summarize your report data by rows and column. And that's where we will pick up in the next chapter.

Chapter 13
Working with Cross-Tabs

In This Chapter

- Inserting Cross-Tabs
- Re-ordering Cross-Tab Data
- Working with Summaries
- Formatting Cross-Tabs
- Analyzing Cross-Tab Data
- Cross-Tabs and Sections
- Charting Cross-Tab Data

INTRODUCTION

Cross-tabs provide an easy way to add complex data summarization and analysis to your report. As the main feature of a report, or as a supporting object within an existing report, they provide an at-a-glance view of information in your database. In this chapter, you will learn how to create, format, and manipulate cross-tabs, adding instant analysis and summarization to your reports.

Cross-tabs, like the one shown over the page in Figure 13.1, are composed of rows and columns of summarized data and can either be placed in the structure of existing reports or can be the main focus and content of their own reports.

With Crystal Reports 2008, there are also a number of new cross-tab features that make it even easier to summarize your report data. We'll be having a look at these features a little later in the chapter, but for now we need to look at some basic cross-tab concepts.

Customer by Country/Type

	Bricks & Mortar	Hybrid	Online	Wholesaler
Australia	$5,684.28	$0.00	$9,327.41	$20,545.26
Canada	$0.00	$10,259.63	$0.00	$0.00
England	$18,938.68	$0.00	$0.00	$0.00
Mexico	$0.00	$9,274.40	$5,566.38	$0.00
USA	$394,360.32	$396,486.43	$267,406.32	$182,412.91

Figure 13.1 A typical cross-tab summarizing domestic and international sales

UNDERSTANDING CROSS-TABS

Cross-tabs look similar to a spreadsheet, but it is important to remember that cross-tabs are database driven. In the Design view of your report, a cross-tab presents a simple row-and-column display. When viewed in the preview of your report, these rows and columns are filled with the data you have requested. The size of the cross-tab varies based on the number of records returned.

To facilitate the creation of reports specifically for use with cross-tabs, Crystal Reports includes a Cross-Tab Expert that walks you through the process of creating a cross-tab report. In addition, you can insert cross-tabs as an element of an existing report.

From the Insert menu, click Insert > Cross Tab. A blank cross-tab is attached to your mouse pointer. Click to place the cross-tab on your report, then right-click the cross-tab and select Cross-tab Expert to open the dialog box as shown in Figure 13.2 on the next page.

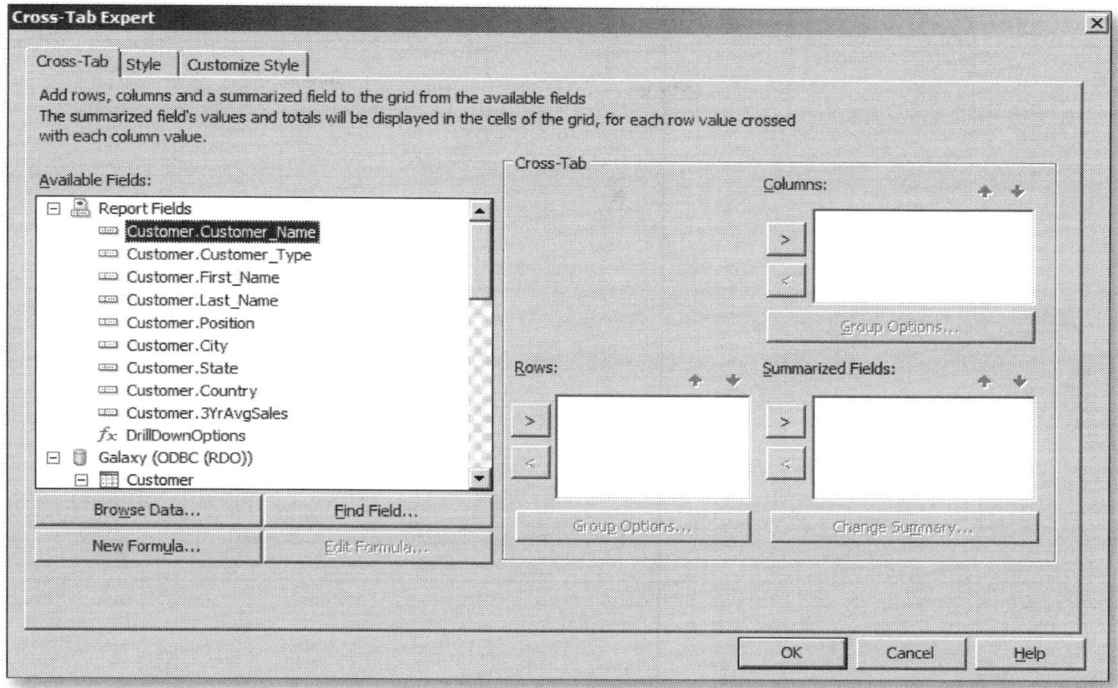

Figure 13.2 The Cross-Tab Expert

CREATING CROSS-TABS

To create a cross-tab, you need three elements: rows, columns, and summarized fields. For a basic cross-tab, you need only one of each. To create new rows, columns, or summarized fields, you can drag a field from the Available Fields box to the corresponding box for rows, columns, or summarized fields.

You can also highlight a field in the Available Fields box and click the arrow button next to text boxes for the Rows, Columns, and Summarized Fields.

To remove a field from a cross-tab, highlight the field and click the left arrow button beside the text box to remove it from the list of selected fields.

If you are curious about the contents of a particular field, you can view some of the values held in the field by highlighting the field name and clicking the Browse button at the bottom-right corner of the dialog box. This action will return about 200 records at a time for you to look through, but keep in mind that on large databases, this process may take some time.

With cross-tabs, you can add as many fields for the columns, rows, and summarized fields as required. When your cross-tab is printed, the fields you specified for the columns and rows will be used to create the cross-tab, and the summarized field you specified will be calculated at the intersection of the columns and rows. For example, in the cross-tab shown in Figure 13.3, the fields for Product Class and Product Name have been used for the rows, with an Order Date field used for the columns. A Quantity value appears in the summarized field.

Sales by Quarter

		01/2009	04/2009	07/2009	10/2009
Action Figures & Collectibles	Animal Farmhouse	$49,575.50	$45,370.50	$18,313.50	$81,722.00
	Dirt Boss Decepticon Action	$33,084.15	$38,716.35	$44,214.50	$59,283.16
	Total	$82,659.65	$84,086.85	$62,528.00	$141,005.16
Apparel	Baby Sleeper Combo	$24,096.10	$12,631.86	$5,204.00	$29,203.80
	Super Heros T-Shirt	$3,679.20	$3,153.60	$1,576.80	$6,832.80
	Total	$27,775.30	$15,785.46	$6,780.80	$36,036.60
Arts & Crafts / Activities	Dashboard Adventures	$11,737.50	$27,469.00	$23,766.60	$35,597.30
	My Little Helper	$13,625.00	$28,724.00	$31,610.25	$15,657.00
	My Little Princess Jewelry M	$21,404.30	$18,232.30	$16,913.35	$54,082.50
	Paint by Numberz - Mona Li	$15,812.00	$27,496.75	$23,502.00	$43,026.95
	Play-All-Day Play Center	$7,770.15	$14,875.00	$34,340.50	$31,916.50

Figure 13.3 A typical cross-tab, summarizing orders over a number of quarters

The value highlighted in Figure 13.3 is the summarized quantity of Guardian "U" Locks ordered in the second quarter of the year (QTR2). By default, the summary operator used for numeric fields is Sum. A little later, you'll learn how to highlight and change the summary operator for summarized fields, but for now it is important to remember that the summarized field is calculated by Crystal Reports at the intersection of rows and columns.

To help you understand how you can create your own cross-tabs, we are going to create a new cross-tab and insert it into an existing report. To do so, use the following steps:

1. Open Crystal Reports and from the book download files, open the QUARTERLYSALES.RPT report.
2. Switch to the Design view of your report, and click Insert > Cross Tab. Then click to place a blank cross-tab in the Report Header in your report.
3. Right-click the cross-tab, and select Cross-Tab Expert from the shortcut menu.
4. Drag the Product Class and Product Name fields from the list of Available Fields to the Rows box.
5. Drag the Invoice Date field from the list of Available Fields to the Columns box.
6. Drag the Qty field from the list of Available Fields to the Summarized Fields box. The Cross-Tab Expert should now look like the one shown in Figure 13.4, over the page.
7. Click OK to return to your report.

If you would like to follow along with the step-by-step instructions in this chapter, this book includes downloadable files that are available from www.kuiperpublishing.com. Also, check out "Setting up the Samples" section in the front of this book for instructions on how to configure the data sources, reports, etc.

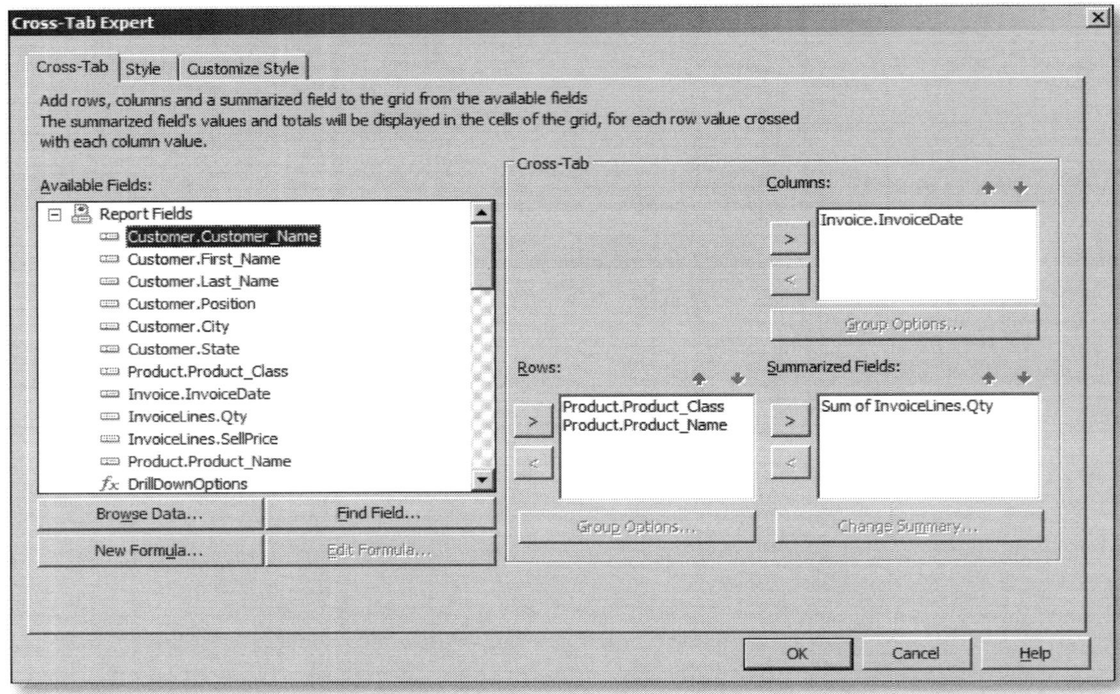

Figure 13.4 The Cross-Tab Expert with all of the required fields selected

When you preview your report, the data will be filled into the cross-tab and shown using the rows, columns, and summarized fields we selected earlier.

It's important to note that when adding a cross-tab to your reports, Crystal Reports allows you to insert a cross-tab into only a report or group header or footer. If you insert your cross-tab into the report header or footer, the cross-tab will display the entire data set you have requested, showing all records. If you place a cross-tab in a group header or footer, your cross-tab will display only the records that relate to that one particular group, as shown in Figure 13.5 on the next page.

Figure 13.5 A cross-tab placed in a group header will be filtered by the group

One of the most common mistakes when placing a cross-tab is trying to place it in the wrong section. If you are inserting a report in the Preview tab, keep an eye on the cursor as you move across sections of your report. Where the "circle with a line through it" symbol appears, you can't add your cross-tab to that section. A good practice is to put it into the Report Header until you figure out where you want to place it on your report or use the Design view to place the cross-tab.

CUSTOMIZING GROUPING OPTIONS

As with reports, you can control the way information is displayed in a cross-tab through the use of grouping. You can apply the same grouping concepts to your cross-tab by using the shortcut menu shown on the next page in Figure 13.6 to edit your cross-tab, then highlight the row or column field you want to change and click the Group Options button.

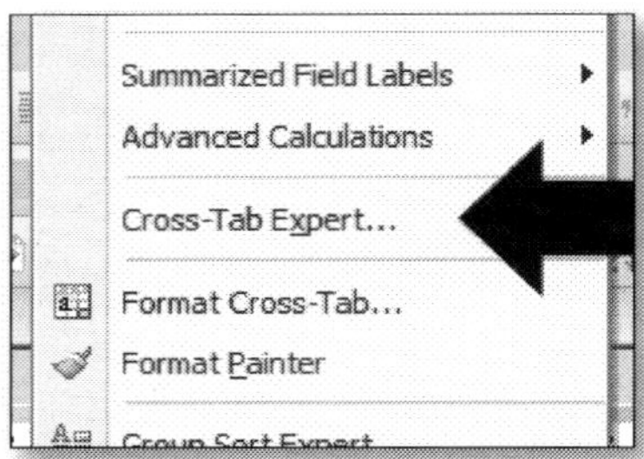

Figure 13.6 To edit a cross-tab, right-click in the top-left corner of the screen, and select Cross-Tab Expert

The Cross-Tab Group Options dialog box, shown in Figure 13.7, opens. Here, you can select a sort order from the following options:

- Ascending
- Descending
- Specified Order

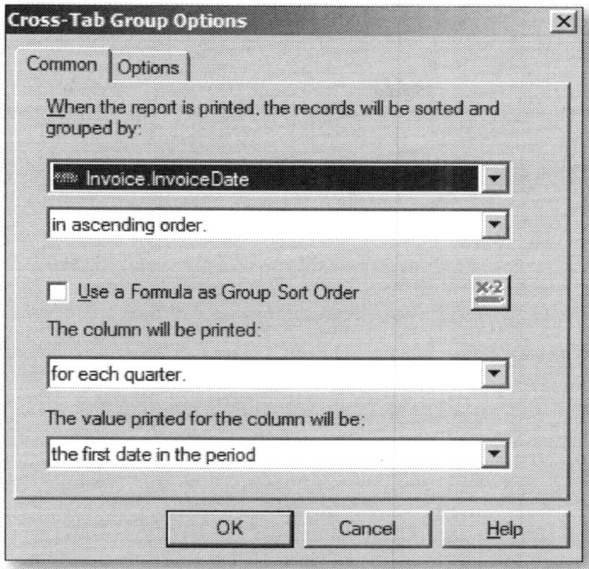

Figure 13.7 A number of grouping options are available for use with cross-tabs

When you use these options to select the sort order, remember that this sort order affects only the items in a single cross-tab. If you were using a cross-tab as the summary page for a long, detailed report, you would also need to change the report sort order so the cross-tab and report would reflect the same sorting.

RE-ORDERING GROUPED DATA

Ascending and descending order is easy to work with, but what about when you want to re-order the data itself, not just sort it? The functionality that allows you to create your own groups is called Specified Grouping. The Specified Order option with cross-tab groups works just as it does for other specified groupings you insert into your report. You need to name each group and then specify the criteria for each.

You might use specified grouping with a cross-tab if you have a particular product grouping not represented in the database. You could use a specified grouping to create separate groups and to establish your own criteria (for example, in our fictional bike company, all the gloves, helmets, and so on could be grouped as Personal Accessories, and saddles and other items could be grouped together as Spare Parts).

Once you have selected Specified Order, a second tab should appear in the Cross-Tab Group Options dialog box, as shown in Figure 13.8.

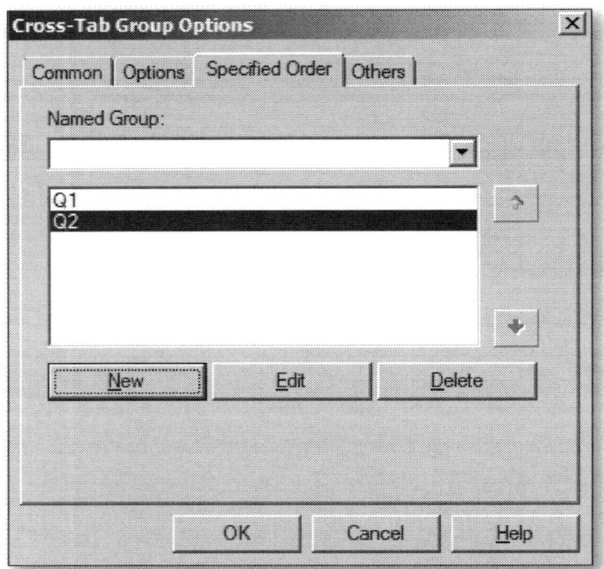

Figure 13.8 The Specified Order tab

The next step is to define a group name, and then you need to specify the group criteria. Start by typing all of the group names you want to create first, pressing the ENTER key after each. This process will build a list of group names.

Once you have all of the group names defined, you can highlight each and click the Edit button to specify the criteria. To establish the criteria for records to be added to your group, use the drop-down menu to select an operator and values.

These are the same operators that were used with record selection.

You can add criteria by clicking the New tab and using the operators to specify additional selection criteria, which are evaluated with an `Or` operator between the criteria that you have specified.

Make sure that you delete any groups you may have added by accident. Even if the criterion is set to Any Value, it can still affect report performance.

After you have entered a single group, another tab appears with options for records that fall outside the criteria that you specify. By default, all of the leftover records are placed in their own group, labeled Others, shown in Figure 13.9. You can change the name of this group by simply editing the name on the Others tab. You can also choose to discard all of the other records or to leave them in their own groups.

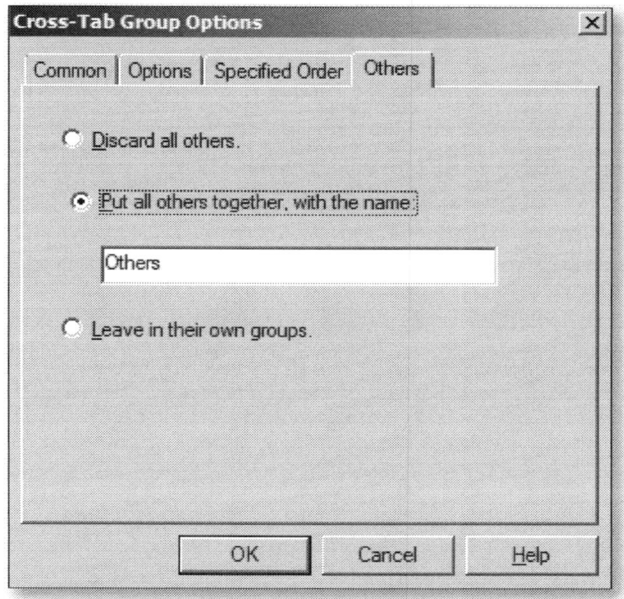

Figure 13.9 Options for handling other records

After you have defined your specified groups and criteria and have reviewed the settings for other records, click OK to accept the changes to the group options. Your specified grouping is now reflected in your cross-tab, as shown below in Figure 13.10.

		Q1	Q2	Q3	Q4
Action Figures & Collectibles	Animal Farmhouse	3419	3129	1263	5636
	Dirt Boss Decepticon Action	2284	3063	3461	4296
	Total	5703	6192	4724	9932
Apparel	Baby Sleeper Combo	1827	1017	392	2192
	Super Heros T-Shirt	336	288	144	624
	Total	2163	1305	536	2816
Arts & Crafts / Activities	Dashboard Adventures	939	2022	1807	2610
	My Little Helper	1090	1852	2367	1026

Figure 13.10 A cross-tab with specified grouping

An example report named XTABGROUP.RPT with a cross-tab with specified grouping is available in the downloadable files that are available from www.kuiperpublishing.com.

CUSTOMIZING THE GROUP NAME FIELD

Another handy option for formatting groups is the ability to select the Group Name field that appears in your cross-tab, using the dialog box shown in Figure 13.11.

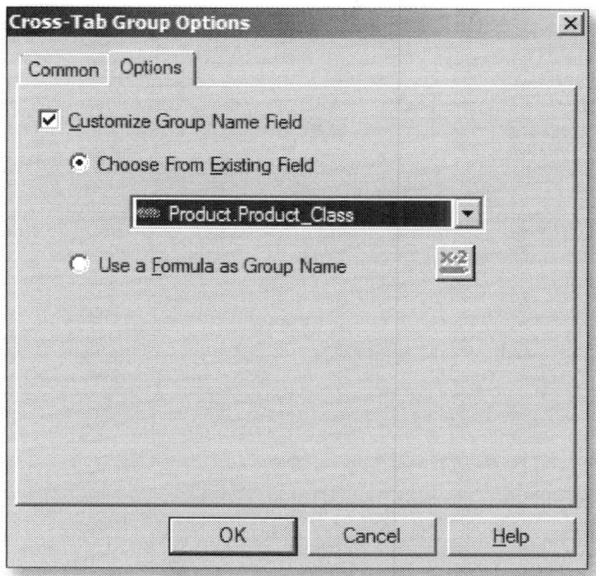

Figure 13.11 You can also set some of the same grouping options that are available for groups inserted into your report

For selecting the group name, you can customize the group name either by selecting a database field or by using a formula. If your database has a table that contains the name you wish to use, you could specify the name of the field that holds the proper name.

For example, if you were creating a cross-tab using the field Product ID, you could customize the group name and select the field Product Name so that the correct product name would appear instead of a numeric Product ID.

If the names of your group are not stored in one of your database tables, you can always create a formula to use as a group name. For example, in an Order Summary report, a cross-tab could be created using the Order Date field. The grouping options could then be set to For every quarter, which would produce some pretty ugly group names using the dates.

A much better solution is to use a formula as the group name. To create a formula for this purpose, click the X+2 button beside Use a Formula as Group Name. The Formula Workshop opens and allows you to enter a formula to be used to display as the group name. From our example, we want to display a quarter number based on the month of the Order Date, so our formula looks like the following:

```
If Month({Purchases.Order Date}) in [1,2,3] then
"QTR1" else
If Month({Purchases.Order Date}) in [4,5,6] then
"QTR2" else
If Month({Purchases.Order Date}) in [7,8,9] then
"QTR3" else
If Month({Purchases.Order Date}) in [10,11,13] then
"QTR4"
```

You can create any formula you like, as long as it is designed to return a string as a group name.

WORKING WITH SUMMARY FIELDS

Summary fields are used to summarize the information contained in your cross-tab. All of the standard Crystal Reports summary fields are available for use. By default, Crystal Reports uses Sum for numeric fields you have inserted into your summarized fields and Count for any string or other type of field. You can change the summary type by highlighting a summarized field and clicking the Change Summary button.

There are a number of summary fields available for use within Cross-tabs, including:

- Sum
- Average
- Minimum
- Maximum
- Count
- Distinct Count
- Correlation
- Covariance
- Weighted Average
- Median
- Pth Percentile
- Nth Largest
- Nth Smallest
- Mode
- Nth Most Frequent
- Variance
- Standard Deviation
- Population Variance
- Population Standard Deviation

In addition to these standard summary field types, some summary types that are specific to cross-tabs follow:

- Percent Of Sum
- Percent Of Average
- Percent Of Maximum
- Percent Of Minimum
- Percent Of Count
- Percent Of Distinct Count

Some of these functions may require that you specify an additional field or value, such as the one shown below in Figure 13.12, which shows the summarized field as a percentage of the total sum.

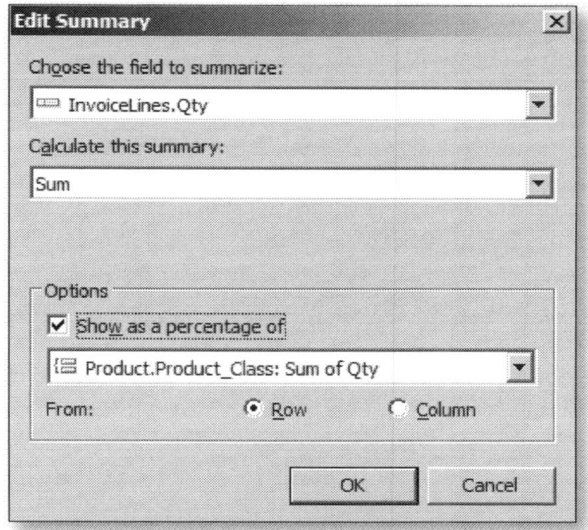

Figure 13.12 Some summary fields may require an additional field or value to be specified

INSERTING FORMULA FIELDS

In addition to database fields, you can also use formula fields in your cross-tabs. Using the Crystal Reports formula language, you can add complex calculations and analysis to your cross-tab. To insert a formula field into a cross-tab, click the New Formula button on the right side of the Format Cross-Tab dialog box.

To edit an existing cross-tab, right-click the cross-tab object and select Format Cross-Tab from the shortcut menu.

As always, the first thing you will need to do is enter a name for the formula you want to create. Click OK to open the Crystal Reports Formula Editor, and enter your formula text, as shown in Figure 13.13.

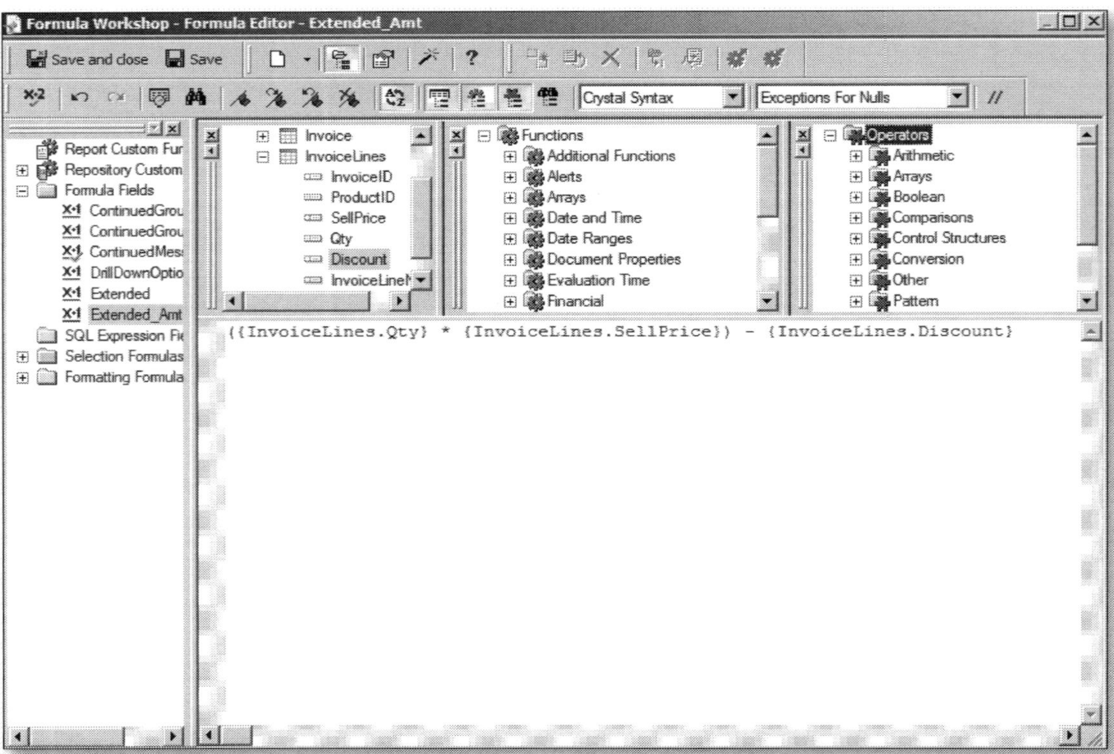

Figure 13.13 You can create a formula using the Cross-Tab Expert

Using the Formula Editor, enter the text for your formula. When you have finished editing your formula, click Save and then the Close button at the top-left corner of the Formula Editor to close your formula and return to the Format Cross-Tab dialog box.

Your formula should now appear in the section marked Available Fields, shown under the Report Fields node. You can then drag the formula to the Row, Column, or Summarized Field box to add it to your cross-tab. When your report is previewed and the cross-tab created, this formula will be evaluated and displayed, as shown in Figure 13.14.

		01/2009	04/2009	07/2009
Action Figures & Collectibles	Animal Farmhouse	$49,575.50	$45,370.50	$18,313.50
	Dirt Boss Decepticon Action	$33,084.15	$38,716.35	$44,214.50
	Total	$82,659.65	$84,086.85	$62,528.00
Apparel	Baby Sleeper Combo	$24,096.10	$12,631.86	$5,204.00
	Super Heros T-Shirt	$3,679.20	$3,153.60	$1,576.80
	Total	$27,775.30	$15,785.46	$6,780.80
Arts & Crafts / Activities	Dashboard Adventures	$11,737.50	$27,469.00	$23,766.60

Figure 13.14 An example of a cross-tab with a formula field inserted

FORMATTING CROSS-TABS

Now that you understand how to insert a cross-tab into your report and control its structure, it's time to take a look at formatting your cross-tab.

CHANGING FIELD, COLUMN, & ROW SIZE

If you have worked with Excel and resized columns, you may try to resize columns in your cross-tabs by resizing the actual column. In Crystal Reports, the column size is determined by the actual fields within the column itself. Just like Excel, column widths are set for the entire column at once (regardless of whether the column has summarized fields, totals, or both).

You can resize a field by clicking the field to select it and then using the handles to change the width or height of the field. You can also right-click the field and select Size and Position from the shortcut menu to open the dialog box shown in Figure 13.15, which gives you precise control over the field's size.

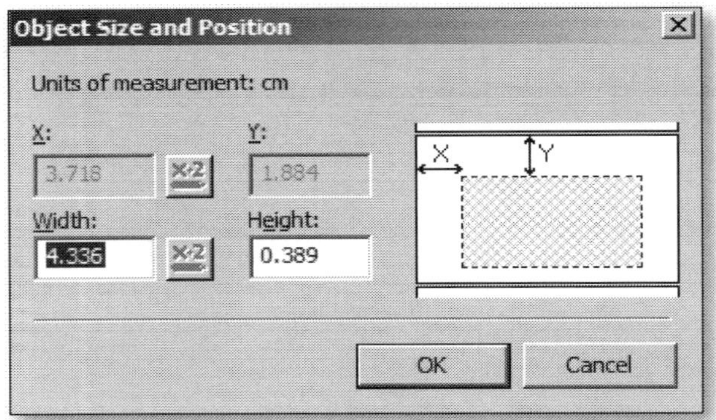

Figure 13.15 The Object Size and Position dialog box

When you have finished resizing your fields, columns, or rows, click anywhere outside the cross-tab to deselect the object with which you were working.

APPLYING A PRE-FORMATTED STYLE

Because the majority of time spent working with cross-tabs is for formatting them, Crystal Reports includes a number of pre-formatted styles for which the formatting attributes have been set for you. This eliminates the need to individually change each of the formatting settings and also provides a way to apply uniform formatting to cross-tabs that may appear on different reports.

To apply a pre-formatted style to your cross-tab object, insert or edit an existing cross-tab and, in the Format Cross-Tab dialog box, click the Style tab to open the dialog box shown in Figure 13.16.

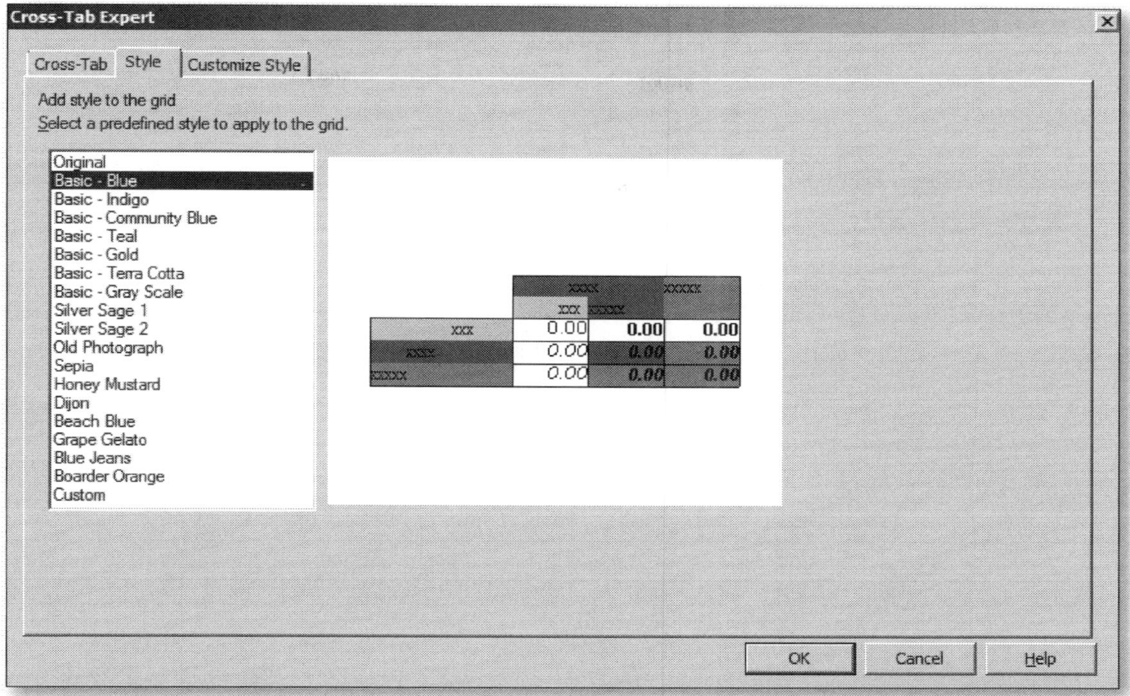

Figure 13.16 You can apply a pre-formatted style to your cross-tab or OLAP grid

To see a preview of the style's formatting, click the style. The pane on the right side of the dialog box will display a preview of a sample cross-tab with the formatting attributes applied. When you have found the style you want to apply, click OK to return to your report and see the style applied to your cross-tab.

One of the most-often asked questions is, "Can I add my own style to this list?" At this time, Crystal Reports does not support adding custom styles. You can, however, use the styles as a starting point for your own formatting, which may make things a little easier.

CUSTOMIZING A STYLE

If you don't particularly care for some of the formatting attributes of a particular style, you can customize the style using the Customize Style tab, also found in the Cross-Tab Expert and shown below in Figure 13.17.

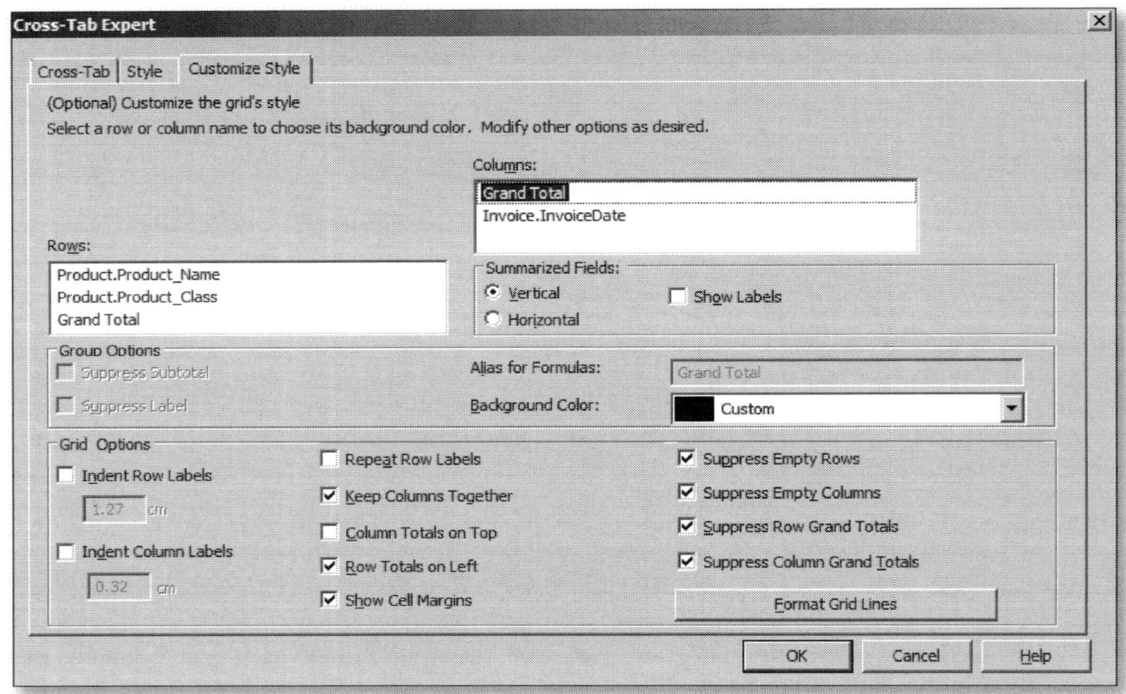

Figure 13.17 You can further customize a pre-formatted style using the Cross-Tab Expert

To change the formatting of a particular row or column, highlight the row or column from the lists provided, and then use the drop-down box immediately below to change the background color and set other options, including the showing of labels, aliases for formula fields, and so on.

On the bottom half of the dialog box are the options for formatting the grid that appears around your cross-tab by default. Table 13.1 on the following page summarizes the gridline options available on the Customize Style tab.

Table 13.1 Cross-Tab Style Options

Option	Description
Show Cell Margins	Displays the internal cell margins in your cross-tab
Indent Row Labels	For each row, indents the labels that appear on the left side of the cross-tab
Repeat Row Labels	Repeats row labels on any new pages
Keep Columns Together	Attempts to keep all columns together on the same page
Row Totals on Top	Moves the row totals from their default location at the bottom of the cross-tab to the top
Column Totals on Left	Moves the column totals from their default location on the right side of the cross-tab to the left
Suppress Empty Rows	Suppresses any empty rows in the cross-tab
Suppress Empty Columns	Suppresses any empty columns in the cross-tab
Suppress Row Grand Totals	Suppresses the grand totals that would appear by default at the bottom of the cross-tab
Suppress Column Grand Totals	Suppresses the grand totals that would appear by default on the right side of the cross-tab

You can also control the grid lines that appear in your cross-tab by clicking the Format Grid Lines button to open the dialog box shown in Figure 13.18 over the page.

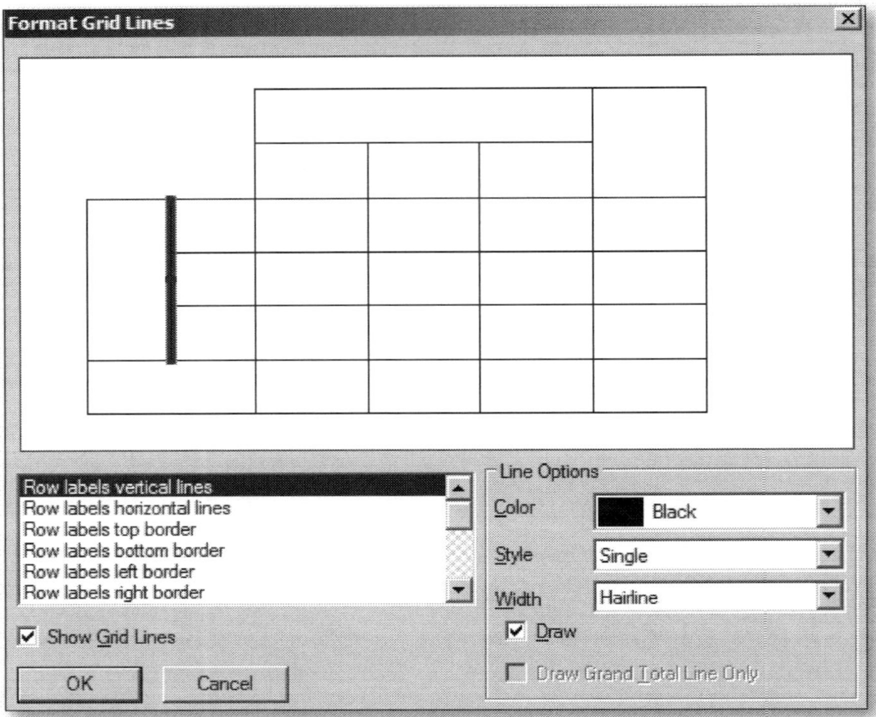

Figure 13.18 You can format individual grid lines as well

Using this dialog box, select the part of your cross-tab or OLAP grid that you want to format and choose the line color, style, and width. The following grid lines are available:

- Row Labels Vertical Lines
- Row Labels Horizontal Lines
- Row Labels Top Border
- Row Labels Bottom Border
- Row Labels Left Border
- Row Labels Right Border
- Column Labels Vertical Lines
- Column Labels Horizontal Lines
- Column Labels Top Border
- Column Labels Bottom Border
- Column Labels Left Border
- Column Labels Right Border
- Cells Vertical Lines
- Cells Horizontal Lines
- Cells Bottom Border
- Cells Right Border

When you have finished setting the grid lines for your cross-tab or OLAP grid, click OK to return to your report design. The changes you have made should be reflected in your cross-tab.

ANALYZING CROSS-TAB DATA

In addition to summarizing data, the data in cross-tabs can also be manipulated to help analyze the data presented and find information quickly. In the following sections we are going to look at some of the tips and techniques to help you analyze the data that is summarized in your cross-tab.

USING THE HIGHLIGHTING EXPERT

The Highlighting Expert can be used to quickly highlight exceptions or abnormal values in a cross-tab. To use the Highlighting Expert, use the following steps:

If you would like to follow along with the step-by-step instructions in this chapter, this book includes downloadable files that are available from www.kuiperpublishing.com. Also, check out "Setting up the Samples" section in the front of this book for instructions on how to configure the data sources, reports, etc.

1. Open Crystal Reports and from the book download files, open the XTABHIGHLIGHT. RPT report.
2. Right-click the field in your cross-tab that you want to highlight. In this example, right-click the Quantity field. From the shortcut menu, select the Highlighting Expert to open the dialog box shown in Figure 13.19 over the page.
3. Click the New button on the left side of the dialog box.
4. Next, in the Item Editor section on the right side of the dialog box, select an operator from the pull-down Value Is list. In the box immediately below the operator, enter the criteria to specify when the highlighting should occur. For this report, select Is greater than and then enter 20000.
5. Now select the font color, background, and border that will be triggered when this criteria is met. For this report, select a background color of Yellow. If you want to enter multiple criteria, click New Item in the Item List on the left.

To change the order of precedence for highlighting criteria, use the up and down arrows.

6. After you have entered the criteria and formatting options, click OK to exit the Highlighting Expert.

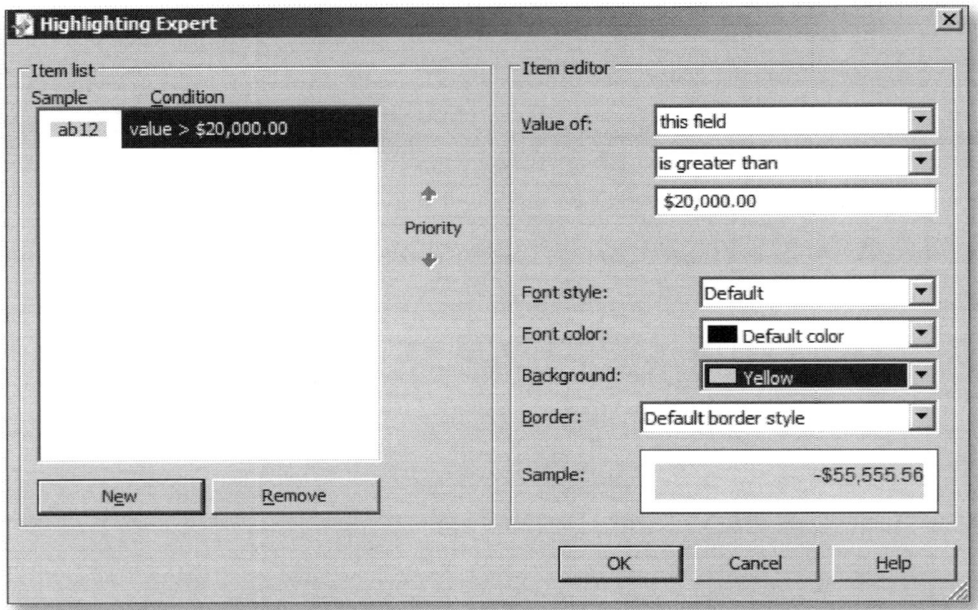

Figure 13.19 The cross-tab Highlighting Expert

When you preview your report, the field you originally selected in your cross-tab should reflect the options set in the Highlighting Expert.

CHANGING CROSS-TAB ORIENTATION

Sometimes when creating cross-tabs, we don't have a good indication of what the resulting summarized data is going to look like. To make cross-tabs a bit easier to read and to bring at least one of its features into line with Microsoft Excel, you can pivot cross-tabs by right-clicking the cross-tab and selecting Pivot Cross-Tab from the shortcut menu.

This action will cause the rows and columns in the cross-tab to be switched, so a report with different countries across the top in columns would then have the countries showing as rows.

This feature is available only at design time — when users view the report through any of the export formats or through one of Crystal Report's Web delivery methods, this feature is not available. You would have to pivot the cross-tab to the desired configuration before distributing or publishing your report.

CROSS-TAB USES AND LIMITATIONS

Traditionally, Crystal Reports developers have either loved or hated cross-tabs. One of the main reasons that report designers are sometimes unhappy with cross-tabs is that they look at a cross-tab and see rows and columns, similar to those in an Excel spreadsheet, and they expect the same functionality. However, a cross-tab and a spreadsheet are not the same.

For starters, a cross-tab's rows and columns all share the same height and width. For example, you can't show 10 columns of data and have one column wider than the rest. Another key point is that any formatting you apply to a field in a cross-tab will be applied to every field in the cross-tab grid.

Over the different versions of Crystal Reports, improvements have been made to fix problems with page breaks, formatting, gridlines, row and column suppression, and so on that really do make cross-tabs a viable analysis and presentation option, but you must understand and work within the framework presented.

If you do need some of the specialized formatting features found in Excel and other spreadsheet programs, Crystal Reports does include a number of export formats that allow you to export to a spreadsheet file. What you do with the data at that point is up to you. The advantage of using a cross-tab is that even though it may not have all of the formatting functions of a spreadsheet, the information presented is data driven and does not require manual updating as a spreadsheet would.

SUMMARY

Subreports are a handy way to show data from two or more disparate sources in the same report. Offering the flexibility of being either related or unrelated to the underlying report data, subreports can often make impossible reports seem possible. Another handy trick you learned in this chapter was the use of cross-tabs to display data in summarized rows and columns, presenting information clearly and concisely. In the next chapter, we'll take that concept even further, with details on how to use charts and geographic mapping effectively within your reports.

Chapter 14
Charting and Mapping

In This Chapter

- Adding Charts to Your Reports
- Inserting an Advanced Chart
- Inserting a Group Chart
- Inserting a Cross-Tab Chart
- Formatting Charts
- Advanced Chart Formatting

INTRODUCTION

In recent years, organizations have seen an explosion in the amount of data that they must collect. With the majority of business applications and systems tied to a back end database, an organization can have millions of rows of records that need to be condensed, analyzed, and displayed. Crystal Reports is the obvious solution for this problem, because even a single report can distill millions of rows of information into a clear, concise format.

You have already seen some of the ways that Crystal Reports can turn raw data into information, using summary and drill-down reports, running totals, summaries, and more. Until now, all of the summary methods we have looked at have been text-based and provided exact numbers and totals that reflect the underlying data in a report.

In this chapter, we are going to explore the different ways you can display data visually, providing report results at a glance using graphs and charts. Using these graphics you can distill millions of rows of data into a concise graph or chart that is easy to read and understand, like some of the examples shown in Figure 14.1 over the page.

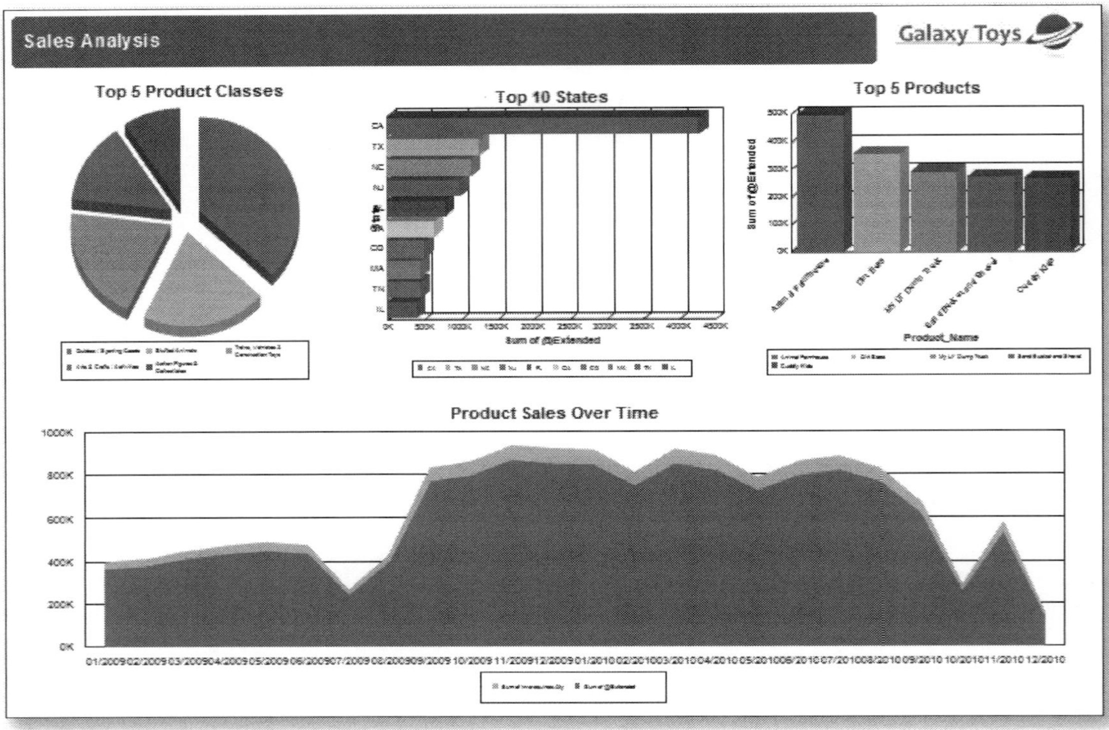

Figure 14.1 An example of some of the chart types Crystal Reports supports

Driving these charts is a sophisticated graphing engine that provides support for a wide range of business and scientific charts, with more than 40 templates to select from. These charts can be customized to meet your specific needs, and numerous formatting attributes are available to achieve the desired result.

Using charting in your reports means that you can convey the meaning of a report and the data it represents at a glance, as shown in some of the examples in Figure 14.2.

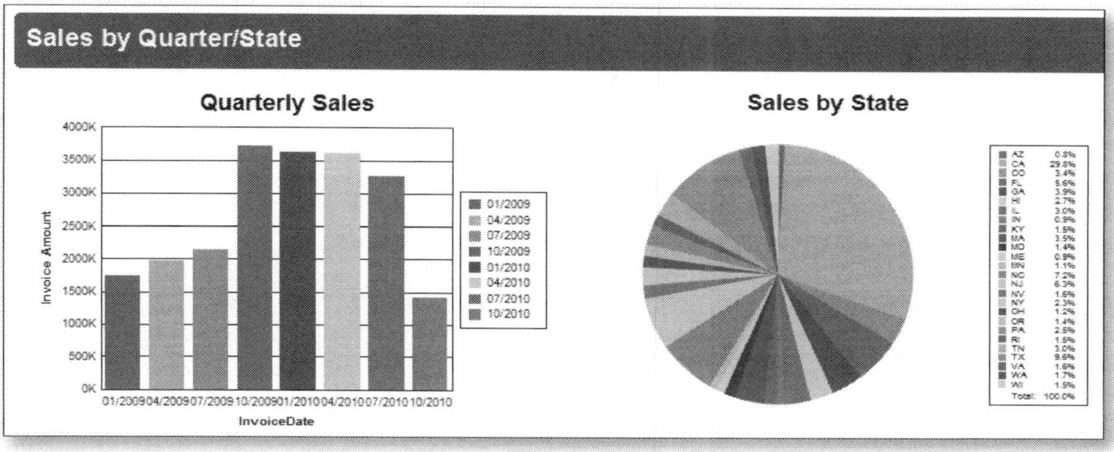

Figure 14.2 An example of side-by-side charts embedded into a report

In this chapter, we will look at charting in detail, starting with a look at the different chart layouts that are available.

ADDING CHARTS TO YOUR REPORTS

Crystal Reports supports a number of different chart formats, including bar charts, pie charts, and line charts. These formats may determine the look and feel, but Crystal Reports charts generally fall into four distinct types or layouts: advanced, group, cross-tab, and OLAP grid, as shown in Figure 14.3.

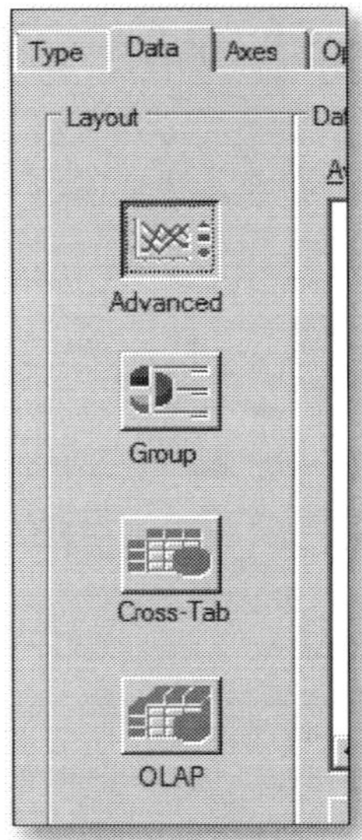

Figure 14.3 Crystal Reports chart types

ADVANCED CHARTS

Advanced charts work like the charts you may remember from high school mathematics where you plot a chart or graph based on x and y values. Crystal Reports' implementation of this type of charting is a little more sophisticated; you can specify summary fields to be generated, perform complex analysis, and control how the information is grouped when displayed on your graph.

GROUP CHARTS

Group charts are commonly used in reports and can be used wherever you have inserted a group into your report and created a summary field based on that group (Sum, Average, and so on). A group chart can appear once, representing the data in the entire report, or you can present one chart for each group.

Group charts can be used to create a drill-through effect, where you start with a graph of the highest level of data (the whole report) and then drill down through the different groups, with a graph displayed for each group, all the way down to the details.

An example of a drill-through report (DRILLTHROUGH.RPT) is included in the book download files in the Chapter 14 folder.

Keep in mind that the difference between advanced and group charts is that advanced charts don't require a group or summary field to be inserted into your report. Frequently, advanced charts are used with formula fields and can be used to create complex graphs that would not normally have been possible with just a group graph.

CROSS-TAB CHARTS

Cross-tabs are used to display summarized rows and columns of information in your report, similar in presentation to an Excel spreadsheet. You can create a chart directly from cross-tab data, providing a way to visualize the data that has been summarized in the cross-tab grid, as shown in Figure 14.4.

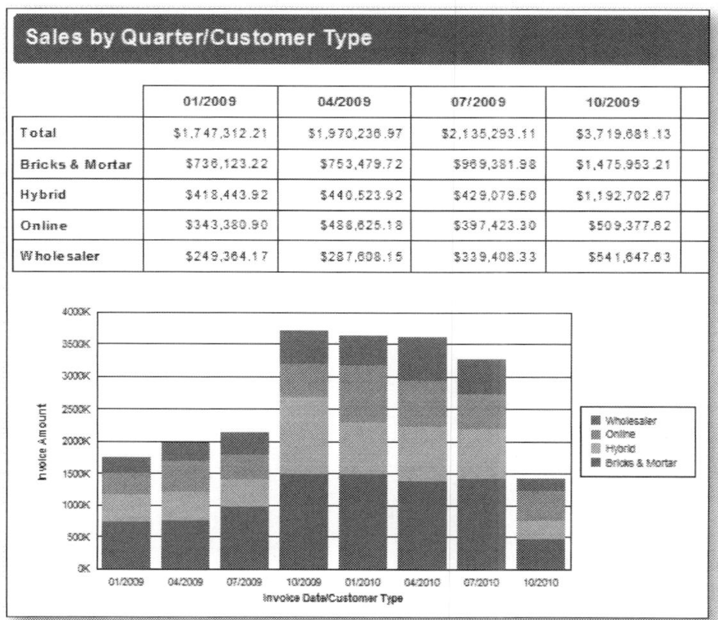

Sales by Quarter/Customer Type

	01/2009	04/2009	07/2009	10/2009
Total	$1,747,312.21	$1,970,236.97	$2,135,293.11	$3,719,681.13
Bricks & Mortar	$736,123.22	$753,479.72	$969,381.98	$1,475,953.21
Hybrid	$418,443.92	$440,523.92	$429,079.50	$1,192,702.67
Online	$343,380.90	$488,625.18	$397,423.30	$509,377.62
Wholesaler	$249,364.17	$287,608.15	$339,408.33	$541,647.63

Figure 14.4 A cross-tab grid and chart

OLAP GRID CHARTS

OLAP grids are similar to cross-tabs, except that OLAP grids display multidimensional OLAP data, as opposed to cross-tabs, which display relational data. Other than that distinction, they behave in the same manner and share most formatting attributes.

INSERTING AN ADVANCED CHART

Advanced charts are drawn based on x and y values, commonly referred to in Crystal Reports as an On change of field and Values to be displayed. In the following steps, we create an advanced chart using the Order Listing report from the book download files as a starting point.

1. Open Crystal Reports, and open the ORDERLISTING.RPT report from the book download files, as shown in Figure 14.5.

Order Listing

Print Date: Tuesday, June 1, 2010

Invoice ID	InvoiceDate	Customer Name	Amount
3800	01/01/2009	Dancing Bear Toys, Ltd	$11,829.60
3801	01/01/2009	Puzzle Master	$2,870.35
3802	01/01/2009		$2,808.00
3803	01/01/2009	Legion Toys	$125.00
3804	01/01/2009		$1,800.00
3805	01/04/2009	Turtle's Nest	$1,800.00
3806	01/04/2009	Red Wagon Toys	$1,800.00
3807	01/04/2009	abraKIDabra Toys Inc	$139.50
3808	01/04/2009	Toys and Stuff	$145.00
3809	01/04/2009	Hobby Time	$125.00
3810	01/07/2009		$145.00
3811	01/07/2009		$2,088.00
3812	01/07/2009	Kidz Stuff Toyz	$900.00
3813	01/07/2009	Black Forest Books & Toys	$3,750.00

Figure 14.5 The Order Listing report provides a good starting point for creating an advanced graph

2. Switch to the Design view of your report, and then click Insert > Chart. Click to place the chart in your Report Header section and to open the Chart Expert dialog box shown in Figure 14.6.

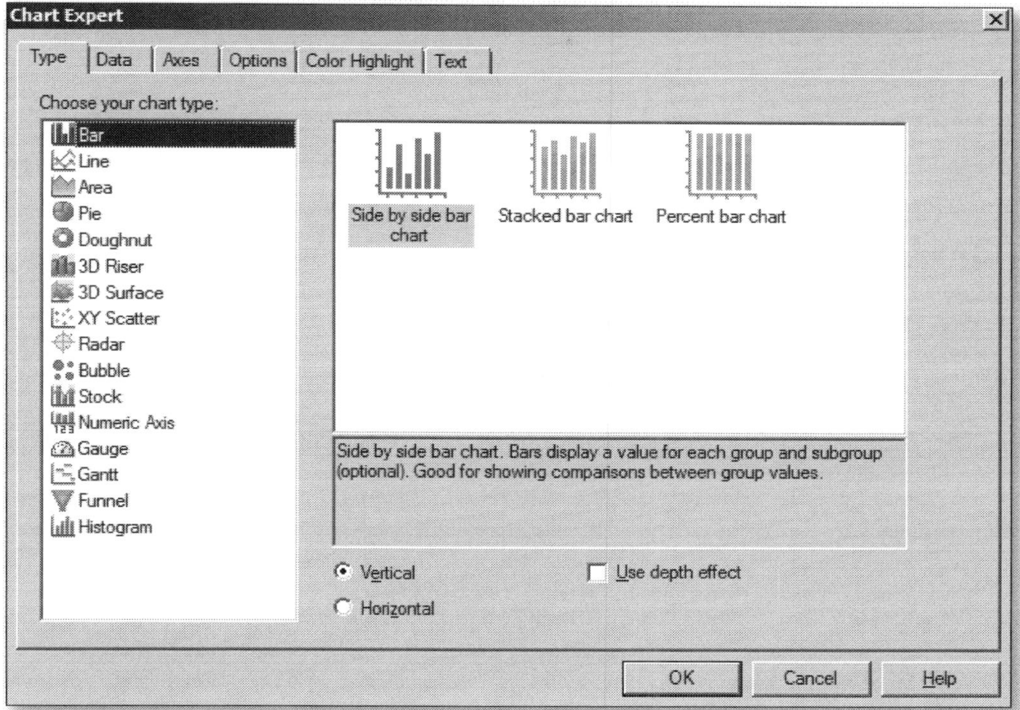

Figure 14.6 The Chart Expert dialog box

3. Click the Type tab, select a Bar chart type, and then click the first icon on the right side to select a Side-by-side chart.
4. Click the Data tab at the top of the dialog box. The dialog box shown in Figure 14.7 on the following page opens.

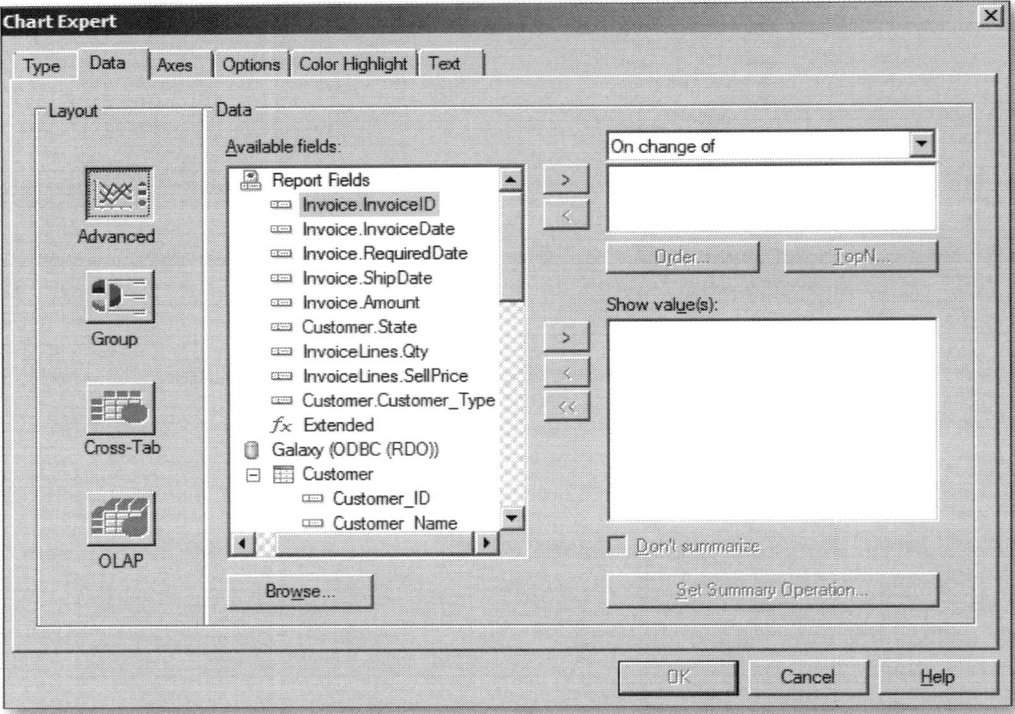

Figure 14.7 The options for an advanced chart

In this example, we are going to create an advanced chart that displays the amounts of each order, sorted by the order date. This chart will be displayed in your report header and will appear on the first page of the report. By default, the placement will be set to One per report and Header, but you could as easily change this placement to put the chart in the report footer.

5. First, click to select the Orders.OrderDate field, and then click the right-arrow button near the top of the dialog box to move it to the On change of field, which is the equivalent of an x-axis value (if you remember your high school algebra).

When working with this *x*-axis value, you have three options for how this field will be used:

On Change Of — When the value contained in the field changes, a new bar, pie piece, and so on will be generated.

For Each Record — A new bar, pie piece, and so on will be created for each record in the database.

For All Records — One bar, pie piece, and so on will be created for all of the records in the database.

6. Because the chart in this example calls for one bar to be shown for each order date, leave the drop-down list to read On Change Of.
7. Next, click to select the Orders.Order Amount field, and then click the right-arrow button closer to the bottom of the dialog box to move it to the Show values field, which could also be considered the y values for your chart.

> **You'll notice that you can choose multiple x-axis values in an advanced graph. This allows you the flexibility you need to create complex graphs that require multiple x-axis values. But be warned that not all of the different graph types support multiple x axes.**

8. Click the Text tab at the top to set the text labels for your chart. As with other graph types, Crystal Reports creates text labels for your chart by default, but you can change and format this text as you see fit.
9. To override the default values, uncheck the Auto-Text boxes, and enter the text you want to see on your chart.
10. To change the formatting attributes associated with this text, select an item (title, subtitle, and so on) from the list at the bottom-right corner of the dialog box. Click the Font button to change these attributes.
11. Click OK at the bottom of the Chart Expert to insert your advanced chart into your report, which will now look like Figure 14.8 on the following page.

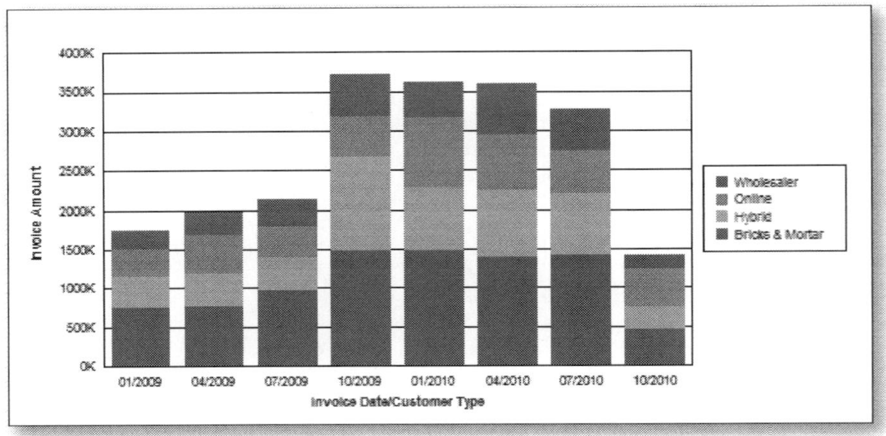

Figure 14.8 Your finished report with an advanced chart inserted

CHANGING THE SORT ORDER

Unfortunately, the chart we just created doesn't really tell us much about the underlying data, because there are too many bars or data points for us to make any sense of the chart. Some reorganization of the chart data would definitely come in handy, and we can change the sort order of this chart to make it a bit more readable.

On the Data tab of the Chart Expert, you may have noticed a button for Order, as shown in Figure 14.9. This button is grayed out until you actually select one of the On change of or x-axis fields.

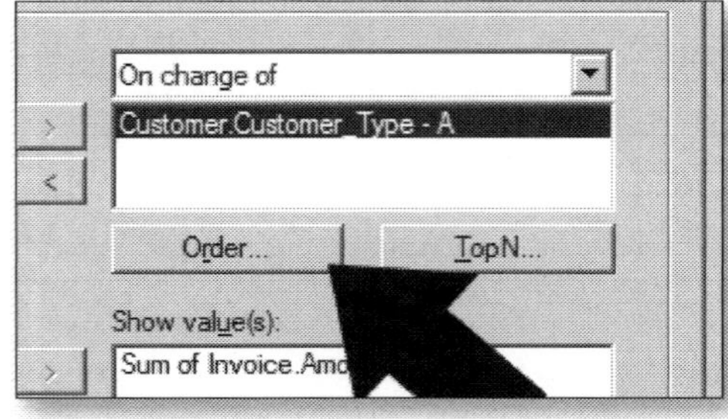

Figure 14.9 Data options for your chart

You can use this button to order and summarize the data that appears on your chart by using the following steps:

1. From the Order Listing report we have been working on, right-click the advanced chart we just created, and select Chart Expert from the shortcut menu to open the Chart Expert.
2. Click the Data tab.
3. Locate the Orders.Order Date field in the text box below the drop-down box labeled On Change Of, and click to select it.
4. Click the Order button to open the Sort Order dialog box shown in Figure 14.10.

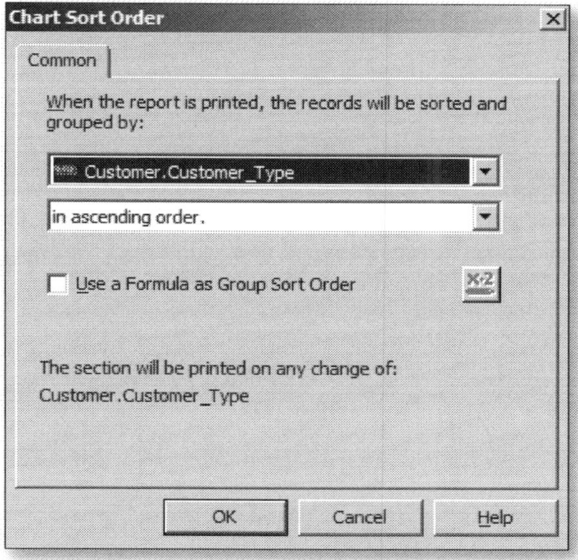

Figure 14.10 Chart-sorting options

The following four sorting options are available:

Ascending — A to Z, 0 to 9, and so on.

Descending — Z to A, 9 to 0, and so on.

Specified — Similar to specified grouping option with groups inserted into your report. This option is for naming a group and specifying the criteria for the values that should be included in that group.

Original — The original order of the data.

5. By default, Ascending is selected. This is the sort order we want, so leave the dialog box as shown.

Because we have created an advanced chart on a date field (that is, Invoice Date), a third drop-down box appears with grouping options specifically for use with date fields (for each day, week, month, and so on).

6. To make this chart a bit more readable, change this drop-down list to read For each Month, and click OK.
7. Click OK again from the Chart Expert to return to the Report Designer. Your report should now look similar to the one shown in Figure 14.11.

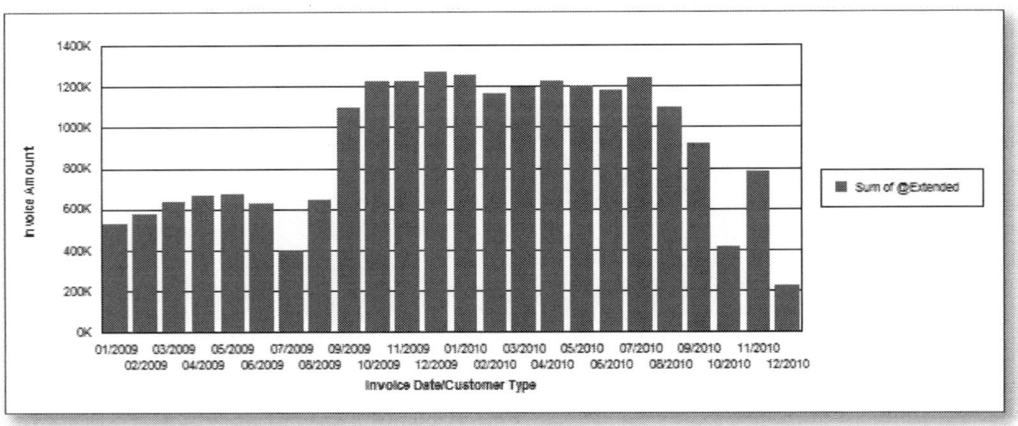

Figure 14.11 The same report with a monthly grouping

APPLYING TOPN/SORT ALL ANALYSIS

TopN and sorting are popular analysis methods in reports because they draw attention to trends in the data. For example, the N in TopN stands for a number, so you could create a Top 5 report to see your top five suppliers, according to how much you purchased from each. In addition, the Sort All functionality allows you to put groups in order by their totals or summary fields. You could take that same supplier report and use Sort All to put the suppliers in order from the supplier with whom you spend the most money to the supplier with whom you spend the least.

In addition to using this functionality in your report, you can also apply this functionality to your charts. To apply TopN/Sort All analysis to your chart, use the following steps:

1. Right-click your existing chart, and select Chart Expert.
2. In the Chart Expert, click the Data tab.
3. Locate the field that controls sort order in the text box below the pull-down box labeled On Change Of, and click to select it.
4. With the field highlighted, the TopN button is enabled. Click to open the TopN/Sort dialog box.

   Three types of analysis are available using this feature, and each has its own parameters:

   TopN — Enter a value for N to determine the TopN values based on the *y* field you have selected.

   BottomN — Enter a value for N to determine the Bottom N values based on the *y* field you have selected.

   Sort All — Choose this option to sort all data items in either ascending or descending order based on the *y* value.

   With TopN and BottomN, you must also determine what to do with the values that are not included in your N sample. Just as with TopN/BottomN analysis on a report, you can discard the other values, keep them in a group called Other, or simply leave them in their own groups — it's up to you.

5. When you have finished setting the TopN/Sorting options, click OK to accept your changes and return to your report's Design or Preview view. Your graph should now reflect the analysis options you have selected.

CHANGING THE SUMMARY OPERATION

Another report feature that can be used in charts is summary operators, which we looked at in Chapter 4. By default, an advanced chart type will use a Sum summary operator, but you can change this to any of the other summary types that Crystal Reports supports. To change the summary operator, use the following steps:

1. Right-click your chart, select Chart Expert, then click the Data tab.
2. Click the summarized field you want to work with in the Show Values dialog box at the bottom-right corner of the dialog box.

3. With the field highlighted, the Set Summary Operation button is enabled. Click it to open the dialog box shown in Figure 14.12.

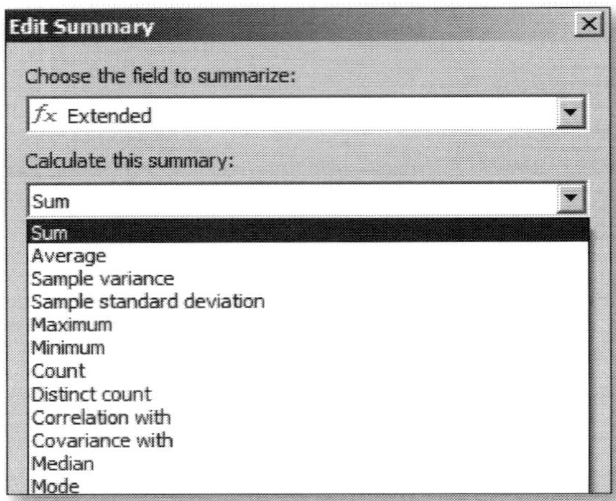

Figure 14.12 Summary operators available for use in your chart

A number of summary operators appear in the drop-down list, depending on the field type. The most popular follow:

Sum — Provides a sum of the contents of a numeric or currency field.

Average — Provides a simple average of a numeric or currency field (that is, the values in the field are all added together and divided by the total number of values).

Minimum — Determines the smallest value present in a database field; for use with number, currency, string, and date fields.

Maximum — Determines the largest value present in a database field; for use with number, currency, string, and date fields.

Count — Provides a count of the values present in a database field; for use with all types of fields.

Distinct Count — Similar to Count, except any duplicate values are counted only once.

4. Click OK to accept your summary type change and return to the Chart Expert.
5. When you have finished editing your chart options, click OK to accept your changes and return to your report's Design or Preview view.

INSERTING A GROUP CHART

Now let's move on to the most common type of graph; the group graph. To use a group graph, you will need two things in your report. The first is a group. This group can be created from a database field, formula field, and so on. The second requirement for a group graph is that you have some sort of summary field inserted onto your report.

Although this field is most frequently a summary on a numeric field, it can also be a summary on another type of field, such as a count of customers. (See Chapters 3 and 4 for information on inserting both groups and summary fields.) Again, to make things a bit easier, we are going to use a report from the book download files as a starting point for our chart.

To insert a group chart into this report, use the following steps:

1. Open Crystal Reports, and open the SALESBYSTATE.RPT report from the book download files, as shown in Figure 14.13.

Sales by State

Invoice ID	Invoice Date	Required Date	Ship Date	Customer Name
AZ				
4271	26-Oct-09	27-Oct-2009	26-Oct-2009	Black Forest Books & Toys
4603	28-Apr-10	01-May-2010	28-Apr-2010	Black Forest Books & Toys
4538	20-Mar-10	27-Mar-2010	24-Mar-2010	Black Forest Books & Toys
4303	30-Oct-09	02-Nov-2009	30-Oct-2009	Black Forest Books & Toys
4170	27-Aug-09	29-Aug-2009	26-Aug-2009	Black Forest Books & Toys
4023	13-May-09	22-May-2009	13-May-2009	Black Forest Books & Toys
3868	9-Feb-09	11-Feb-2009	09-Feb-2009	Black Forest Books & Toys
3813	7-Jan-09	16-Jan-2009	07-Jan-2009	Black Forest Books & Toys
4316	22-Nov-09	29-Nov-2009	22-Nov-2009	Black Forest Books & Toys
CA				
4707	25-Jul-10	03-Aug-2010	25-Jul-2010	Toy Hut

Figure 14.13 A typical customer report with a group and summary inserted

2. Switch to the Design view of your report, click Insert > Chart, and click in the Report Header section to place your chart. Your chart will appear automatically with the default settings in place. To change these settings, right-click the chart and select the Chart Expert.

The first step in customizing a group graph is selecting the type of graph you want to create.

3. Referring to the descriptions of the different types of graphs, click the Type tab and select a chart type from the list on the left, then click an image of the specific type of graph you want to create.
4. The next step is selecting where the data will come from. Click the Data tab to open the dialog box shown in Figure 14.14. From the layouts on the left side of the page, select Group.

If you don't have a group inserted into your report, this option will be unavailable.

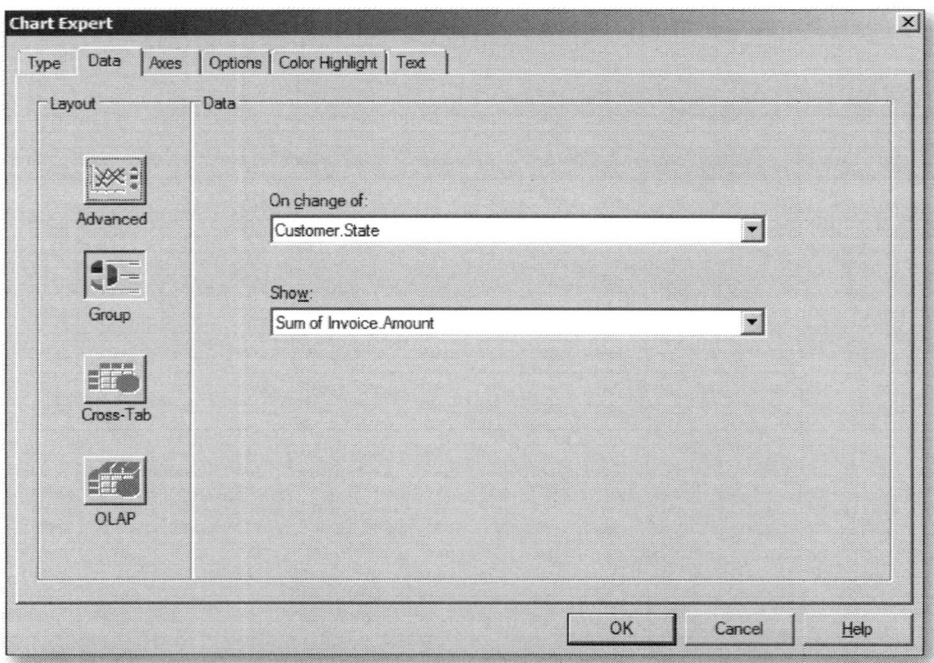

Figure 14.14 You need to select the source of your graph data

5. You need to specify where you want your new group chart placed. If you have only one group and summary in your report, this task is easy; your only options will be to place the graph in either the Report Header or Report Footer section. If you have multiple groups and summaries, you can place the graph in the group headers.
6. In the middle of the page, you need to select On Change Of and Show values. These two options correspond directly to the groups and summaries you have inserted onto your report. You should see all of the groups you have inserted in the On Change Of drop-down list, and all of the summaries should appear in the Show drop-down list.
7. The last step in creating a group graph is setting the text (that is, title, axis labels, and so on) that will appear on your graph. Click the Text tab to display the dialog box shown in Figure 14.15.

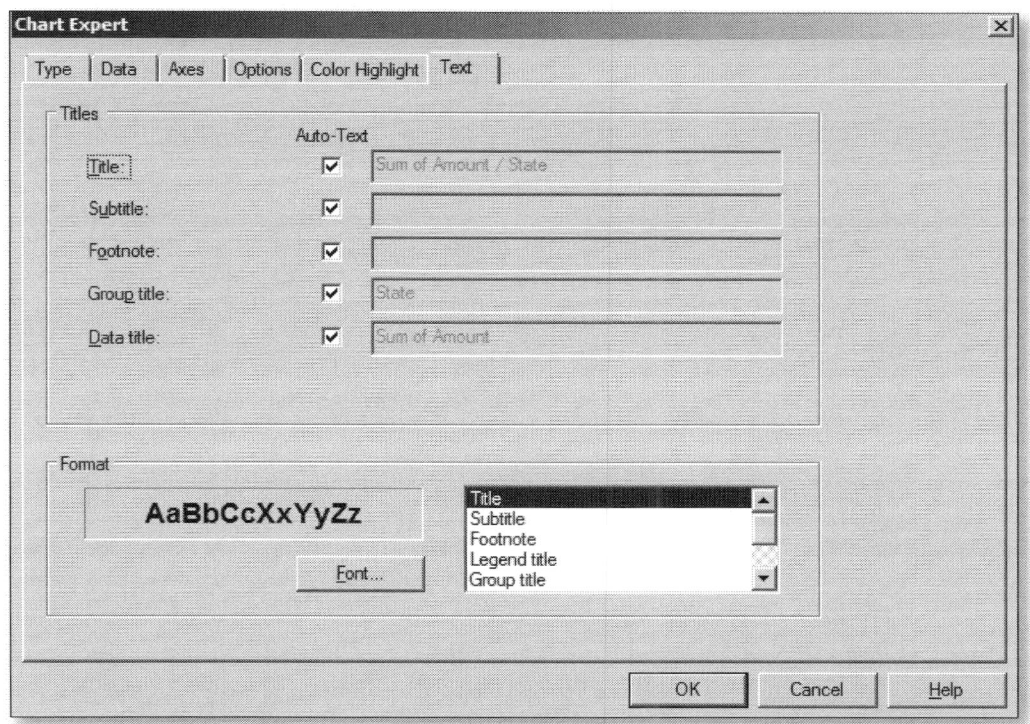

Figure 14.15 You can enter or edit the text for your chart

By default, Crystal Reports creates text labels for your graph. To override these values, uncheck the Auto-Text checkbox found in the Text tab and enter the text you want to see on your graph. To change the formatting attributes associated with this text, select an item from the list at the bottom-right corner of the dialog box, and click the Font button to change the attributes.

8. Now click OK. Your new chart options are applied to the chart you added to your report.

A little later in the chapter, we will look at some of the formatting options for charts, but now it's time to get a little practice with the other graph types.

INSERTING A CROSS-TAB CHART

Cross-tabs are special objects that can be inserted into your report to provide complex summary and analysis features (for more on cross-tabs, see Chapter 13). To take those capabilities even further, you can insert a graph based on the data in a cross-tab.

First, you need to make sure you've added a cross-tab to your report's design and that it is working correctly. You can open a good example of a cross-tab report from the book download files for use in this section.

To add a chart to a cross-tab report, use the following steps:

1. Open Crystal Reports, and open the CROSSTAB.RPT report from the book download files.
   a. Preview the cross-tab to make sure data appear in the rows, columns, and summarized fields.
   b. Switch back to the Design view of your report, click Insert > Chart, click to place your chart, and from the Chart Expert, select a chart type.

By default, Crystal Reports sets chart options automatically, reducing the number of formatting options required. If you want more control over your report, you can turn off this setting by unchecking Automatically Set Chart Options in the Chart Expert. Additional tabs will appear in the Chart Expert dialog box.

2. Click the Data tab to progress to the next step of the Chart Expert. In the dialog box that appears, click the Cross-Tab layout type, and select the placement for your chart.

The options for Cross-Tab layout are set by the cross-tabs you have inserted into your report and their locations. If your report contains only one cross-tab, you can place the graph only in the report header or footer. If you have placed a cross-tab in the group header or footer, you can place a chart alongside it.

3. Using the combo boxes in the middle of the Data page (see Figure 14.16), specify how the chart will be printed, including how its components will be broken out (On Change Of), how they will be split within those breakouts (Subdivided By), and what values will be shown (Show).

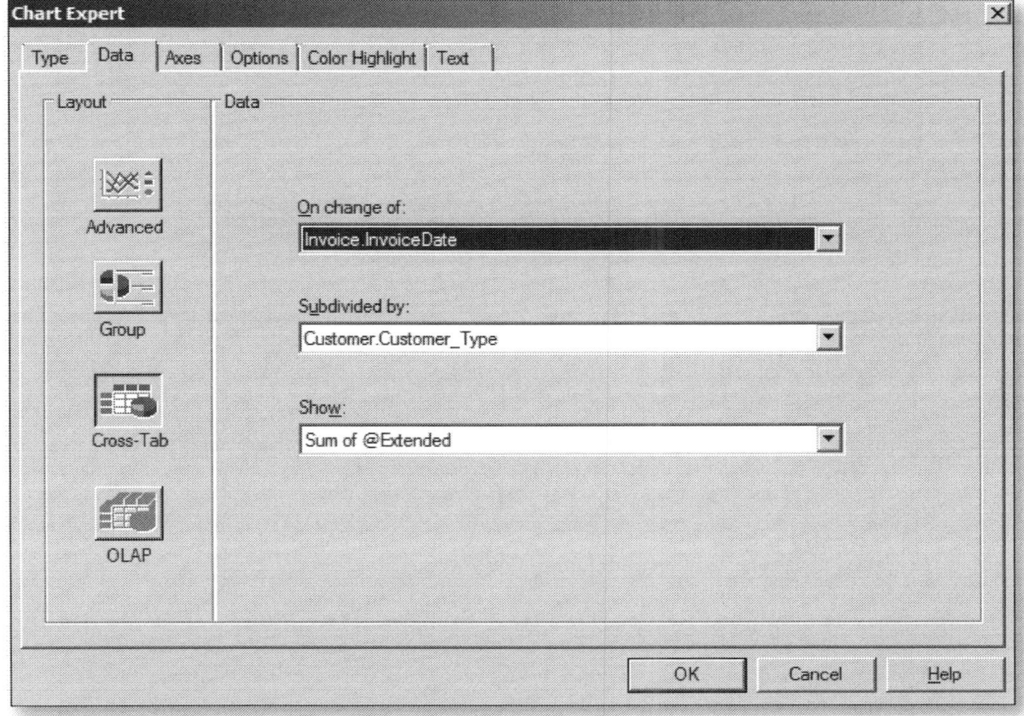

Figure 14.16 Cross-tab chart options

4. Click the Text tab to advance to the final step of the Chart Expert, and choose the text that will appear on your chart and the format of each text object. By default, Crystal Reports will choose the text, but you can override this by unchecking the Auto-Text box and typing your own text in the box provided.
5. When you have finished editing your chart's text, click OK to finish.

Crystal Reports will show a placeholder graph on the Design tab, so your data may not appear immediately. To see your own data, preview your report.

Your report should now look like the one shown in Figure 14.17.

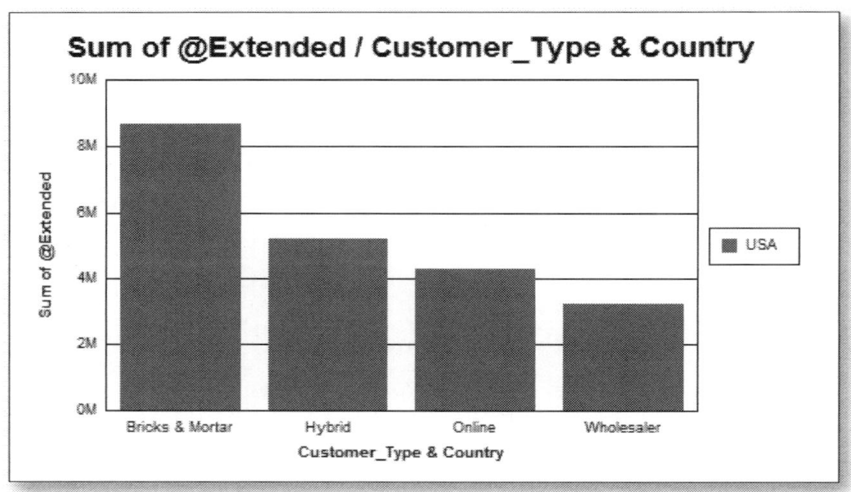

Figure 14.17 The finished report with the cross-tab and chart side-by-side

FORMATTING CHARTS

Regardless of what type of graph or chart you have created, some common formatting options can be applied to all. Most of the formatting options are available through shortcut menus that appear when you right-click your graph or chart.

MOVING CHARTS

Charts can be placed almost anywhere on your report, depending on the type of chart you have created. To change a chart's position within a section, click the graph, and drag and drop the chart to its new position.

To move a chart between sections (for example, between the page header and a group header), use the Chart Expert. To move a chart, use the following steps:

1. Right-click the chart, select Chart Expert, and then click the Data tab.
2. Using the pull-down box at the top of the page, select a new section for your chart. Use the radio buttons to specify whether the chart should be included in the header or footer of that section.

The options available at this point depend on what type of graph, groups, and so on you have inserted. All of your available options will be shown.

3. Click OK to accept your changes. Your chart should now be in the section you have specified.

CHANGING A GRAPH FROM COLOR TO BLACK AND WHITE

By default, Crystal Reports charts appear in full color. However, you can choose to use black-and-white shading and patterns instead, for better visibility when printing on a monotone printer.

To switch a chart to black-and-white mode, use the following steps:

1. Right-click the graph you want to work with, and click Format > Chart Expert.
2. Click the Options tab to open the dialog box shown in Figure 14.18.

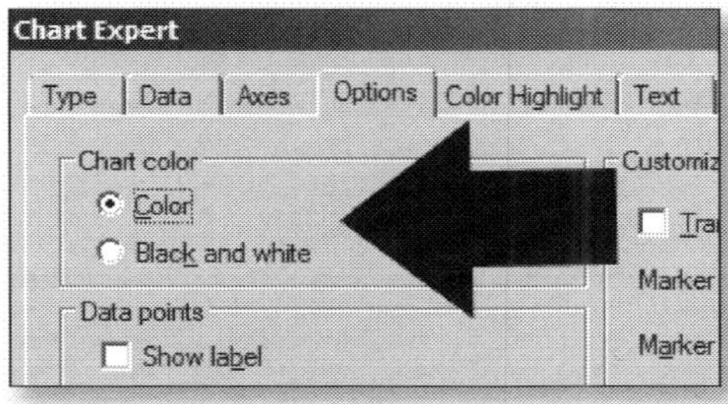

Figure 14.18 A number of chart options are available for you to customize from within the Expert

3. Click the radio button marked Black and White, and then click OK to accept your changes. Your graph will now be recolored using only black and grayscale solids and patterns.

SHOWING A GRAPH LEGEND

Crystal Reports generates a legend for the charts and graphs you install automatically. To work with legend options found in the Chart Expert, use the following steps:

1. Right-click the chart you want to work with, and click Format > Chart Expert.
2. On the Type tab of the Chart Expert, turn off Automatically Set Chart Options, and click the Options tab that appears.
3. Locate the section marked Legend, and select the checkbox to enable the legend.
4. Use the pull-down list next to this setting to choose the placement of the legend (right, left, or bottom).
5. Click OK to accept your changes.

CONTROLLING CHART GRID LINES & SCALE FOR BAR & AREA GRAPHS

Grid lines are an easy way to add value to your chart or graph and provide an instant reference to the grid's dimensions and values. To control the grid lines and scale within your graph, use the following steps:

1. Right-click a grid line in the chart, and select Format Grid Lines. A dialog box with multiple tabs, shown in Figure 14.19, opens.

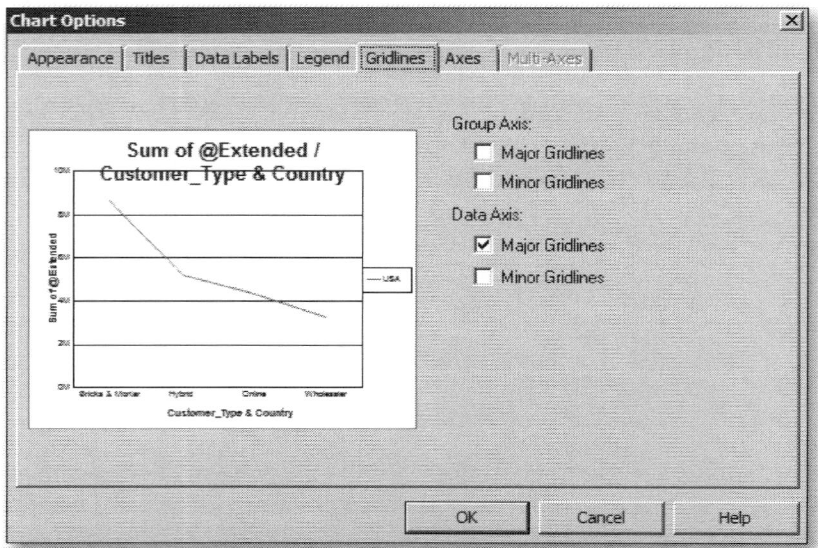

Figure 14.19 You can set grid formats using these options

2. Click either the Scales or Layout tabs.
3. On the Layout tab, choose one of the grid line options for this axis, which include the following:
   - Show Gridlines
   - Grid Style (regular, grids and ticks, inner ticks, outer ticks, spanning ticks)
   - Draw Custom Line As
4. Click the Scales tab to set the options for scaling, which include the following:
   - Use Logarithmic Scale
   - Always Include Zero
   - Use Manual Settings for Minimum Value
   - User Manual Settings for Maximum Value
5. When you have finished setting grid and scale options, click OK to accept your changes.

CHANGING THE GRAPH TYPE

You can apply a number of different chart types to your report data. In addition to the standard types of bar, pie, and so on, some chart types suit statistical data, radar and bubble graphs, three-dimensional graphs, and more. To change the graph type, use the following steps:

1. Right-click the chart you want to change, and select Load Template from the shortcut menu.
2. From the Gallery tab, select a new chart type by clicking an item on the list.

The preview image and options on the right side of the page will change according to the graph type you choose.

3. For custom chart types or any templates you may have saved, click the User Defined category. In this dialog box, click to select a category. A preview of all of the charts in that category appears on the right side of the page.

Any templates you have saved can be found in the category called User Defined.

4. Click a chart preview image to select it.
5. Click OK to accept your change and return to your report's design or preview.

SETTING CHART TITLES AND TEXT

As mentioned earlier, chart titles and text are set by default by Crystal Reports. To create your own chart titles and text for your graph, use the following steps:

1. Right-click your graph or chart, select Chart Options from the shortcut menu, then click on the Titles tab.
2. Enter the text you want to appear for the title, subtitle, and so on.

> **Changing the text that appears in this dialog box is the equivalent of unchecking the Auto-Text option in the Chart Expert and entering text.**

3. When you have finished entering your text, click OK to accept your changes and return to your report's Design or Preview view.

ADVANCED FORMATTING

Beyond the basic options, Crystal Reports offers more granular control over your charts and their data, which you can access by right-clicking a chart and choosing Chart Options. Here are some of the more advanced options you can apply.

CONTROLLING FONT SIZE & COLOR

One of the most common formatting tasks you'll perform is changing the font size and color. To do so, use the following steps:

1. Locate the element of your graph that you want to change, right-click it, and select Format XXX to open the Formatting dialog box.
2. Using the dialog box shown in Figure 14.20 on the next page, select the font, size, style, and color for the object you selected.
3. When you are finished, click OK to return to the report's Design or Preview view.

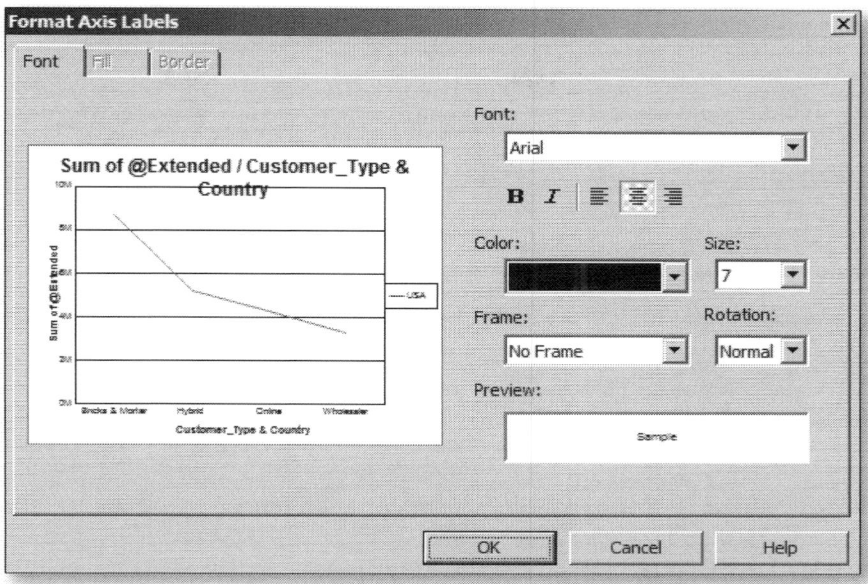

Figure 14.20 You can control the attributes of a specific element of your graph or chart

If you need a hand getting your graph or chart organized, you can use the Auto Arrange feature to have Crystal Reports do the arranging for you. To auto-arrange your chart contents, right-click anywhere on the graph, and select Auto Arrange Chart from the shortcut menu. Your graph contents will be rearranged, and the automatic changes should be reflected when you preview the report.

RESIZING A CHART

Charts and graphs may be just one element of a Crystal Report, and trying to get all of the elements to fit in the allotted space can be tricky. You can resize your chart or graph body so it fits within your report. To resize a chart, click to select the chart you want to resize. A blue handle should appear on each side of your chart. Drag the blue handles to resize your chart.

Alternatively, right-click the graph and select Object Size and Position from the shortcut menu. This option will allow you to control the size and position of your graph or chart precisely within the Design or Preview tab.

SUMMARY

Regardless of whether you have 10 records or 10,000, charts and graphs are an easy way to consolidate and present information in a report and can add visual impact. Often reports will combine a variety of presentation methods, including some of the techniques we have looked at in previous chapters, combining charts and graphs with cross-tabs, summary information and drill-downs into more detailed data. You will put all of these techniques and more to good use, as you integrate reports into your own applications, which is where we pick up in the next chapter.

Chapter 15
Exporting Reports

In This chapter

- Exporting to Excel
- Excel Exporting Tricks
- Exporting Formatted Reports
- Exporting to Word
- Exporting to HTML
- Exporting to XML
- Exporting to Text
- Exporting to a BusinessObjects Server platform

INTRODUCTION

The majority of this book has been devoted to creating information-rich, presentation quality reports from your own data, but up until this point we haven't yet talked about how to distribute those reports to other users.

In addition to distributing the Crystal Reports file (.RPT) you have created, Crystal Reports includes 18 different export formats, covering a wide range of applications and uses, including formats for use with word processing programs, spreadsheet applications, and just about everything in between. These export formats are listed on the following page.

Crystal Reports	Record Style – Columns with Spaces
HTML 3.2	Record Style – Columns without spaces
HTML 4.0	Report Definition
Microsoft Excel (97-2003)	Rich Text Format
Microsoft Excel (97-2003) Data Only	Separated Values (CSV)
Microsoft Word (97-2003)	Tab Separated Values (TTX)
Microsoft Word (97-2003) Editable	Text
ODBC	XML
PDF	XML (Legacy)

With any of these formats, you are extracting the content and format of the report at that point in time — users won't be able to refresh the data to get the latest results.

If your users need to refresh report data, you will need to look at purchasing a copy of Crystal Reports for these users (which can get expensive and difficult to manage) or alternately you can look at a server-based solution like Crystal Reports Server.

Crystal Reports Server is a web-based framework that provides a portal where you can publish reports, as well as a server architecture that provides the ability to refresh, schedule and distribute reports in a secure manner, just using your web browser.

We'll be looking at some Crystal Reports Server basics a little later in Chapter 18, including how to publish your report to this platform.

The same technology that underpins Crystal Reports Server is also available as BusinessObjects Edge for mid-sized companies and BusinessObjects Enterprise for larger organizations, as well as Business Intelligence OnDemand (www.ondemand.com/bi) which is a hosted version of the server technology behind these products.

If you are thinking of distributing your Crystal Reports to a wider audience, I recommend investing in one of the server-based platforms listed above, rather than getting caught in the cycle of refreshing and distributing reports yourself (as nobody made you the "Report Monkey", even though you may be getting paid bananas!).

SELECTING A DISTRIBUTION METHOD

Before we actually get into the technical details of how to distribute and export reports, we need to look at how to select the right distribution method and format. One of the first questions you need to ask is "Will this report need to be distributed with exactly the same format I have created?"

If the answer is "Yes", you will probably want to consider using one of Crystal Report's WYSIWYG formats. WYSIWYG is an acronym for "what you see is what you get" and these formats will try to faithfully recreate the report's layout in the exported file.

However, given that there are a number of users who want to take the report data and perform further analysis and manipulation (most likely using Excel), there are some export formats that are better suited to this "data only" approach.

Since this is the most popular method, we are going to start our look at export formats with this "Data Only" approach in mind. A little later in the chapter, we will look at some of the other formatted export formats, for when you want to keep the actual design and formatting of your report intact.

EXPORTING TO EXCEL

The most popular export format by far is Microsoft Excel. For Crystal Reports developers, there are two export options available — Excel and Excel "Data Only". The difference between the two is that with the standard Excel export, Crystal Reports will attempt to faithfully recreate your report layout in the Excel file itself. In the example on the following page in Figure 15.1, a formatted Crystal Report has been exported to the standard Excel format.

Employee by Department — Galaxy Toys

Departments: Marketing
Finance
Management

Last Name	First Name	Position	Hire Date	Salary	FTE	WorkStatus	Insurance	Other Benefits	Retirement	Total Package
Admin Department										
Clark	Raymond	General Admin	10/18/2009	$68,500.00	1.00	Full Time	$6,342.00	$941.00	$1,208.00	$76,991.00
Johnson	Lisa	Admin Assistant	10/19/2010	$80,500.00	1.00	Full Time	$8,392.00	$318.00	$3,428.00	$92,638.00
Gagnon	Alice	Office Manager	12/17/2010	$77,000.00	1.00	Full Time	$7,014.00	$781.00	$6,616.00	$91,411.00
	Average Salary:	**$75,333.33**	**Total Salary:**	**$226,000.00**						
Finance Department										
Coolidge	Barry	Chief Financial Officer	06/04/2008	$180,000.00	1.00	Full Time	$8,269.00	$319.00	$5,651.00	$194,239.00
White	Lorie	Accounts Payable	11/02/2010	$45,000.00	0.75	Part Time	$1,186.00	$805.00	$1,698.00	$48,689.00
Williams	Nathan	Accounts Recievable	11/19/2010	$90,000.00	1.00	Full Time	$8,324.00	$749.00	$966.00	$100,039.00
	Average Salary:	**$105,000.00**	**Total Salary:**	**$315,000.00**						
IT Department										
Wilson	Peter	CIO	11/02/2010	$160,000.00	1.00	Full Time	$9,038.00	$463.00	$4,496.00	$173,997.00
	Average Salary:	**$160,000.00**	**Total Salary:**	**$160,000.00**						
Management Department										

This report is provided as part of the downloadable files for the book "***Mastering Crystal Reports 2008***". For more information, visit http://www.kuiperpublishing.com

KUIPER

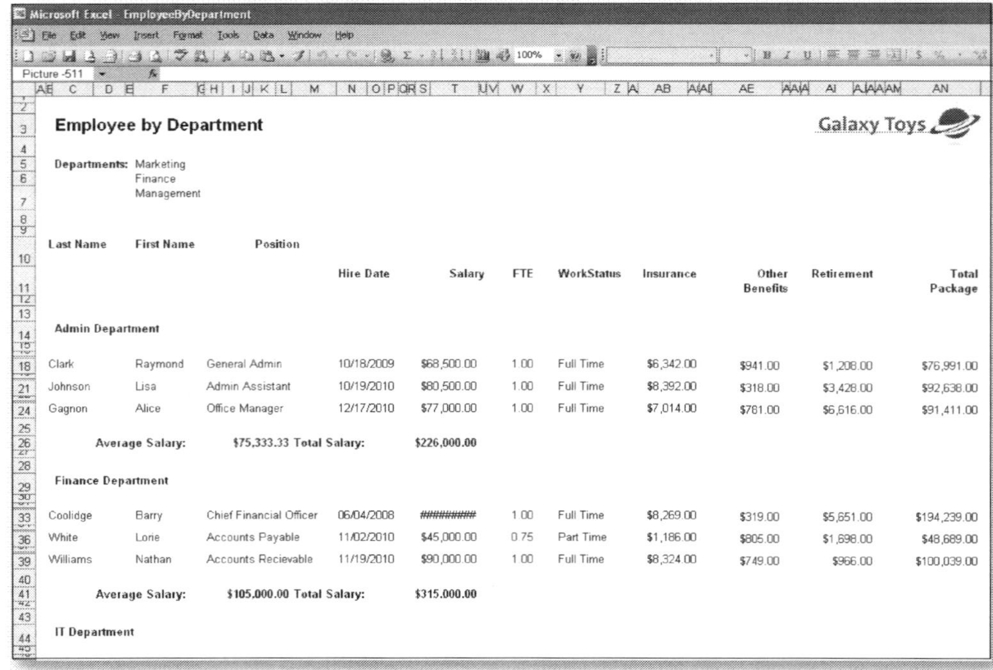

Microsoft Excel - EmployeeByDepartment

Employee by Department — Galaxy Toys

Departments: Marketing
Finance
Management

Last Name	First Name	Position	Hire Date	Salary	FTE	WorkStatus	Insurance	Other Benefits	Retirement	Total Package
Admin Department										
Clark	Raymond	General Admin	10/18/2009	$68,500.00	1.00	Full Time	$6,342.00	$941.00	$1,208.00	$76,991.00
Johnson	Lisa	Admin Assistant	10/19/2010	$80,500.00	1.00	Full Time	$8,392.00	$318.00	$3,428.00	$92,638.00
Gagnon	Alice	Office Manager	12/17/2010	$77,000.00	1.00	Full Time	$7,014.00	$781.00	$6,616.00	$91,411.00
	Average Salary:	**$75,333.33 Total Salary:**		**$226,000.00**						
Finance Department										
Coolidge	Barry	Chief Financial Officer	06/04/2008	#########	1.00	Full Time	$8,269.00	$319.00	$5,651.00	$194,239.00
White	Lorie	Accounts Payable	11/02/2010	$45,000.00	0.75	Part Time	$1,186.00	$805.00	$1,698.00	$48,689.00
Williams	Nathan	Accounts Recievable	11/19/2010	$90,000.00	1.00	Full Time	$8,324.00	$749.00	$966.00	$100,039.00
	Average Salary:	**$105,000.00 Total Salary:**		**$315,000.00**						
IT Department										

Figure 15.1 A formatted report (above) and the resulting exported file (below)

Using this export format, Crystal Reports will try to represent your report design using the features within Excel — for example, you may see merged cells, centering across a selection and even padded spaces used to help Crystal Reports mimic the report format in an Excel spreadsheet.

However, the only problem with this is that once the report data is exported, most users will want to do something further with it. For example, they may want to add additional columns or copy and paste the data into another spreadsheet, etc. So for most Excel users, they would much rather have their data exported using the Excel "Data Only" option, which will place their data into nice, orderly columns, as shown in Figure 15.2.

Microsoft Excel - EmployeeByDepartment-DATAONLY

A8 Finance Department

	A	B	C	D	E	F	G	H	I	J	K
1	Employee by Department	Departments:	Marketing□Finance□Management								
2	Last Name	First Name	Position	Hire Date	Salary	FTE	WorkStatus	Insurance	Other Benefits	Retirement	Total Package
3	Admin Department										
4	Clark	Raymond	General Admin	10/18/2009	$68,500.00	1.00	Full Time	$6,342.00	$941.00	$1,208.00	$76,991.00
5	Johnson	Lisa	Admin Assistant	10/19/2010	$80,500.00	1.00	Full Time	$8,392.00	$318.00	$3,428.00	$92,638.00
6	Gagnon	Alice	Office Manager	12/17/2010	$77,000.00	1.00	Full Time	$7,014.00	$781.00	$6,616.00	$91,411.00
7	Average Salary:	$75,333.33	Total Salary:	$226,000.00							
8	Finance Department										
9	Coolidge	Barry	Chief Financial Officer	06/04/2008	$180,000.00	1.00	Full Time	$8,269.00	$319.00	$5,651.00	$194,239.00
10	White	Lorie	Accounts Payable	11/02/2010	$45,000.00	0.75	Part Time	$1,186.00	$805.00	$1,698.00	$48,689.00
11	Williams	Nathan	Accounts Recievable	11/19/2010	$90,000.00	1.00	Full Time	$8,324.00	$749.00	$966.00	$100,039.00
12	Average Salary:	$105,000.00	Total Salary:	$315,000.00							
13	IT Department										
14	Wilson	Peter	CIO	11/02/2010	$160,000.00	1.00	Full Time	$9,038.00	$463.00	$4,496.00	$173,997.00
15	Average Salary:	$160,000.00	Total Salary:	$160,000.00							
16	Management Department										
17	Smith	Jane	CEO	03/20/2006	$210,000.00	1.00	Full Time	$5,655.00	$84.00	$7,214.00	$222,953.00
18	Average Salary:	$210,000.00	Total Salary:	$210,000.00							
19	Marketing Department										
20	King	Leanne	Marketing Manager	10/09/2006	$112,000.00	1.00	Full Time	$6,704.00	$552.00	$8,792.00	$128,048.00
21	Average Salary:	$112,000.00	Total Salary:	$112,000.00							
22	Sales Department										
23	Lee	John	Account Executive	12/15/2007	$65,000.00	1.00	Full Time	$2,446.00	$250.00	$1,191.00	$68,887.00
24	Roy	Michael	Account Executive	05/04/2009	$98,000.00	1.00	Full Time	$8,770.00	$823.00	$1,208.00	$108,801.00
25	McGraw	Colin	Account Executive	06/21/2007	$86,000.00	1.00	Full Time	$3,024.00	$875.00	$6,534.00	$96,433.00
26	Long	Alex	Account Executive	04/21/2007	$45,999.00	1.00	Full Time	$299.00	$345.00	$7,057.00	$53,700.00
27	Martin	Bob	Account Executive	02/16/2009	$86,250.00	1.00	Full Time	$5,224.00	$517.00	$1,100.00	$93,091.00
28	Cunningham	Randall	National Sales Manager	05/04/2007	$120,000.00	1.00	Full Time	$2,270.00	$166.00	$9,295.00	$131,731.00
29	Brown	David	Account Manager	05/23/2006	$54,555.00	1.00	Full Time	$7,588.00	$172.00	$5,837.00	$68,152.00
30	Average Salary:	$79,400.57	Total Salary:	$555,804.00							
31	Warehouse Department										
32	Foster	Isabella	Warehouse Stockman	06/18/2010	$72,000.00	1.00	Full Time	$4,457.00	$749.00	$6,592.00	$83,798.00
33	Phillips	Aleigha	Warehouse Manager	01/28/2009	$86,500.00	1.00	Full Time	$4,104.00	$911.00	$2,944.00	$94,459.00
34	Long	Rodney	Warehouse Stockman	07/07/2007	$65,000.00	1.00	Full Time	$2,586.00	$956.00	$3,571.00	$72,113.00
35	Sanders	Madison	Warehouse Stockman	11/23/2007	$65,000.00	1.00	Full Time	$9,373.00	$573.00	$6,373.00	$81,319.00
36	Ross	Olivia	Warehouse Stockman	08/04/2008	$65,000.00	1.00	Full Time	$2,195.00	$650.00	$4,435.00	$72,280.00
37	Morales	John	Inventory Control	01/31/2011	$75,000.00	1.00	Full Time	$8,605.00	$933.00	$827.00	$85,365.00

Figure 15.2 A report exported to the Excel "Data Only" format

Even with the "Data Only" option, successfully exporting your data to Excel can be tricky, especially with complex reports that may be heavily formatted, but still require the ability to export to this format.

In the following sections, we are going to look at some of your Excel "Data Only" exporting options, as well as some of the guidelines for getting that perfect Excel export every time.

SETTING EXCEL FORMAT OPTIONS

When you select File > Export > Export Report and select the "Microsoft Excel (97–2003) Data Only" export format, you will get a dialog like the one shown below in Figure 15.3.

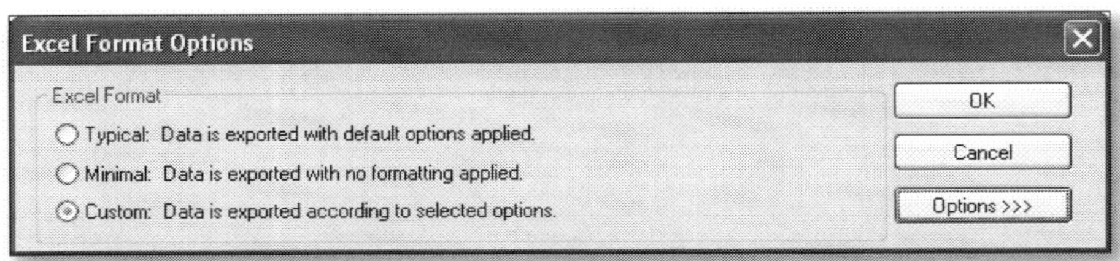

Figure 15.3 The Excel "Data Only" export format

This dialog is used to control the Excel exporting options and there are three choices to be made for the Excel format you want to use:

Typical — Where the data in your report is exported using the default options applied

Minimal — Where the data in your report is exported with no formatting applied

Custom — Where you choose the options that are applied to the exported file format.

So using our report from before, let's look at each of these options to see their effect on the resulting Excel output.

Using our example report, if you were to select the "Typical" option and export your report, it would look like the file shown in Figure 15.4.

Microsoft Excel - EmployeeListing-TYPICAL

	A	B	C	D	E	F	G	H	I	J
1	Employee List									
2	Last Name	First Name	Position	Hire Date	Salary	FTE	Work Status	Insurance	Other Benefits	Retirement
3	Clark	Raymond	General Admin	10/18/2009	$68,500.00	1.00	Full Time	$6,342.00	$941.00	$1,208
4	Johnson	Lisa	Admin Assistant	10/19/2010	$80,500.00	1.00	Full Time	$8,392.00	$318.00	$3,428
5	Gagnon	Alice	Office Manager	12/17/2010	$77,000.00	1.00	Full Time	$7,014.00	$781.00	$6,616
6	Coolidge	Barry	Chief Financial Officer	06/04/2008	$180,000.00	1.00	Full Time	$8,269.00	$319.00	$5,651
7	White	Lorie	Accounts Payable	11/02/2010	$45,000.00	0.75	Part Time	$1,186.00	$805.00	$1,698
8	Williams	Nathan	Accounts Recievable	11/19/2010	$90,000.00	1.00	Full Time	$8,324.00	$749.00	$966
9	Wilson	Peter	CIO	11/02/2010	$160,000.00	1.00	Full Time	$9,038.00	$463.00	$4,496
10	Smith	Jane	CEO	03/20/2006	$210,000.00	1.00	Full Time	$5,655.00	$84.00	$7,214
11	King	Leanne	Marketing Manager	10/09/2006	$112,000.00	1.00	Full Time	$6,704.00	$552.00	$8,792
12	Lee	John	Account Executive	12/15/2007	$65,000.00	1.00	Full Time	$2,446.00	$250.00	$1,191
13	Roy	Michael	Account Executive	05/04/2009	$98,000.00	1.00	Full Time	$8,770.00	$823.00	$1,208
14	McGraw	Colin	Account Executive	06/21/2007	$86,000.00	1.00	Full Time	$3,024.00	$875.00	$6,534
15	Long	Alex	Account Executive	04/21/2007	$45,999.00	1.00	Full Time	$299.00	$345.00	$7,057
16	Martin	Bob	Account Executive	02/16/2009	$86,250.00	1.00	Full Time	$5,224.00	$517.00	$1,100
17	Cunningham	Randall	National Sales Manager	05/04/2007	$120,000.00	1.00	Full Time	$2,270.00	$166.00	$9,295
18	Brown	David	Account Manager	05/23/2006	$54,555.00	1.00	Full Time	$7,588.00	$172.00	$5,837
19	Foster	Isabella	Warehouse Stockman	06/18/2010	$72,000.00	1.00	Full Time	$4,457.00	$749.00	$6,592
20	Phillips	Aleigha	Warehouse Manager	01/28/2009	$86,500.00	1.00	Full Time	$4,104.00	$911.00	$2,944
21	Long	Rodney	Warehouse Stockman	07/07/2007	$65,000.00	1.00	Full Time	$2,586.00	$956.00	$3,571
22	Sanders	Madison	Warehouse Stockman	11/23/2007	$65,000.00	1.00	Full Time	$9,373.00	$573.00	$6,373
23	Ross	Olivia	Warehouse Stockman	08/04/2008	$65,000.00	1.00	Full Time	$2,195.00	$650.00	$4,435
24	Morales	John	Inventory Control	01/31/2011	$75,000.00	1.00	Full Time	$8,605.00	$933.00	$827
25	Powell	Tom	Purchasing Specialist	12/17/2010	$65,000.00	1.00	Full Time	$1,049.00	$288.00	$6,301
26	Sullivan	John	Purchasing Specialist	05/26/2010	$71,000.00	0.50	Part Time	$5,441.00	$647.00	$6,711
27	Russell	Hailey	Purchasing Specialist	05/26/2011	$65,000.00	1.00	Full Time	$8,434.00	$128.00	$9,472
28	Ortiz	Abigail	Inventory Control	06/09/2011	$45,000.00	0.75	Part Time	$3,036.00	$331.00	$9,943
29	Jenkins	Kaitlyn	Logistics Coordinator	06/26/2011	$78,090.00	1.00	Full Time	$5,502.00	$263.00	$2,871
30	Gutiérrez	James	Logistics Coordinator	07/07/2007	$90,900.00	0.00	Seperated	$3,802.00	$885.00	$870
31	Perry	Edward	Shipping Specialist	01/31/2010	$65,000.00	1.00	Full Time	$2,782.00	$500.00	$7,864
32	Myers	Sophia	Inventory Control	04/02/2010	$65,000.00	1.00	Full Time	$4,444.00	$168.00	$9,468
33	GRAND TOTAL			$2,552,294.00	28.00		$156,365.00	$16,142.00	$150,533.00	$2,875,324

Figure 15.4 Excel export using the "Typical" option

Note that this export option uses the Details section to set the column width in your spreadsheet and that it exports both the page header and page footer as well. If you were to perform the same action and this time selected the "Minimal" option, your report would look a little more spare, with just the data represented in columns, as shown over the page in Figure 15.5.

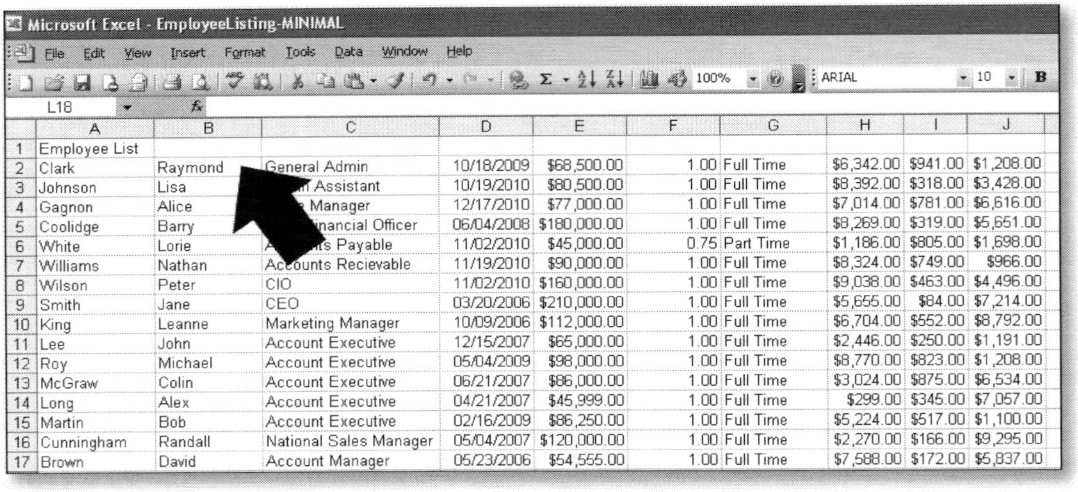

Microsoft Excel - EmployeeListing-MINIMAL

	A	B	C	D	E	F	G	H	I	J
1	Employee List									
2	Clark	Raymond	General Admin	10/18/2009	$68,500.00	1.00	Full Time	$6,342.00	$941.00	$1,208.00
3	Johnson	Lisa	n Assistant	10/19/2010	$80,500.00	1.00	Full Time	$8,392.00	$318.00	$3,428.00
4	Gagnon	Alice	Manager	12/17/2010	$77,000.00	1.00	Full Time	$7,014.00	$781.00	$6,616.00
5	Coolidge	Barry	inancial Officer	06/04/2008	$180,000.00	1.00	Full Time	$8,269.00	$319.00	$5,651.00
6	White	Lorie	A ts Payable	11/02/2010	$45,000.00	0.75	Part Time	$1,186.00	$805.00	$1,698.00
7	Williams	Nathan	Accounts Recievable	11/19/2010	$90,000.00	1.00	Full Time	$8,324.00	$749.00	$966.00
8	Wilson	Peter	CIO	11/02/2010	$160,000.00	1.00	Full Time	$9,038.00	$463.00	$4,496.00
9	Smith	Jane	CEO	03/20/2006	$210,000.00	1.00	Full Time	$5,655.00	$84.00	$7,214.00
10	King	Leanne	Marketing Manager	10/09/2006	$112,000.00	1.00	Full Time	$6,704.00	$552.00	$8,792.00
11	Lee	John	Account Executive	12/15/2007	$65,000.00	1.00	Full Time	$2,446.00	$250.00	$1,191.00
12	Roy	Michael	Account Executive	05/04/2009	$98,000.00	1.00	Full Time	$8,770.00	$823.00	$1,208.00
13	McGraw	Colin	Account Executive	06/21/2007	$86,000.00	1.00	Full Time	$3,024.00	$875.00	$6,534.00
14	Long	Alex	Account Executive	04/21/2007	$45,999.00	1.00	Full Time	$299.00	$345.00	$7,057.00
15	Martin	Bob	Account Executive	02/16/2009	$86,250.00	1.00	Full Time	$5,224.00	$517.00	$1,100.00
16	Cunningham	Randall	National Sales Manager	05/04/2007	$120,000.00	1.00	Full Time	$2,270.00	$166.00	$9,295.00
17	Brown	David	Account Manager	05/23/2006	$54,555.00	1.00	Full Time	$7,588.00	$172.00	$5,837.00

Figure 15.5 Excel export using the "Minimal" option

In addition to "Typical" and "Minimal", you also have the ability to control the export settings using the "Custom" option. Using this option, you select what your resulting Excel column widths are based on (details, report footer, etc.) as well as which elements of the report to export (page headers/footers), etc. You can set these attributes by clicking on the "Options" button to open the dialog shown in Figure 15.6 on the facing page.

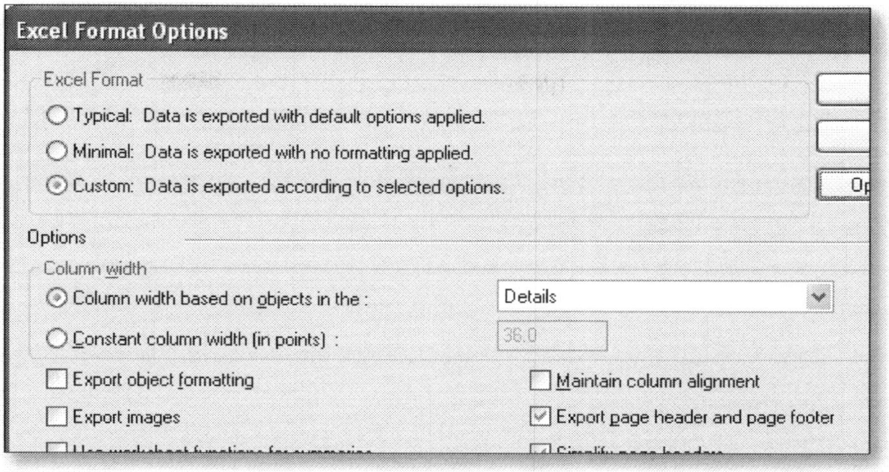

Figure 15.6 Excel export custom options

As a report developer, you probably don't want to set these settings each time you export the report. Luckily, in Crystal Reports 2008, we can set the defaults for our export formats by selecting File > Export > Report Export Options, then selecting the "Microsoft Excel (97-2003) Data Only" format and clicking OK to open the dialog shown in Figure 15.7.

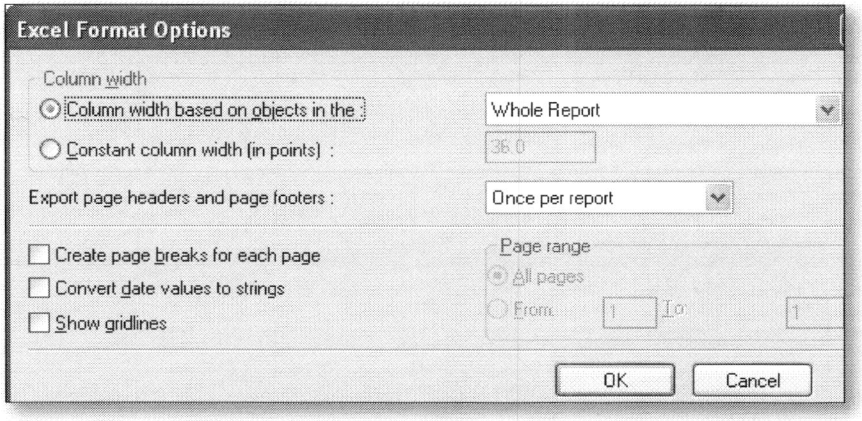

Figure 15.7 Excel saved export options

Using this dialog you can set your default settings for this export format, so they are remembered each time the report is exported.

DESIGNING A REPORT FOR EXCEL EXPORT

In training courses, I always tell students that designing reports for export to Excel is a bit of a "black art", but it is definitely a skill you can learn. Over the years, there have been a number of white papers dedicated to getting the perfect report design for export to Excel. The tips below come from my own experience of trial and error, but they are tried and true. It is by no means an exhaustive list and no doubt every Crystal Reports developer has found their own tricks to get a data-perfect Excel export every time.

Keeping It Simple

If you are designing a report specifically for Excel export, my first advice is to keep it simple. Avoid applying a lot of formatting, logos, headers, etc. to your report. This is especially true if you are providing the report solely for use with Excel, or to format for import or use in another application. In the example below in Figure 15.8, the report has been stripped down to just the detail fields and field headings, with all of the extraneous formatting removed.

Last Name	First Name	Position	Hire Date	Salary	FTE	Work Status	Insurance
Clark	Raymond	General Admin	10/18/2009	$68,500.00	1.00	Full Time	$6,342.00
Johnson	Lisa	Admin Assistant	10/19/2010	$80,500.00	1.00	Full Time	$8,392.00
Gagnon	Alice	Office Manager	12/17/2010	$77,000.00	1.00	Full Time	$7,014.00
Coolidge	Barry	Chief Financial Officer	06/04/2008	$180,000.00	1.00	Full Time	$8,269.00
White	Lorie	Accounts Payable	11/02/2010	$45,000.00	0.75	Part Time	$1,186.00
Williams	Nathan	Accounts Recievable	11/19/2010	$90,000.00	1.00	Full Time	$8,324.00
Wilson	Peter	CIO	11/02/2010	$160,000.00	1.00	Full Time	$9,038.00
Smith	Jane	CEO	03/20/2006	$210,000.00	1.00	Full Time	$5,655.00
King	Leanne	Marketing Manager	10/09/2006	$112,000.00	1.00	Full Time	$6,704.00
Lee	John	Account Executive	12/15/2007	$65,000.00	1.00	Full Time	$2,446.00
Roy	Michael	Account Executive	05/04/2009	$98,000.00	1.00	Full Time	$8,770.00
McGraw	Colin	Account Executive	06/21/2007	$86,000.00	1.00	Full Time	$3,024.00
Long	Alex	Account Executive	04/21/2007	$45,999.00	1.00	Full Time	$299.00
Martin	Bob	Account Executive	02/16/2009	$86,250.00	1.00	Full Time	$5,224.00
Cunningham	Randall	National Sales Manager	05/04/2007	$120,000.00	1.00	Full Time	$2,270.00
Brown	David	Account Manager	05/23/2006	$54,555.00	1.00	Full Time	$7,588.00
Foster	Isabella	Warehouse Stockman	06/18/2010	$72,000.00	1.00	Full Time	$4,457.00
Phillips	Aleigha	Warehouse Manager	01/28/2009	$86,500.00	1.00	Full Time	$4,104.00

Figure 15.8 A stripped down report format

This report should export nicely to Excel (regardless of which Excel format you choose) as there is no formatting, logos, etc. to get in the way.

Get In Line

When Crystal Reports exports your report, it will try to determine which fields to place into each Excel column. You can make that decision a lot easier by making all of your field and field-heading objects the same size, and also lining them up with guidelines, as shown in Figure 15.9.

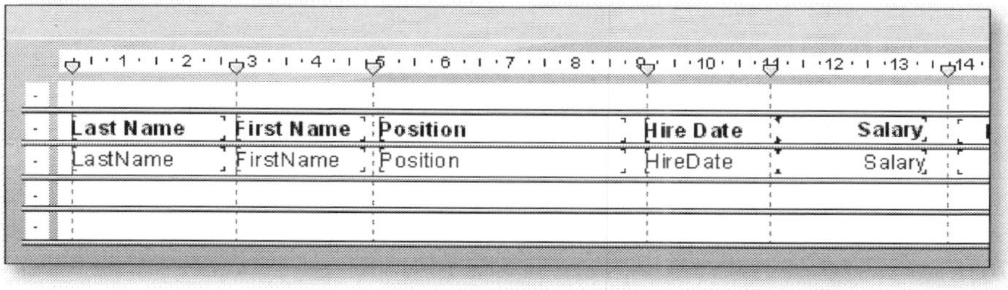

Figure 15.9 Aligning objects to Guidelines

While some Crystal Reports developers prefer NOT to use guidelines in their reports, this is one instance where they are essential. You can create a new guideline by clicking within the ruler, and then align objects by dragging them close to the guideline until they "snap" onto it (believe me, this is an acquired skill — it may take you a couple of go's to get the hang of it).

To get the fields and field headings the same size, use CTRL-CLICK to select both objects and then select Format > Make Same Size and choose either Width, Height or Both.

If you really want to get precise with the alignment, you can get it down to the pixel, inch or centimeter by selecting Format > Size and Position and then entering both the size and position down to the fraction. Often, if you have a "wayward" field that looks out of place in the Excel export, you can attribute it to being slightly off in the alignment to the rest of the fields. You will have to go into the Size and Position dialog and tweak the settings to get it "just right" (it's no wonder my hairline is receding — I am usually pulling it out at this point).

That same Format menu is also handy when moving objects around, as you can use the Format > Align function to align Tops, Middles, Lefts, etc. Also, don't forget that in Design mode, you can use the arrow keys to "bump" objects around, pixel by pixel if you have to.

Size Isn't Everything

Another handy trick when formatting a report for export is to make the fields smaller — and in this case, size does matter. If you are creating a report specifically for export to Excel, chances are you don't care about how this looks in a formatted Crystal Reports. By making the fields smaller, you can add more fields to the report canvas and line them up with their field headings. When the file is exported (as shown in Figure 15.10) the columns will have the correct data and you can then resize the columns in Excel.

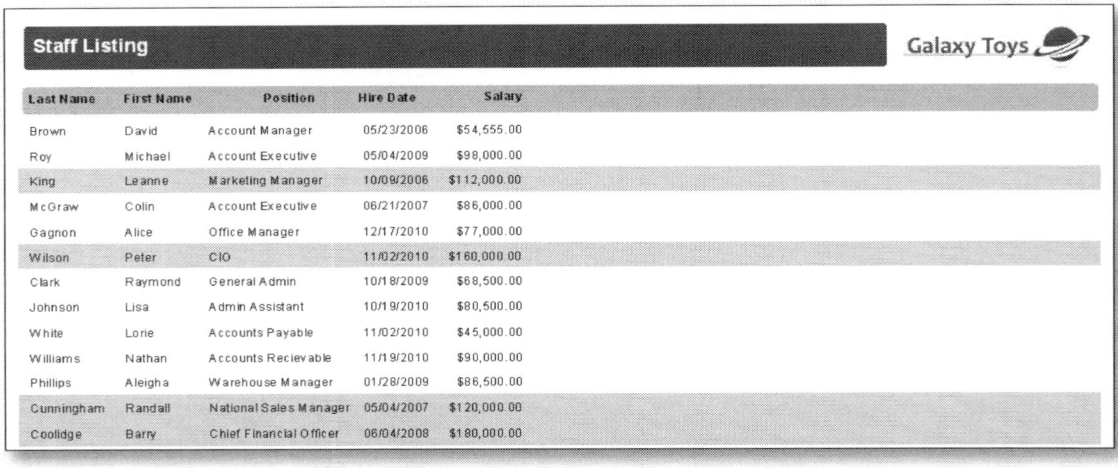

Staff Listing

Galaxy Toys

Last Name	First Name	Position	Hire Date	Salary
Brown	David	Account Manager	05/23/2006	$54,555.00
Roy	Michael	Account Executive	05/04/2009	$98,000.00
King	Leanne	Marketing Manager	10/09/2006	$112,000.00
McGraw	Colin	Account Executive	06/21/2007	$86,000.00
Gagnon	Alice	Office Manager	12/17/2010	$77,000.00
Wilson	Peter	CIO	11/02/2010	$160,000.00
Clark	Raymond	General Admin	10/18/2009	$68,500.00
Johnson	Lisa	Admin Assistant	10/19/2010	$80,500.00
White	Lorie	Accounts Payable	11/02/2010	$45,000.00
Williams	Nathan	Accounts Recievable	11/19/2010	$90,000.00
Phillips	Aleigha	Warehouse Manager	01/28/2009	$86,500.00
Cunningham	Randall	National Sales Manager	05/04/2007	$120,000.00
Coolidge	Barry	Chief Financial Officer	06/04/2008	$180,000.00

Figure 15.10 A "before" and "after" with smaller columns

EXPORTING FORMATTED REPORTS

Now that we have had a look at how to get the data out of your Crystal Report, we need to look at how to distribute formatted reports to users. Considering so much time developing a report is spent formatting your reports, it's no wonder we want to show off the report just as it has been designed. Using Crystal Reports, you have a few options for getting a true WYSIWYG copy of your report, and they are outlined below.

REPORT WITH SAVED DATA

As you have been creating reports, you have been using the Crystal Reports 2008 designer, which is a licensed product. But if you want to distribute your reports with saved data, to allow users to interact with the report, drill-down, etc. you can distribute your reports using the free Crystal Reports 2008 viewer, which is available for both the Windows and Mac platform.

Using Crystal Reports Viewer, you could save your report with data and distribute the report file with the saved data inside. The Save Report with Data option is actually a setting that Crystal Reports turns on by default, and it provides an easy method for distributing your report to other Crystal Reports users without having to give them access to your data source.

In addition, if you are using Crystal Reports Server (or any of the other BusinessObjects server products) you can schedule reports to be distributed to users in .RPT format.

To save your report with data, verify that the option Save Data with Report is checked by selecting File > Save Data with Report, as shown in Figure 15.11.

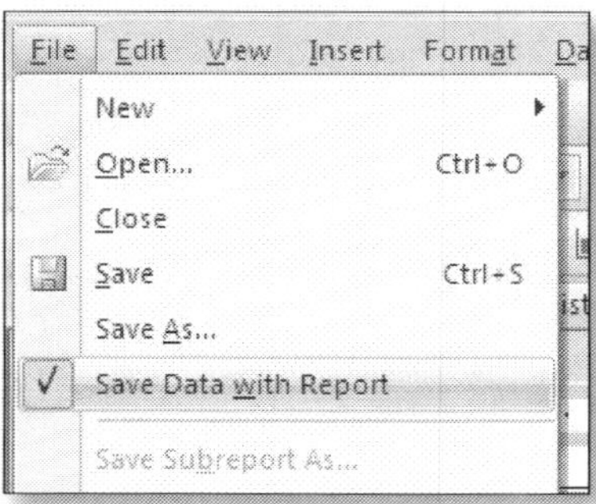

Figure 15.11 You can save your report with data for distribution to other Crystal Reports users

From that point, all you need to do is refresh your report data (using F5) and save your report using File > Save, specifying a file name and location for the report. Once you have saved the .rpt file, you can send it to other Crystal Reports users, and they can view and print both the report and the data contained within.

EXPORTING TO AN ACROBAT PDF FILE

Another true "WYSIWYG" format available for Crystal Reports is Adobe's Portable Document Format, which is one of the most widely supported document formats.

To export your report to Acrobat format, select File > Export to open the Export dialog box, shown in Figure 15.12.

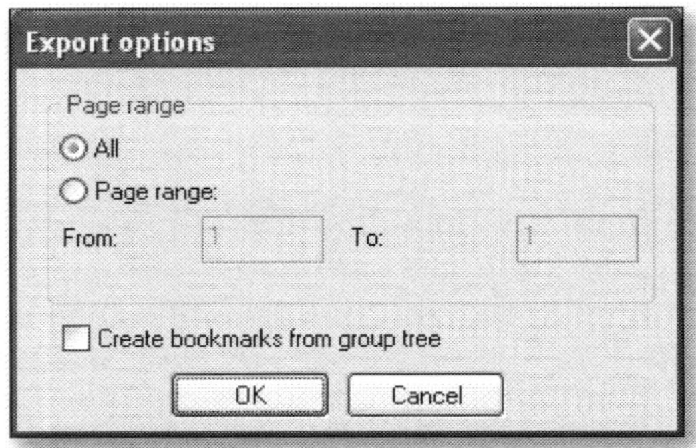

Figure 15.12 The Crystal Reports Export dialog box

In the Export dialog box, select the format Acrobat Format (PDF) and the destination Disk File and click OK. In the Export Options dialog box that appears, enter the page range to be exported and click OK to proceed.

At this point, you will need to enter a file name and location for the exported file and click OK. A progress dialog box, shown in Figure 15.13, will appear while the records are being exported, and you will be returned to your report's design or preview page when the export is finished.

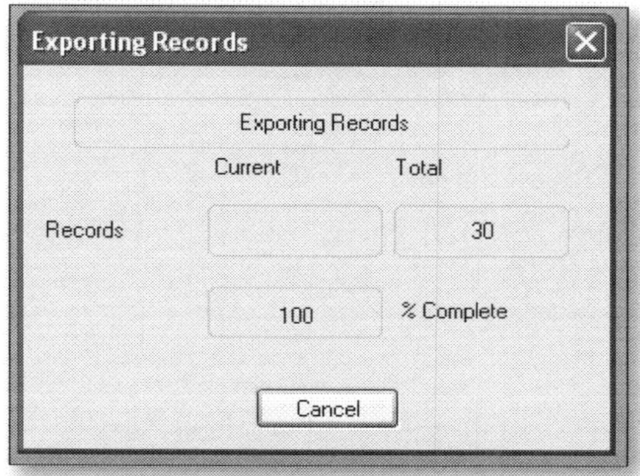

Figure 15.13 A progress dialog box appears while the report is being exported

If you have installed the Adobe Acrobat Reader before-hand, you should be able to locate the file you just exported and double-click to view the report.

As with the majority of export formats, you will not be able to use any of the native Crystal Reports features such as drill-down and record selection when you export to PDF.

EXPORTING TO WORD & RTF

In previous versions of Crystal Reports, there are actually two different file export formats available for Microsoft Word. The first, "Microsoft Word (97-2003)", was originally designed as an export format specifically for Word. This export format tried to reflect the design and content of your report within a Word document, with mixed results.

A much better choice was to export your report using the Rich Text Format export option. A significant amount of development work has been put into the RTF format export to ensure that most formatting options from your report design will be exported.

To export to either Word option, you will need to select File > Export. Then in the Export dialog box, select your export format and a destination of Disk File. Once you click OK, a second dialog box will open and prompt you for a page range to export. Once you have selected either "All Pages" or entered a page range, your report will be exported.

When you view the resulting .DOC or .RTF file, you will notice that the majority of your report's features have been translated into the export format, as shown in Figure 15.14.

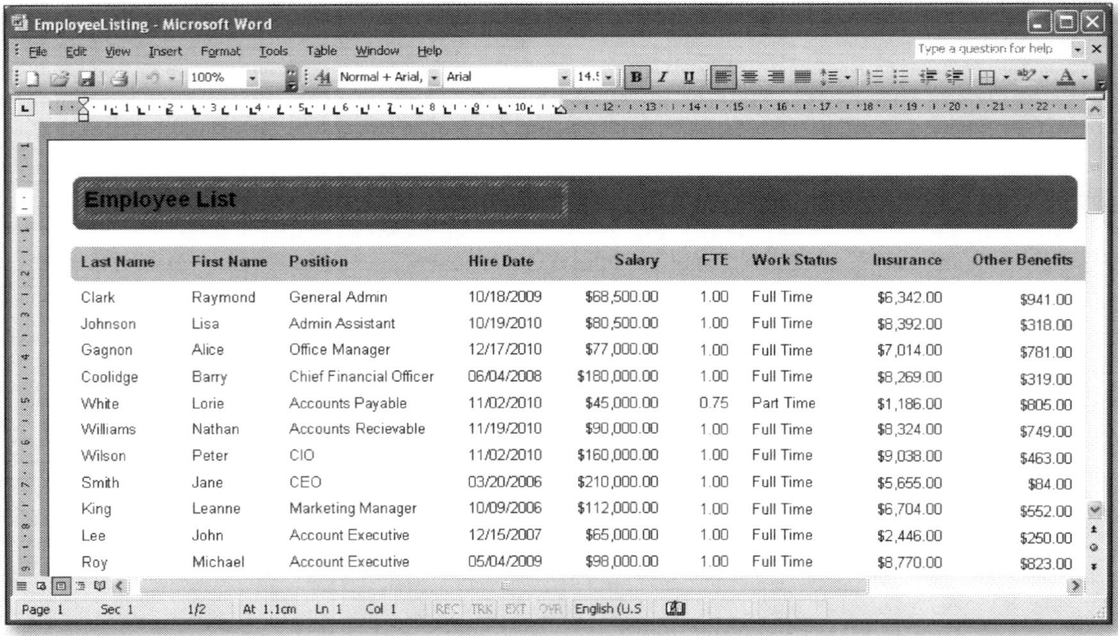

Last Name	First Name	Position	Hire Date	Salary	FTE	Work Status	Insurance	Other Benefits
Clark	Raymond	General Admin	10/18/2009	$68,500.00	1.00	Full Time	$6,342.00	$941.00
Johnson	Lisa	Admin Assistant	10/19/2010	$80,500.00	1.00	Full Time	$8,392.00	$318.00
Gagnon	Alice	Office Manager	12/17/2010	$77,000.00	1.00	Full Time	$7,014.00	$781.00
Coolidge	Barry	Chief Financial Officer	06/04/2008	$180,000.00	1.00	Full Time	$8,269.00	$319.00
White	Lorie	Accounts Payable	11/02/2010	$45,000.00	0.75	Part Time	$1,186.00	$805.00
Williams	Nathan	Accounts Recievable	11/19/2010	$90,000.00	1.00	Full Time	$8,324.00	$749.00
Wilson	Peter	CIO	11/02/2010	$160,000.00	1.00	Full Time	$9,038.00	$463.00
Smith	Jane	CEO	03/20/2006	$210,000.00	1.00	Full Time	$5,655.00	$84.00
King	Leanne	Marketing Manager	10/09/2006	$112,000.00	1.00	Full Time	$6,704.00	$552.00
Lee	John	Account Executive	12/15/2007	$65,000.00	1.00	Full Time	$2,446.00	$250.00
Roy	Michael	Account Executive	05/04/2009	$98,000.00	1.00	Full Time	$8,770.00	$823.00

Figure 15.14 An example of a report exported to Microsoft Word

To keep the approximate page layout, these export formats use extensive use of frames within the document to position text precisely where it would have appeared on the report.

Also, the export formats don't make good use of the features within Word — for example, if your page header includes a graphic, that graphic will be repeated on every page of your Word document (instead of being inserted once with Word's own header/footer functionality) and can drastically increase the size of the file.

Likewise, if you wanted to change the page headings and your report was 60 pages, you would have to change the headings on each page individually.

So, the best practice for working with users who need a report exported to Word or RTF format is to do all of your report formatting using Crystal Reports and make exporting the file your final step when you are ready to distribute the file. If your users do need to change the format of the document afterwards, consider giving the report to them in Excel or one of the text file formats, where they can apply all of the formatting they want very easily.

EXPORTING TO HTML

Crystal Reports includes export formats for both HTML 3.2 and HTML 4.0 (DHTML), allowing you to export your report for viewing in a Web browser without any additional plug-ins. The HTML 4.0 (DHTML) format is the better of the two as it can translate more of the report's design into HTML using some of the features introduced with HTML 4.0, as shown in Figure 15.15.

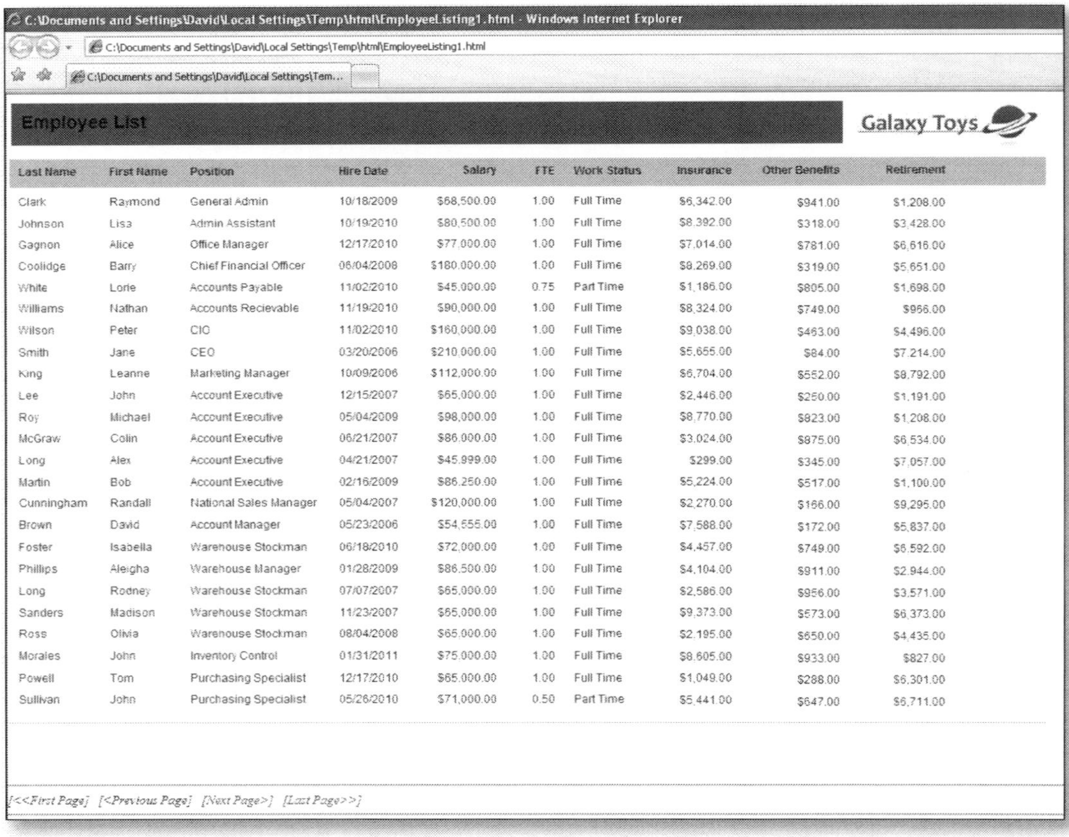

C:\Documents and Settings\David\Local Settings\Temp\html\EmployeeListing1.html - Windows Internet Explorer

Employee List

Galaxy Toys

Last Name	First Name	Position	Hire Date	Salary	FTE	Work Status	Insurance	Other Benefits	Retirement
Clark	Raymond	General Admin	10/18/2009	$68,500.00	1.00	Full Time	$6,342.00	$941.00	$1,208.00
Johnson	Lisa	Admin Assistant	10/19/2010	$80,500.00	1.00	Full Time	$8,392.00	$318.00	$3,428.00
Gagnon	Alice	Office Manager	12/17/2010	$77,000.00	1.00	Full Time	$7,014.00	$781.00	$6,616.00
Coolidge	Barry	Chief Financial Officer	06/04/2008	$180,000.00	1.00	Full Time	$8,269.00	$319.00	$5,651.00
White	Lorie	Accounts Payable	11/02/2010	$45,000.00	0.75	Part Time	$1,186.00	$805.00	$1,698.00
Williams	Nathan	Accounts Recievable	11/19/2010	$90,000.00	1.00	Full Time	$8,324.00	$749.00	$966.00
Wilson	Peter	CIO	11/02/2010	$160,000.00	1.00	Full Time	$9,038.00	$463.00	$4,496.00
Smith	Jane	CEO	03/20/2006	$210,000.00	1.00	Full Time	$5,655.00	$84.00	$7,214.00
King	Leanne	Marketing Manager	10/09/2006	$112,000.00	1.00	Full Time	$6,704.00	$552.00	$8,792.00
Lee	John	Account Executive	12/15/2007	$65,000.00	1.00	Full Time	$2,446.00	$250.00	$1,191.00
Roy	Michael	Account Executive	05/04/2009	$98,000.00	1.00	Full Time	$8,770.00	$823.00	$1,208.00
McGraw	Colin	Account Executive	06/21/2007	$86,000.00	1.00	Full Time	$3,024.00	$875.00	$6,534.00
Long	Alex	Account Executive	04/21/2007	$45,999.00	1.00	Full Time	$299.00	$345.00	$7,057.00
Martin	Bob	Account Executive	02/16/2009	$86,250.00	1.00	Full Time	$5,224.00	$517.00	$1,100.00
Cunningham	Randall	National Sales Manager	05/04/2007	$120,000.00	1.00	Full Time	$2,270.00	$166.00	$9,295.00
Brown	David	Account Manager	05/23/2006	$54,555.00	1.00	Full Time	$7,588.00	$172.00	$5,837.00
Foster	Isabella	Warehouse Stockman	06/18/2010	$72,000.00	1.00	Full Time	$4,457.00	$749.00	$6,592.00
Phillips	Aleigha	Warehouse Manager	01/28/2009	$86,500.00	1.00	Full Time	$4,104.00	$911.00	$2,944.00
Long	Rodney	Warehouse Stockman	07/07/2007	$65,000.00	1.00	Full Time	$2,586.00	$956.00	$3,571.00
Sanders	Madison	Warehouse Stockman	11/23/2007	$65,000.00	1.00	Full Time	$9,373.00	$573.00	$6,373.00
Ross	Olivia	Warehouse Stockman	08/04/2008	$65,000.00	1.00	Full Time	$2,195.00	$650.00	$4,435.00
Morales	John	Inventory Control	01/31/2011	$75,000.00	1.00	Full Time	$8,605.00	$933.00	$827.00
Powell	Tom	Purchasing Specialist	12/17/2010	$65,000.00	1.00	Full Time	$1,049.00	$288.00	$6,301.00
Sullivan	John	Purchasing Specialist	05/26/2010	$71,000.00	0.50	Part Time	$5,441.00	$647.00	$6,711.00

[<<First Page] [<Previous Page] [Next Page>] [Last Page>>]

Figure 15.15 A Crystal Report exported to HTML 3.2 and HTML 4.0 (DHTML)

With either export format, you can choose whether you want the report displayed as a single page or multiple pages, as well as whether the exported report has built-in page navigation. You can also specify the page range to be exported and the base file name and subdirectory where the files are stored.

Exporting to HTML provides a static view of your report, as all of the pictures, graphs, and maps in your report are created as static JPEG files and stored in the same directory structure as the HTML files generated. If you need a live, dynamic report within a Web browser, you may want to consider publishing your report to Crystal Reports Server, which allows you to view your report in a browser window. This will also give you a live, dynamic view of the report, allowing you to refresh the data contained; drill down into charts, graphs, and summary fields; search for fields, and more.

When you are working with Web reporting, Crystal Reports Server provides a better solution for deploying reports to an Intranet or Internet site, but exporting can also be helpful for one-off exports or where you need to manually edit the HTML before it is published.

To export your report, select File > Export to open the export dialog box. In the export dialog box, select the format HTML 4.0 (DHTML) and the destination Disk File. In the dialog box, shown in Figure 15.16, select a directory name and location for storing your HTML and related files.

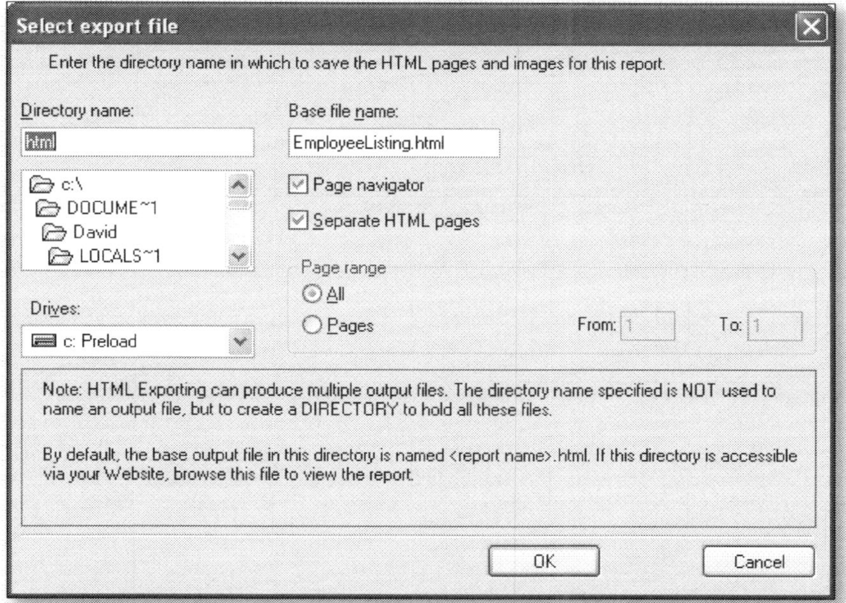

Figure 15.16 You can select a directory name and file location for storing your HTML files

Select a base file name for your report and use the check boxes to specify whether you want a page navigator to appear and whether you want the report exported to a single HTML document.

Finally, select a page range to be exported and click OK to start the export. A progress dialog box should appear while the records are being exported; you will be returned to your report's design or preview page when the export is finished.

From that point, you can view your report using Internet Explorer or Netscape Navigator by locating the folder where your HTML files were created and viewing the base file. At the bottom of each HTML page will be links to navigate between pages (if specified), and you can print each of the report pages independently.

EXPORTING TO TEXT FILES

Exporting using Crystal's text file formats is popular because most word processing, spreadsheet, and database applications can read them. To make life easier, there are a number of text file formats to choose from when exporting your report. Which you use will probably be based on the target application or platform for the exported file. With all of these file formats, there are some report design considerations that need to be taken into account to successfully export your report.

First, if you are considering creating a report for export to text-only format, keep in mind that Crystal Reports will attempt to export all of the elements of the report to the text file. This means that if you have page numbers at the bottom of every page, these numbers could end up in the text file as values, as shown in Figure 15.17.

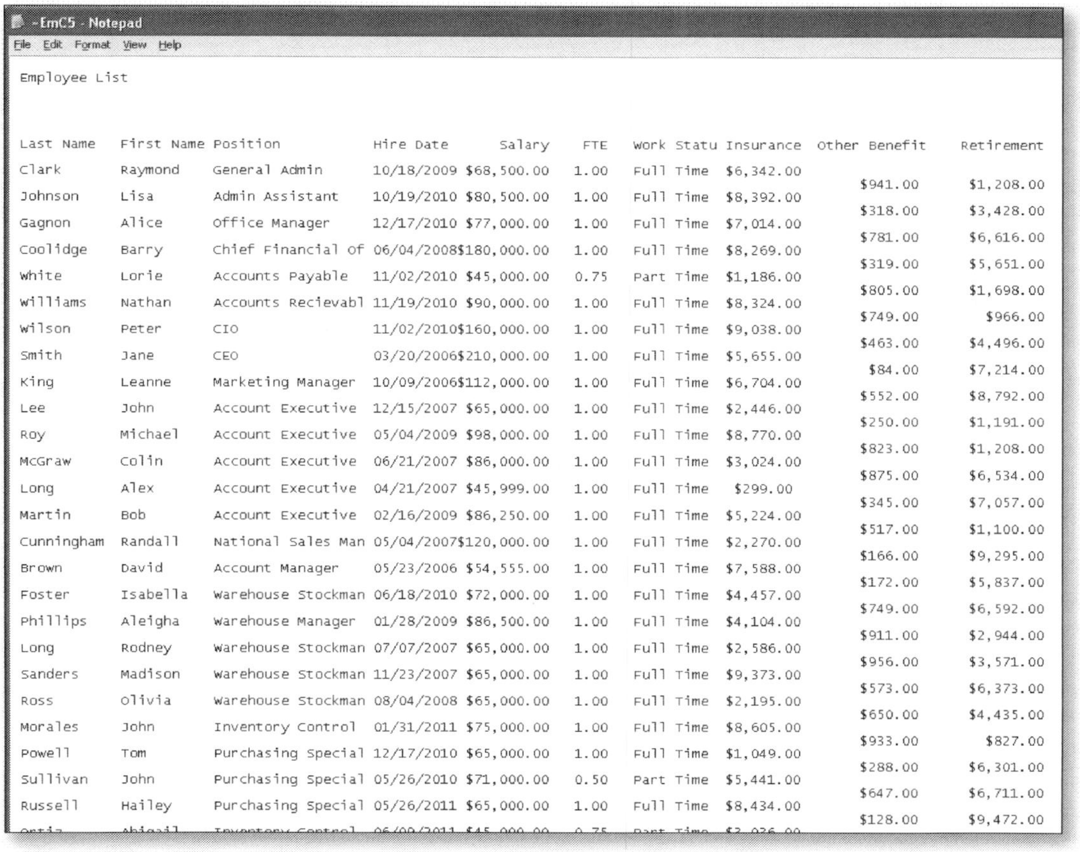

~EmC5 - Notepad

File Edit Format View Help

Employee List

Last Name	First Name	Position	Hire Date	Salary	FTE	Work Statu	Insurance	Other Benefit	Retirement
Clark	Raymond	General Admin	10/18/2009	$68,500.00	1.00	Full Time	$6,342.00	$941.00	$1,208.00
Johnson	Lisa	Admin Assistant	10/19/2010	$80,500.00	1.00	Full Time	$8,392.00	$318.00	$3,428.00
Gagnon	Alice	Office Manager	12/17/2010	$77,000.00	1.00	Full Time	$7,014.00	$781.00	$6,616.00
Coolidge	Barry	Chief Financial Of	06/04/2008	$180,000.00	1.00	Full Time	$8,269.00	$319.00	$5,651.00
White	Lorie	Accounts Payable	11/02/2010	$45,000.00	0.75	Part Time	$1,186.00	$805.00	$1,698.00
Williams	Nathan	Accounts Recievabl	11/19/2010	$90,000.00	1.00	Full Time	$8,324.00	$749.00	$966.00
Wilson	Peter	CIO	11/02/2010	$160,000.00	1.00	Full Time	$9,038.00	$463.00	$4,496.00
Smith	Jane	CEO	03/20/2006	$210,000.00	1.00	Full Time	$5,655.00	$84.00	$7,214.00
King	Leanne	Marketing Manager	10/09/2006	$112,000.00	1.00	Full Time	$6,704.00	$552.00	$8,792.00
Lee	John	Account Executive	12/15/2007	$65,000.00	1.00	Full Time	$2,446.00	$250.00	$1,191.00
Roy	Michael	Account Executive	05/04/2009	$98,000.00	1.00	Full Time	$8,770.00	$823.00	$1,208.00
McGraw	Colin	Account Executive	06/21/2007	$86,000.00	1.00	Full Time	$3,024.00	$875.00	$6,534.00
Long	Alex	Account Executive	04/21/2007	$45,999.00	1.00	Full Time	$299.00	$345.00	$7,057.00
Martin	Bob	Account Executive	02/16/2009	$86,250.00	1.00	Full Time	$5,224.00	$517.00	$1,100.00
Cunningham	Randall	National Sales Man	05/04/2007	$120,000.00	1.00	Full Time	$2,270.00	$166.00	$9,295.00
Brown	David	Account Manager	05/23/2006	$54,555.00	1.00	Full Time	$7,588.00	$172.00	$5,837.00
Foster	Isabella	Warehouse Stockman	06/18/2010	$72,000.00	1.00	Full Time	$4,457.00	$749.00	$6,592.00
Phillips	Aleigha	Warehouse Manager	01/28/2009	$86,500.00	1.00	Full Time	$4,104.00	$911.00	$2,944.00
Long	Rodney	Warehouse Stockman	07/07/2007	$65,000.00	1.00	Full Time	$2,586.00	$956.00	$3,571.00
Sanders	Madison	Warehouse Stockman	11/23/2007	$65,000.00	1.00	Full Time	$9,373.00	$573.00	$6,373.00
Ross	Olivia	Warehouse Stockman	08/04/2008	$65,000.00	1.00	Full Time	$2,195.00	$650.00	$4,435.00
Morales	John	Inventory Control	01/31/2011	$75,000.00	1.00	Full Time	$8,605.00	$933.00	$827.00
Powell	Tom	Purchasing Special	12/17/2010	$65,000.00	1.00	Full Time	$1,049.00	$288.00	$6,301.00
Sullivan	John	Purchasing Special	05/26/2010	$71,000.00	0.50	Part Time	$5,441.00	$647.00	$6,711.00
Russell	Hailey	Purchasing Special	05/26/2011	$65,000.00	1.00	Full Time	$8,434.00	$128.00	$9,472.00

Figure 15.17 A standard text file export without considering the report format

For reports that are specifically created for export to a text file, it is a good idea to remove any unnecessary fields that you may have inserted, including the report title, page numbers, special fields, and comments. Another consideration is that all of the column headings and fields should be relatively the same size and aligned in your report.

Each of the export formats listed here may have additional options that allow you to control the file's contents. Here is a rundown of all of the text file formats, the additional options they present, and the output they return:

Separated Values (.*SV) — Standard CSV file format, with values enclosed in quotation marks and separated by commas. CSV allows you to choose the date and number formats.

Tab-Separated Values (TSV) — Creates a tab-delimited file, with text in quotation marks and values shown as is. TSV allows you to specify the date and number formats.

Record Style (Columns no spaces) (REC) — Presents distinct columns of text with no spaces between them. Also allows you to choose the date and number formats.

Record Style (Columns with spaces) (REC) — Presents distinct columns of text WITH spaces and allows you to choose the date and number formats.

Text (TXT) — A simple text file export representing the report's design as closely as possible.

To export your report to a text file, select File > Export to open the Export dialog box and select from one of the text file formats: Text File, Comma Delimited (CSV), and so on.

Keep in mind that some of these export formats may require additional information, such as how you want the numbers or text to appear.

Select a destination of Disk File and click OK. Enter a file name and location for the exported file and click OK. A progress dialog box should appear while the records are being exported, and you will be returned to your report's design or preview page when the export is finished.

From that point, you should be able to open your text file using Notepad or any other software application that is capable of opening text files. If the resulting file is not what you expected, try changing your report's design (removing page numbers, headings, and so on) to optimize the text file produced.

EXPORTING TO XML

Crystal Reports features a powerful XML export facility that has been enhanced with the Crystal Reports 2008 release. With this release, you can now specify an XML transform that will be used when the XML file is actually generated. By default, Crystal Reports has a default XML format, which you can view by selecting File > Export > Manage XML Formats to open the dialog shown below in Figure 15.18.

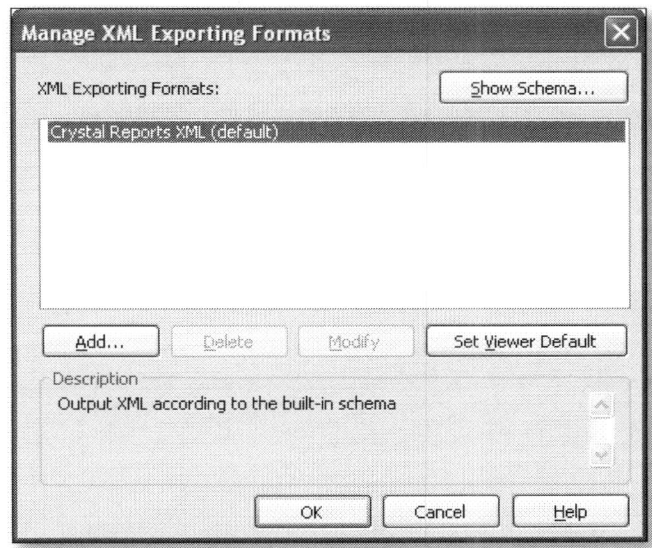

Figure 15.18 Manage XML Exporting Formats

You can view the default schema by clicking the "Show Schema" button — the default schema closely maps all of the objects in your report to the sections in which they are used, and includes the name of the object, location, etc. as well as the field value.

You can add your own schema by clicking the Add button, which will open the dialog shown below in Figure 15.19. This will allow you to import an existing XML transformation to be used when exporting your reports.

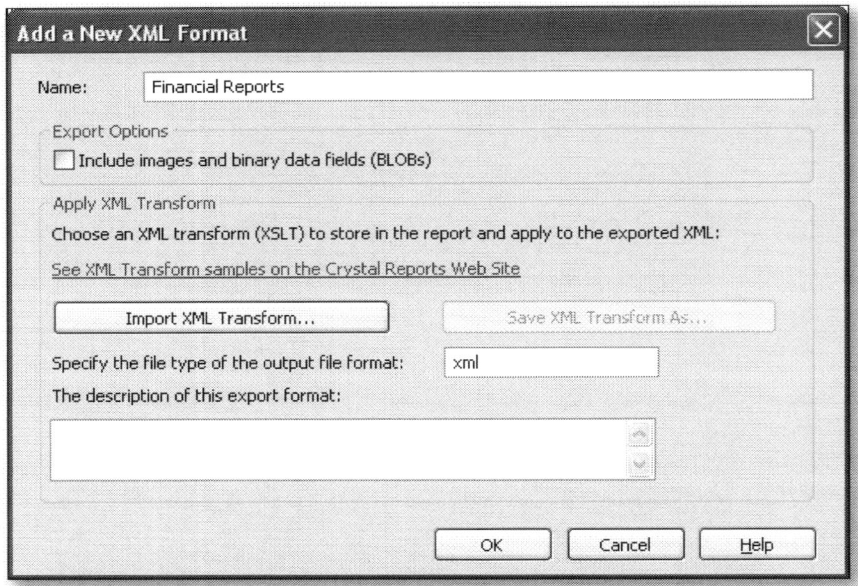

Figure 15.19 Add a New XML Format

To help you get started, there is also a link available on this dialog to some sample XML transforms that you can have a look at to see how they are put together. This link has been shortened here:

http://bit.ly/fPSGd4

Using these as a guide, you can then create your own XML transforms to use with the export process.

Unless you are an XML guru and love Notepad, it is always a good idea to use an XML editor like XMLSpy (www.xmlspy.com) to help you create the transformation.

Once you have added your transformation to the XML Exporting Formats, it will then be available whenever a user exports their report to XML using File > Export > Export Report. Selecting the XML format for the export will open the dialog shown below in Figure 15.20 and allow the user to export their report data as an XML file.

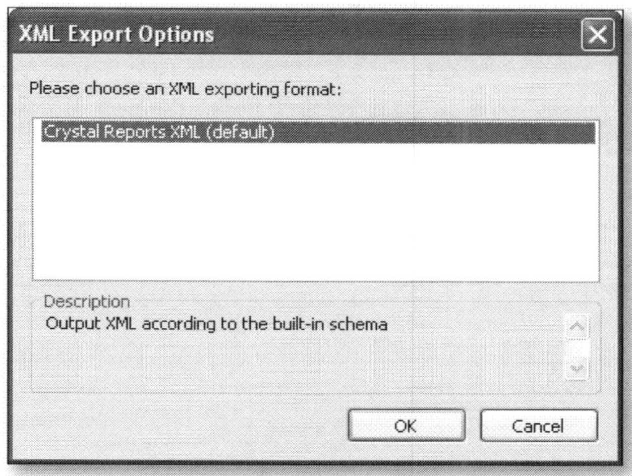

Figure 15.20 XML Export Options

It is important to note that the XML export that was used in previous versions of Crystal Reports is still available — this has been named "XML Legacy" and is still available under File > Export > Export Report. But if you are considering using XML as a data transfer format, you should be using the new format, which gives you more control over the XML report output.

OTHER EXPORT FORMATS

In addition to the formats listed, Crystal Reports also supports some unusual export options that may come in handy. In the following sections we are going to look at some of these exporting options.

EXPORTING YOUR REPORT'S DEFINITION

If you ever need to document a Crystal Report you have created, you will be delighted at how easy it can be. Using the export options within Crystal Reports, you can generate a text file that describes and documents the structure and layout of your Crystal Report.

This file is called the Report Definition and it is one of those hidden gems within Crystal Reports. Shown in Figure 15.21, it contains a detailed description of your report, broken into logical sections. This description includes not only the fields, formulas, groups, and record selection in your report, but it also includes a summary of the formatting options applied to each field, object, or section to give you the precise details of how the report was created.

```
-EmC7 - Notepad
File  Edit  Format  View  Help
        Crystal Report Professional v12.0 (32-bit) - Report Definition

1.0 File Information

        Report File:
        Version: 12.0

2.0 Record Sort Fields

3.0 Group Sort Fields

4.0 Formulas

4.1 Record Selection Formula


4.2 Group Selection Formula


4.3 Other Formulas

5.0 Sectional Information

5.1 Page Header Section
        Visible, Keep Together

        Last Name
                String, Visible, Left Alignment, Top Alignment,
                Keep Together


        First Name
                String, Visible, Left Alignment, Top Alignment,
                Keep Together


        Position
                String, Visible, Left Alignment, Top Alignment,
                Keep Together


        Hire Date
                String, Visible, Left Alignment, Top Alignment,
                Keep Together


        Salary
                String, Visible, Right Alignment, Top Alignment,
                Keep Together


        FTE
                String, Visible, Right Alignment, Top Alignment,
                Keep Together
```

Figure 15.21 A sample report definition file

To export your report's definition, select File > Export to open the Export dialog box. In the Export dialog box, select the Report Definition format and a destination of Disk and click OK. Once you have selected a file location and name, a progress dialog box should appear while the records are being exported. You will be returned to your report's design or preview page when the export is finished.

You can then locate the text file that has been created and open it using Notepad or some other word processing application.

This report definition is for documentation purposes only. It does not control or drive the report's design and cannot be modified to change the report's design. The report definition is also for export only, meaning that you can generate it at any time, but Crystal Reports cannot read a report definition file or create a new report from it.

EXPORTING TO AN ODBC DATABASE

Another option frequently overlooked is the ability to push a report's contents directly back into an ODBC database as a new table. This functionality is available for most ODBC drivers and requires that you have create permission on the database in question. (Your database administrator should be able to help you with this.) When you choose this option, Crystal Reports will ask you for a table name and proceed to create a table using the report's definition and export the data from your report into the same table.

To get started, you will need to configure an ODBC data source name that points to the database where you want to export your report contents.

Before you get started, make sure you verify with your database administrator that you have create and update rights on the database you want you use.

From that point, select File > Export to open the Export dialog box, and in the Export dialog box, select the format ODBC-dsn name (where dsn name is the name of the ODBC data source you have created) and click OK. In the dialog box that appears, shown in Figure 15.22, enter a table name for your exported report.

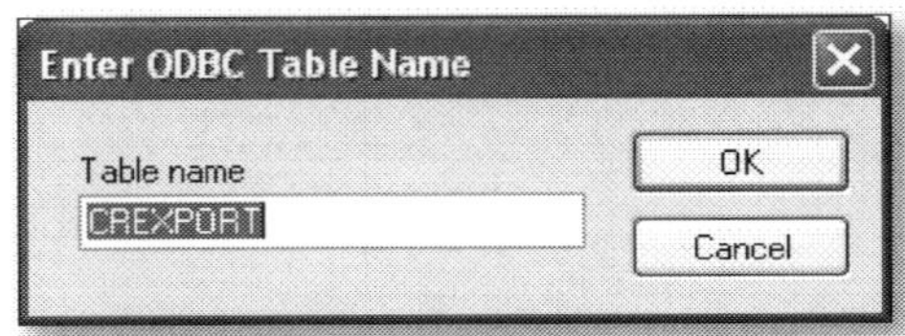

Figure 15.22 You will need to specify a name for your exported table of report information

By default, this table name is CREXPORT, but you can change it to anything you like within your database's naming constraints.

A progress dialog box should appear while the records are being exported, and you will be returned to your report's design or preview page when the export is finished. Your report contents should now be exported to the table you specified in your database. If the results are not as you expected, modify the report design to optimize it for exporting to a database table. The most common culprits are page headings, page numbers, and report titles because these elements are also placed in your database table.

DISTRIBUTING YOUR REPORT

In addition to various file formats, Crystal Reports 9.0 also includes a number of methods for distributing these exported files. These methods range from simply saving a file to disk, to placing a report in a Lotus Notes database or Exchange folder, to sending the exported file as an e-mail attachment. Here are the destinations available:

File Export — You can save an exported report directly to a file, using any of the export formats listed earlier. Once you specify a file name, Crystal Reports adds the appropriate extension to the file name.

Application — If you specify a destination of Application, Crystal Reports will save a temporary file in the format you specify and launch the associated application. For example, if you select Adobe Acrobat (PDF) and specify Application, Crystal Reports will create the PDF file and then immediately launch Acrobat Reader to display the file.

Lotus Domino — You can push an exported report back into a Lotus Notes or Domino database file (NSF) using any of the supported Crystal Reports format. In addition, Crystal Reports can be run from within Lotus Notes or Domino using LotusScript.

Lotus Domino Mail — Reports can be sent as e-mail attachments using Lotus Domino Mail

If you are using cc:Mail 4.0+, note that Lotus switched to a MAPI interface, and the Microsoft Mail (MAPI) option can be used instead.

Microsoft Exchange — A report can be placed in an Exchange public folder and shared with Exchange and Outlook users. These reports can be exported to any format, and when they are launched from within a public folder, the associated application will be launched.

Microsoft Mail (MAPI) — Reports can be sent as e-mail attachments using a simple Microsoft MAPI hook. Mail clients supported include Outlook, Outlook Express, and any other MAPI-compliant mail program.

Crystal Reports Server — Crystal Reports Server is a scalable report scheduling and distribution solution from the makers of Crystal Reports. A copy of Crystal Reports Server Standard is included with every copy of Crystal Reports 9.0 and provides the Web reporting framework for Crystal Reports.

Application Integration — Crystal Reports includes a number of integration methods for popular programming languages, including Visual Basic, Visual C++, Delphi, and Java, allowing you to integrate Crystal Reports design and viewing functionality without your application.

SENDING YOUR REPORT AS AN E-MAIL ATTACHMENT (MAPI)

A simple MAPI hook has been included with Crystal Reports to send an e-mail message and attachment from directly within Crystal Reports. To send your exported report as an attachment, you need to verify that you have a working MAPI mail client (such as Outlook or Outlook Express) installed on the computer from which you want to export your report, and that you can send and receive e-mail successfully using this client.

From that point, select File > Export to open the Export dialog box. In the Export dialog box, select the export format and the destination Microsoft Mail (MAPI) and click OK.

If Microsoft Mail does not appear as an option, your e-mail client may not be correctly installed and configured as the default mail application on your computer. See your e-mail administrator to correct this problem.

You will not be prompted for a file name when sending a file by e-mail. The file name will be determined by the report name and will have the correct extension for the format you use (for example, MyReport.xls for an Excel export of a report named MyReport.rpt).

A blank e-mail form will appear, as shown in Figure 15.23, allowing you to enter the e-mail address where this report will be sent as well as a subject and message.

Figure 15.23 Crystal Reports includes a simple MAPI hook to create an e-mail message and attachment from within Crystal Reports

Use the Address button to select e-mail addresses from your address book and the Check Names button to check the names you have manually entered against your address book.

When you have finished, click the Send button to send your e-mail. Your exported report will be attached to the e-mail message you have composed and sent through the default mail client installed on your machine. A progress dialog box should appear while the records are being exported, and you will be returned to your report's design or preview page when the export is finished.

SENDING YOUR REPORT AS AN E-MAIL ATTACHMENT (LOTUS DOMINO)

Crystal Reports can also send your exported report as an e-mail attachment through a connection to Lotus Domino Mail.

Again, just as with the other mail clients, you will need to verify that you have a working Lotus Domino client installed on the computer from which you want to export your report, and that you can send and receive e-mail successfully using this client.

If you are having problems sending an e-mail message with a report as an attachment, make sure that the Domino mail client is correctly installed and configured on your computer and that you can open the client and send e-mails from the same.

To attach and send your report, select File > Export to open the Export dialog box. In the Export dialog box, select the export format and the destination Lotus Domino Mail. A blank e-mail form will appear, allowing you to enter the e-mail address where this report will be sent as well as a subject and message.

Use the Address button to select e-mail addresses from your address book and the Check Names button to check the names you have manually entered against your address book.

When you have finished, click the Send button to send your e-mail. Your exported report will be attached to the e-mail message you have composed and sent through the default mail client installed on your machine.

A progress dialog box should appear while the records are being exported, and you will be returned to your report's design or preview page when the export is finished.

SENDING YOUR REPORT TO AN EXCHANGE PUBLIC FOLDER

Exchange public folders can be used to disseminate a wide range of information to Outlook users, and exported reports can be placed directly into Exchange public folders without leaving the Crystal Reports interface.

Before you get started, you will need to verify that Microsoft Outlook has been installed and configured correctly on your machine. You should be able to open Outlook and view the public folder where you want to place your exported report. Also, check with your Exchange administrator to ensure that you have write access to the Exchange public folder where you want to place your report.

To place a report in an Exchange public folder, select File > Export to open the Export dialog box. In the Export dialog box, select the export format and the destination Exchange Folder and click OK.

A dialog box will open and allow you to choose an Exchange public folder for storing your report. Click to select the folder and OK to accept your selection.

A progress dialog box should appear while the records are being exported, and you will be returned to your report's design or preview page when the export is finished.

SUMMARY

So now that you know one way to distribute your reports, we are next going to look at one of the techniques alluded to in this chapter — publishing a report to Crystal Reports Server. This flexible web-based platform provides the ability to publish and share reports in a secure environment, as well as view reports on demand, schedule reports, distribute via e-mail, burst to multiple users and more.

Chapter 16
Working with Data Sources

In This Chapter

- Accessing Data with Crystal Reports
- Creating Reports Using the Standard Report Wizard
- Saving Your Report
- Working with the Report Designer
- Report Design Environment
- Customizing the Design Environment

INTRODUCTION

Regardless of what type of report you want to create, it all starts with a connection to a data source. One of Crystal Report's strengths is its ability to report from just about any database, from desktop and file-based data structures to data warehouses, relational databases, OLAP data sources, and enterprise resources planning (ERP) systems such as SAP®, Oracle® JD Edwards®, and more.

In this chapter, we will be looking at how you can access these different data sources from Crystal Reports, as well as some of the techniques for working with data sources of all types. We'll deep-dive into how to use the Database Expert to select data sources, as well as link tables within these sources, verify databases and set the data source location.

ACCESSING DATA WITH CRYSTAL REPORTS

Crystal Reports can access data sources in two ways, either through a translation layer (like ODBC) or through a native driver or connection. A native connection from Crystal Reports to your data is accomplished through a specialized interface that is specific to your data source or application.

Over the years, Crystal Reports has teamed up with database, application, and other vendors to create a number of native drivers for PC or file-type databases, relational databases, and ERP systems, including native drivers for Microsoft SQL Server, Oracle, and others.

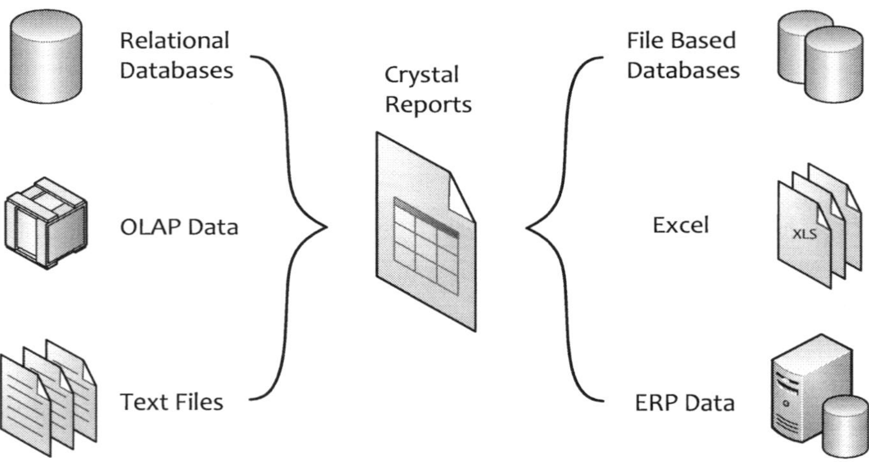

Figure 16.1 Crystal Reports data access methods

Through a native-driver interface, you can access non-traditional data sources as well, such as log files, Microsoft Outlook®, and even the file system itself.

The second data-access method uses a translation layer, which is most often ODBC. ODBC is a common interface for querying relational databases. Regardless of where the data resides, ODBC provides a reliable, stable platform that vendors can use to develop drivers and data-access methods.

Whether you use a native or ODBC connection depends on the data or application you are reporting from and the availability of a native driver or an ODBC for that connection.

In terms of performance, it is always preferable to use a native driver when possible to eliminate the extra layer of translation that ODBC requires.

In addition to commercially available ODBC drivers, Crystal Reports includes drivers free of charge for some of the most popular databases. These drivers are provided by DataDirect and are available for download from the Crystal Reports Start Page.

In the tables below, the different data source Crystal Reports supports are listed — this list was current as of Crystal Reports 2008 SP1 and with each new service pack there may be additional database formats/versions added.

Also, some of the data sources require a separate Integration Kit to be installed in order to access data using the driver mentioned. These integration kits may be licensed separately or require other components to be installed and configured, so contact your BusinessObjects reseller or account executive if you are interested in using these kits.

You can find additional information about these integration kits, how they are installed and configured, what features and functionality they provide, etc from http://help.sap.com

And finally, if you do not see your database or data source listed in the table, remember that Crystal Reports can access data through ODBC, JDBC, OLE DB, etc. and you may need to source your own ODBC driver to use these data sources.

Relational Data Sources	Native	ODBC	OLE DB	JDBC	Data-Direct ODBC 5.3	DAO
Generic JDBC Data Sources				X		
Generic OLE DB Data Sources			X			
Generic ODBC Data Sources		X				
HP Neoview 2.2		X				
HP Neoview 2.3.1		X				

Relational Data Sources	Native	ODBC	OLE DB	JDBC	Data-Direct ODBC 5.3	DAO
IBM DB2/UDB for iSeries 5.3	X	X	X	X	X	
IBM DB2/UDB for iSeries 5.4	X	X		X	X	
IBM DB2/UDB for NT/Unix/Linux 8.2	X	X	X	X	X	
IBM DB2/UDB for NT/Unix 9.1	X			X	X	
IBM DB2/UDB for NT/Unix/Linux 9.5	X	X	X	X	X	
IBM DB2/UDB for zSeries 7	X	X			X	
IBM DB2/UDB for zSeries 8	X	X			X	
IBM Informix Dynamic Server 10.0	X	X			X	
IBM Informix Dynamic Server 9.4	X	X			X	
Ingres 2006		X				
Microsoft Access 2003 (.mdb)		X	X			X
Microsoft Access 2007 (.accdb)		X	X			
Microsoft SQL Server 2000 SP4		X	X	X	X	X
Microsoft SQL Server 2005		X	X	X	X	X
Microsoft SQL Server 2008		X	X	X		
MySQL AB 5.0		X		X		
NCR Teradata V2R6		X				
NCR Teradata V2R6.1		X		X		
NCR Teradata V2R6.2		X		X		
Teradata 12.0		X		X		
Netezza 2.5		X				
Netezza 3.0		X				
Netezza 3.1		X				

Relational Data Sources	Native	ODBC	OLE DB	JDBC	Data-Direct ODBC 5.3	DAO
Netezza NPS Server 4.0		X		X		
Oracle 9.2	X	X	X	X	X	
Oracle 10g R1 (10.1)	X	X	X	X	X	
Oracle 10g R2(10.2)	X	X	X	X	X	
Oracle 11g R1 (11.1)	X	X	X	X	X	
Progress OpenEdge 10.0B		X		X		
PostgreSQL 8.2		X		X		
SAP BusinessOne		X				
Sybase ASE 12.5.x	X	X		X	X	
Sybase ASE 15	X	X		X	X	

OLAP Data Sources	Native	ODBC	OLE DB	JDBC	Data-Direct ODBC 5.3	DAO
Essbase Server 7.1.x	X					
Hyperion System 9	X					
Microsoft SQL Server Analysis Services 2000 SP4	X					
Microsoft SQL Server Analysis Services 2005 SP1	X					
Microsoft SQL Server Analysis Services 2005 SP2	X					
Microsoft SQL Server Analysis Services 2008	X					

Integration Kits	Native	ODBC	OLE DB	JDBC	DataDirect ODBC 5.3	DAO
SAP R/3	X					
SAP BW	X					
JD Edwards EnterpriseOne (ODA Driver)		X				
Oracle eBusiness Suite	X					
PeopleSoft Enterprise	X					
Siebel	X					

Other Data Sources	Native	ODBC	OLE DB	JDBC	DataDirect ODBC 5.3	DAO
Act!	X					
ADO.NET	X					
Borland Database Engine	X					
Microsoft Exchange 5.5	X					
Microsoft Exchange 2000	X					
Microsoft Exchange 2003	X					
Microsoft Excel 2000		X	X			X
Microsoft Excel 2003		X	X			X
Microsoft Excel 2007		X	X			
Field Definitions	X					
File System	X					
Lotus Notes 6		X				
Lotus Notes 7		X				

Other Data Sources	Native	ODBC	OLE DB	JDBC	DataDirect ODBC 5.3	DAO
Microsoft Outlook 2000	X					
Microsoft Outlook 2003	X					
Microsoft Outlook 2007	X					
Pervasive SQL 2000		X				
Pervasive SQL 8		X				
Salesforce.com	X					
Text		X				
XML	X					
Web Services	X					

WORKING WITH RELATIONAL DATABASES

To begin examining how to work with databases, we are going to look at how to connect to a relational database. By far the most popular data-access method for relational data is through a native or ODBC connection to a relational database, as Crystal Reports ships native and ODBC drivers for the most popular RDBMS's.

- IBM DB2®
- IBM Informix®
- Microsoft SQL Server
- Oracle
- Sybase®

Most of these native drivers require that the standard database client be installed and configured before they can be used. For example, if you want to access an Oracle database, you will need the Oracle client installed and configured locally on your computer. Chances are good that you probably already have the correct software installed on your machine. A good rule of thumb is that you should be able to access your data source through the native tools provided with the application — for example, Oracle has SQL*Plus, which is a query tool that you can use to access Oracle data. If you can access your data through SQL*Plus, Crystal Reports should not have a problem.

In addition to using native driver access, you can also access these data sources through an ODBC driver. To access a database through ODBC, you will need to configure the appropriate ODBC driver through the 32-bit ODBC Administrator (accessed through the Windows Control Panel) in order to access the data as shown in Figure 16.2.

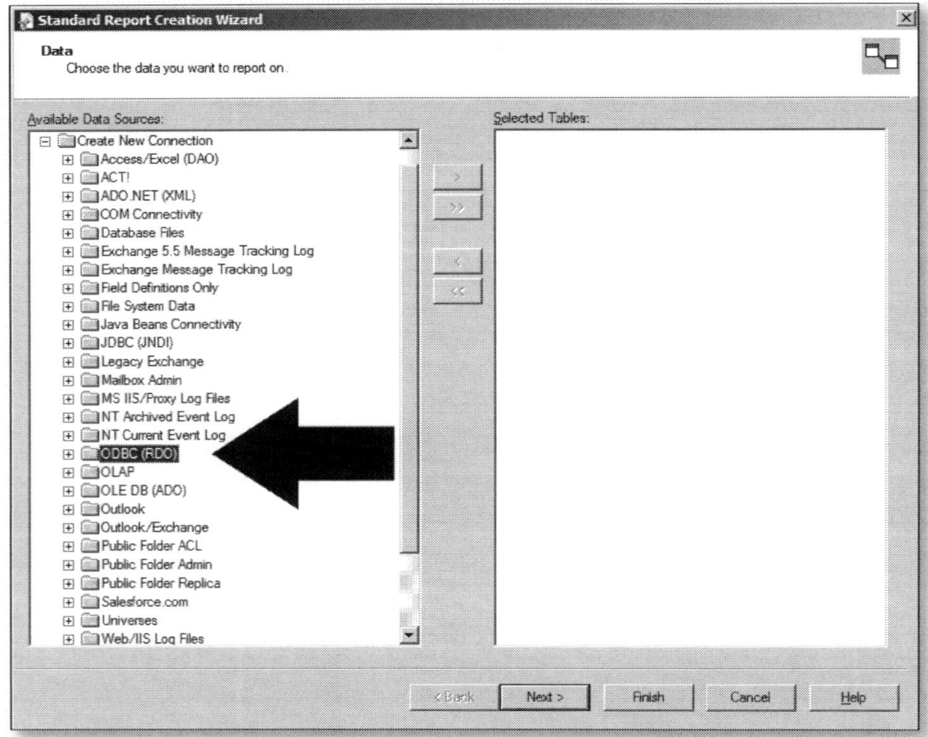

Figure 16.2 Accessing data through ODBC

To see how you can access data from a relational database, follow these steps:

1. Open Crystal Reports, and click File > New > Blank Report.
2. Double-click to expand the folder marked Create New Connection, and then double-click the node marked ODBC (RDO). A list of available data sources appears, as shown in Figure 16.3.

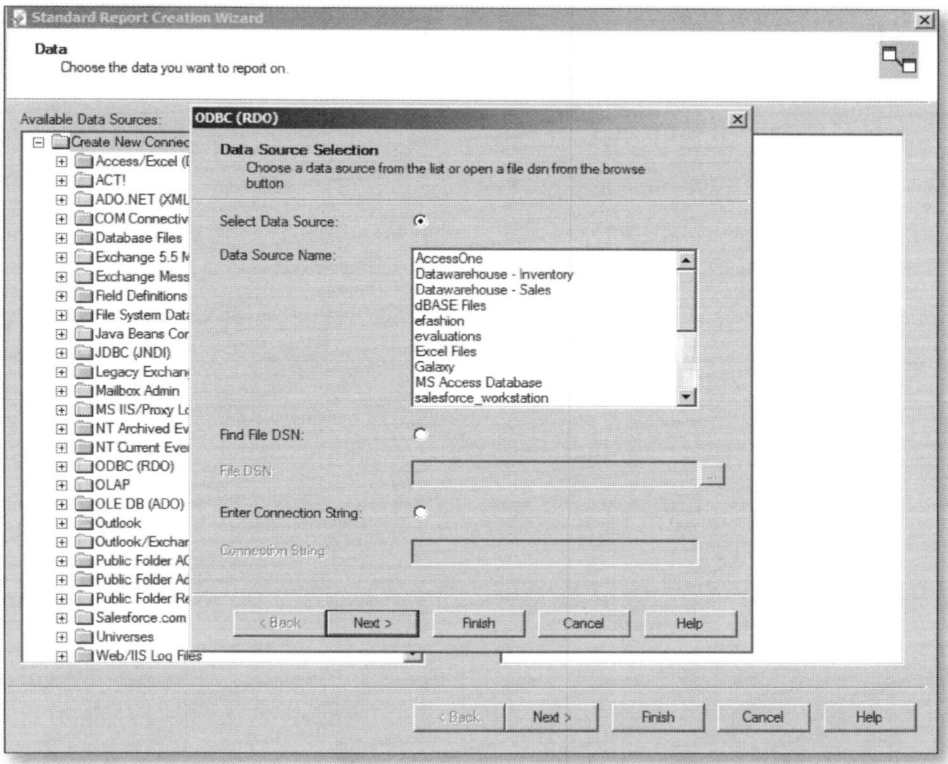

Figure 16.3 A list of available ODBC data sources

3. To select a data source from the list, enter any information the ODBC driver requires — for example, if you are reporting from a secure database, you will need to provide a username and password. For this example, we are using the Galaxy database that is included with the book download files, which does not require a user name or password, so you can just click OK.
4. A list of the available tables, views, stored procedures, and so on within your data source is displayed in the Database Expert. Highlight the table or tables you want to use in your report, and click the right-arrow button to move them to the selected list on the left.
5. When you have finished selecting your table, click OK to return to the Report Design window.

You can now use the tables and fields you selected to build your report as you normally would.

Crystal Reports logs on to your database to retrieve information and perform queries. Frequently, when you close a report and leave Crystal Reports open, the database connection will remain open as well. To log on or off a database server, click Database > Log On or Off Server.

CREATING REPORTS FROM FILE-BASED DATA SOURCES

Crystal Reports can use a direct, native connection to report from file-type databases, including the following:

- Microsoft Excel (.xls)
- XML (.xml)
- Xbase (.dbf, .ndx, .mdx, .bde)
- Paradox® (.db)
- Pervasive® (.ddf)
- Microsoft Access (.mdb)

Through this direct connection, Crystal Reports can extract data without having to submit a query to a database server. To create a report from these types of data sources, you can either create the report from scratch or use one of the report wizards, which we will look at a little later in this chapter.

To create a report based on one of these data sources, use the following steps:

1. Open Crystal Reports, and click File > New > Blank Report. The Database Expert opens.
2. Double-click the folder marked Create New Connection to expand the list of available data sources.
3. Double-click the Database Files folder, and using the standard File > Open dialog box, locate your database file and click OK to select. In this example, select the GALAXY.MDB database file from the book download files.
4. A list of tables will now appear below your data source, as shown in Figure 16.4.

Highlight the table or tables you want to use in your report, and click the right-arrow button to move them to the Selected Tables list on the left. For this report, select the INVOICE table.

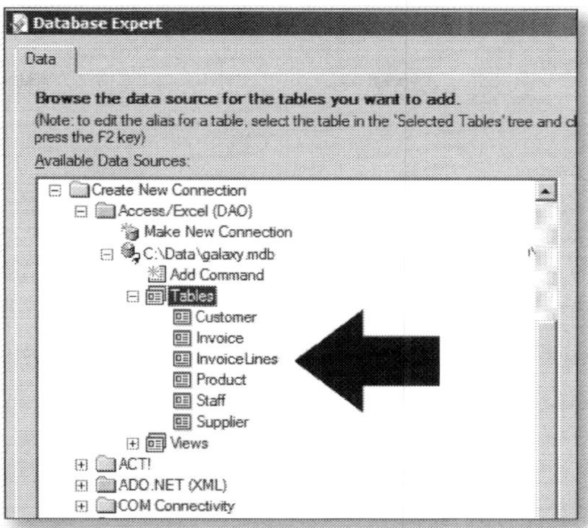

Figure 16.4 Available tables

5. When you have finished selecting your tables, click OK to return to the Report Design window.
6. A new report will be created and displayed in the Report Designer, as shown in Figure 16.5.

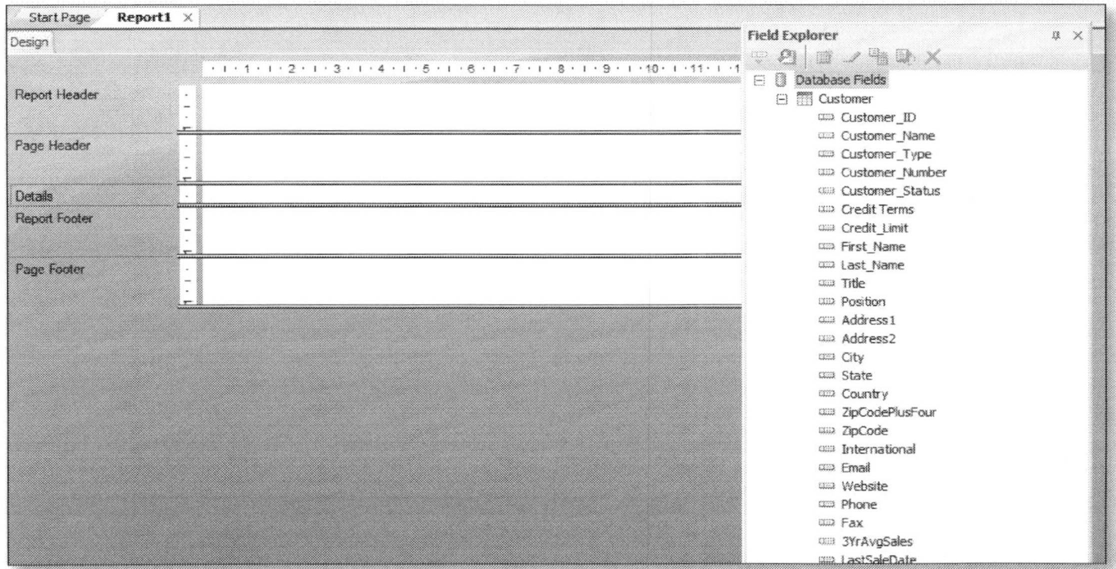

Figure 16.5 Your blank report with the data source listed

7. You can then drag and drop fields from the Field Explorer on to your report canvas, just as you normally would for a report.

A sample report named NATIVECONNECT.RPT created using this method is located in the book download files.

One of the downsides of using this method to access data sources is that you are connecting to a file path and name that may be different, depending on where the report is published and run. Sometimes it is a better option to set up an ODBC driver to these data sources, or alternately put the file on a shared or network drive if you plan to publish the report to Crystal Reports Server (or any of the other BusinessObjects server platforms).

CONNECTING TO OLAP DATA SOURCES

OLAP data (sometimes called multidimensional data) can be accessed through OLE DB for OLAP, a standard interface for accessing OLAP data, or through a number of native OLAP drivers that are included with Crystal Reports, such as:

- SQL Server Analysis Services
- Hyperion® Essbase®
- IBM DB/2
- Holos HDC

OLAP data provides a summary of information that is held in a relational database and can provide insight that millions of rows of data normally do not.

SAP BW is also an OLAP-based data source, but it is covered in its own section a little later in the chapter, along with all of the other SAP data sources.

CONNECTING TO AN OLAP DATA SOURCE

To report from an OLAP data source, use the following steps:

1. Open Crystal Reports, and click File > New > Blank Report. The Database Expert opens.
2. Double-click to expand the folder marked Create New Connection, and then double-click the node marked OLAP. The OLAP Connection Browser opens.
3. To add a new data source, click the Add button. The Connection Properties dialog box appears.

4. From the drop-down list at the top of the dialog box, select your OLAP data source. In this example, we will be using a local cube file (.CUB) .CUB files are a perfect data source for demonstrating OLAP features in Crystal Reports, because they don't require a back-end OLAP server.
5. In the text box provided, enter a caption for your OLAP data source. In this example, enter Sales Information.
6. Use the browse button to browse for your OLAP data source (SALESDATA.CUB). The sample cube is available in the book download files in the DATASOURCES directory.
7. Next, click the Test Connection button to test the connection to your cube. A message box stating Connection Successful opens.
8. To finish the setup, click OK to return to the OLAP Connection Browser, which should now look like the one shown in Figure 16.6.

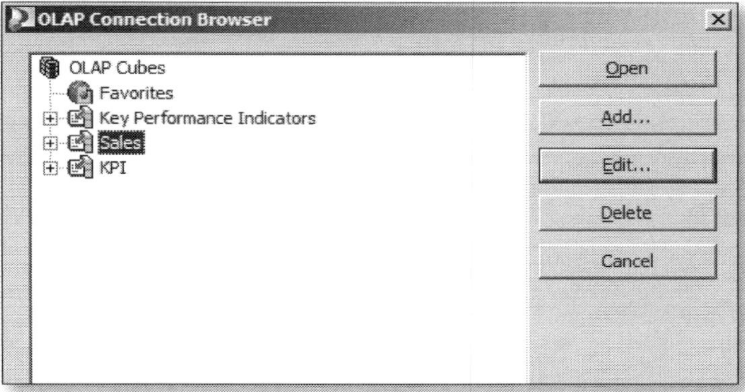

Figure 16.6 OLAP Connection Browser

9. Expand your new data source, and highlight the cube that you want to report from. In this example, expand your data source, select Xtreme, and click OK. You return to the Database Expert, and all of the available cubes in your data source appear in the OLAP folder below the data source name as shown in Figure 16.7 over the page.

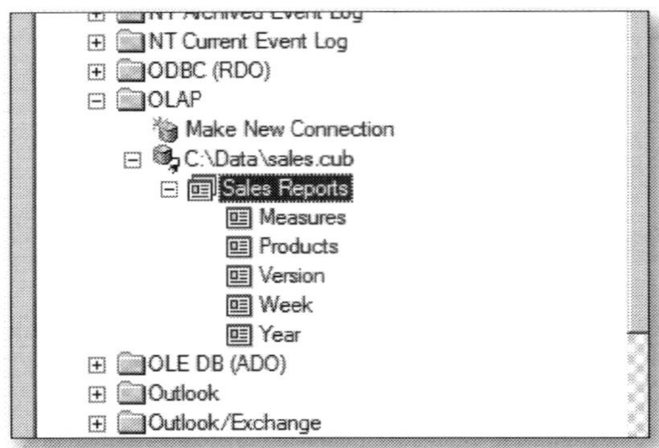

Figure 16.7 The OLAP Connection Browser displaying your new connection

10. Highlight the cube(s) you want to use in your report, and click the right-arrow button to move them to the selected list on the left. For this report, select the CUSTOMER cube.
11. When you have finished selecting your tables, click OK to return to the Report Design window.
12. A list of fields will appear in the Field Explorer on the right side of the Report Designer. You can drag and drop these fields onto your report as you normally would. If the fields look a bit strange (for example, "Monthly, Level 0") don't worry too much about that now.

A sample report named OLAPCONNECT.RPT created using an OLAP data source is located in the book download files in the Chapter 16 folder of the book downloads.

Keep in mind that this setup procedure may be slightly different for the different OLAP data sources that Crystal Reports supports. For example, for the Essbase setup, you will need to enter a server name as well as a user name and password and may need to have the Essbase client software installed and configured on your computer.

WORKING WITH OLAP GRIDS

OLAP grids are very similar to cross-tabs in Crystal Reports — OLAP grids, like the one shown in Figure 16.8, are composed of rows and columns of summarized data and can either be placed in the structure of existing reports or can be the main focus and content of their own reports and are based on OLAP data.

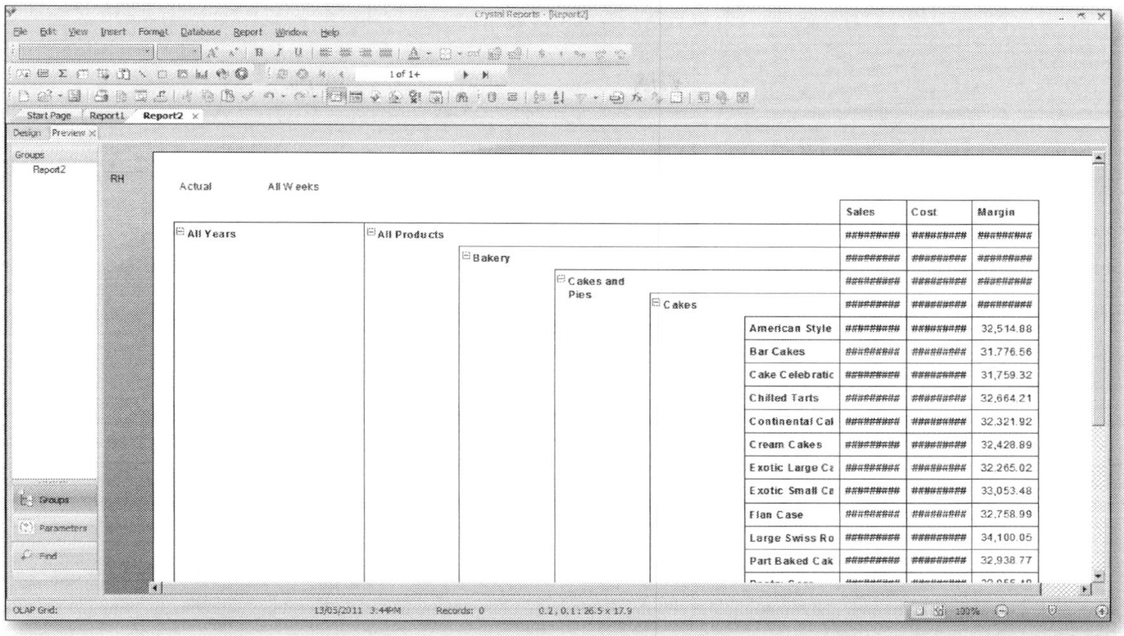

Figure 16.8 A typical OLAP grid summarizing domestic and international sales

OLAP grids look similar to a spreadsheet, but it is important to remember that OLAP grids are driven by the data in the cube. In the Design view of your report, an OLAP grid presents a simple row-and-column display. When viewed in the preview of your report, these rows and columns are filled with the data you have requested. The size of the OLAP grid varies based on the number of records returned.

To facilitate the creation of reports specifically for use with OLAP grids, Crystal Reports includes an OLAP grid Expert that walks you through the process of creating an OLAP grid report. In addition, you can insert OLAP grids as an element of an existing report.

From the Insert menu, click Insert > OLAP Grid. A blank OLAP grid is attached to your mouse pointer. Click to place the OLAP grid on your report, then right-click the cross tab and select OLAP grid Expert to open the dialog box shown in Figure 16.9 on the following page.

To create an OLAP grid, you need three elements: rows, columns, and summarized fields. For a basic OLAP grid, you need only one of each. To create new rows, columns, or summarized fields, you can drag a field from the Available Fields box to the corresponding box for rows, columns, or summarized fields.

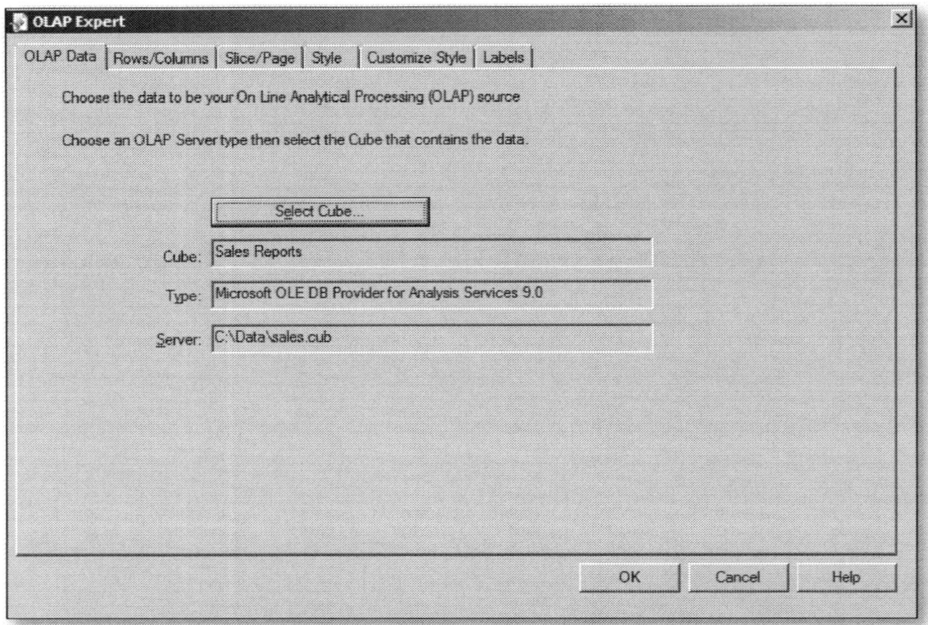

Figure 16.9 The OLAP grid Expert

You can also highlight a field in the Available Fields box and click the arrow button next to text boxes for the Rows, Columns, and Summarized Fields.

To remove a field from an OLAP grid, highlight the field and click the left arrow button beside the text box to remove it from the list of selected fields.

If you are curious about the contents of a particular field, you can view some of the values held in the field by highlighting the field name and clicking the Browse button at the bottom-right corner of the dialog box. This action will return about 200 records at a time for you to look through, but keep in mind that on large databases, this process may take some time.

With OLAP grids, you can add as many fields for the columns, rows, and summarized fields as required. When your OLAP grid is printed, the fields you specified for the columns and rows will be used to create the OLAP grid, and the summarized field you specified will be calculated at the intersection of the columns and rows.

Customizing Grouping Options

As with reports, you can control the way information is displayed in an OLAP grid through the use of grouping. You can apply the same grouping concepts to your OLAP grid by using the shortcut menu shown in Figure 16.10 to edit your OLAP grid, then highlight the row or column field you want to change and click the Group Options button.

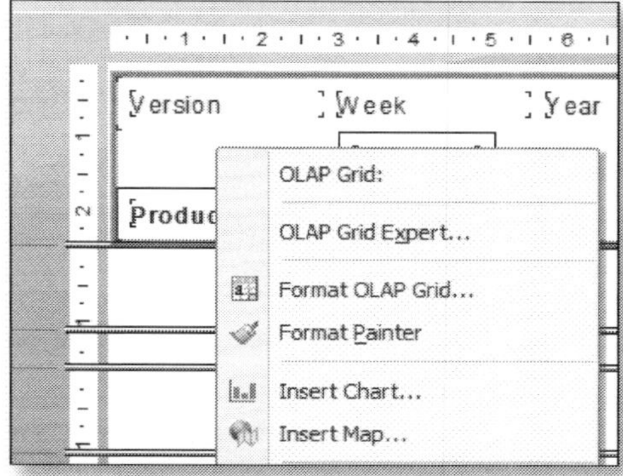

Figure 16.10 To edit an OLAP grid, right-click in the top-left corner of the screen, and select OLAP grid Expert

The OLAP grid Group Options dialog box, shown in Figure 16.11 opens. Here, you can select a sort order from the following options:

- Ascending
- Descending
- Specified Order

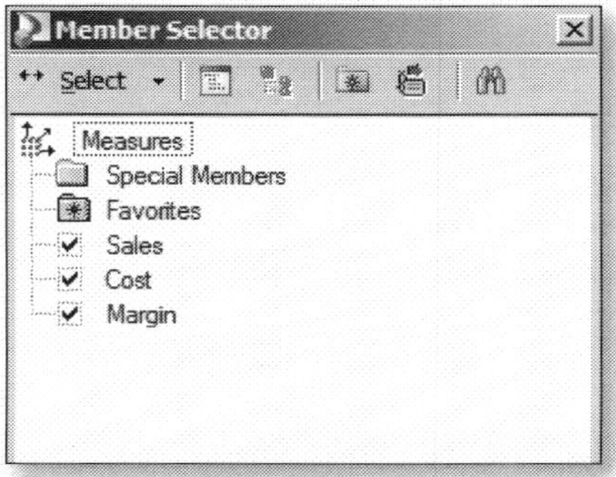

Figure 16.11 A number of grouping options are available for use with OLAP grids

When you use these options to select the sort order, remember that this sort order affects only the items in a single OLAP grid. If you were using an OLAP grid as the summary page for a long, detailed report, you would also need to change the report sort order so the OLAP grid and report would reflect the same sorting.

Reordering Grouped Data

Ascending and descending order is easy to work with, but what about when you want to reorder the data itself, not just sort it? The functionality that allows you to create your own groups is called Specified Grouping. The Specified Order option with OLAP grid groups works just as it does for other specified groupings you insert into your report. You need to name each group and then specify the criteria for each.

You might use specified grouping with an OLAP grid if you have a particular product grouping not represented in the database. You could use a specified grouping to create separate groups and to establish your own criteria (for example, in our fictional bike company, all the gloves, helmets, and so on could be grouped as Personal Accessories, and saddles and other items could be grouped together as Spare Parts).

Once you have selected Specified Order, a second tab should appear in the OLAP grid Group Options dialog box, as shown in Figure 16.12.

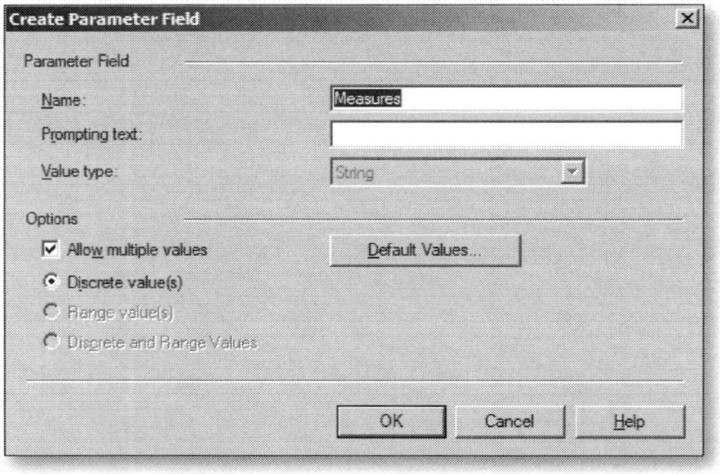

Figure 16.12 The Specified Order tab

The next step is to define a group name, and then you need to specify the group criteria. Start by typing all of the group names you want to create first, pressing the ENTER key after each. This process will build a list of group names.

Once you have all of the group names defined, you can highlight each and click the Edit button to specify the criteria. To establish the criteria for records to be added to your group, use the drop-down menu to select an operator and values.

These are the same operators that were used with record selection.

You can add criteria by clicking the New tab and using the operators to specify additional selection criteria, which are evaluated with an `Or` operator between the criteria that you have specified.

Make sure that you delete any groups you may have added by accident. Even if the criterion is set to Any Value, it can still affect report performance.

After you have entered a single group, another tab appears with options for records that fall outside of the criteria that you specify. By default, all of the leftover records are placed in their own group, labeled Others, shown in Figure 16.13. You can change the name of this group by simply editing the name on the Others tab. You can also choose to discard all of the other records or to leave them in their own groups.

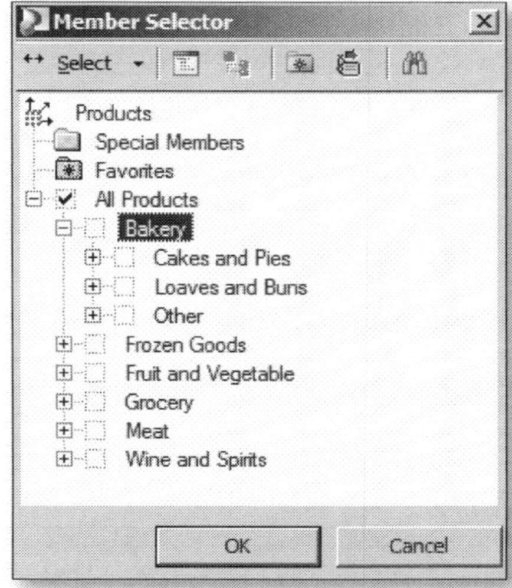

Figure 16.13 Options for handling other records

After you have defined your specified groups and criteria and have reviewed the settings for other records, click OK to accept the changes to the group options. Your specified grouping is now reflected in your OLAP grid, as shown in Figure 16.14.

Actual | All Weeks | All Years

		Sales	Cost	Margin
⊟ All Products		170,888,357.64	139,298,880.85	31,589,476.80
	⊞ Bakery	10,510,721.83	8,505,040.99	2,005,680.83
	⊞ Frozen Goods	5,371,773.20	4,725,072.66	646,700.54
	⊞ Fruit and Vege	33,436,399.40	26,261,723.40	7,174,676.02
	⊞ Grocery	76,376,769.52	63,676,754.16	12,700,015.30
	⊞ Meat	32,338,893.35	26,554,678.06	5,784,215.34
	⊞ Wine and Spir	12,853,800.35	9,575,611.58	3,278,188.77

Figure 16.14 An OLAP grid with specified grouping

FORMATTING OLAP GRIDS

Now that you understand how to insert an OLAP grid into your report and control its structure, it's time to take a look at formatting your OLAP grid.

Changing Field, Column, and Row Size

If you have worked with Excel and resized columns, you may try to resize columns in your OLAP grids by resizing the actual column. In Crystal Reports, the column size is determined by the actual fields within the column itself. Just like Excel, column widths are set for the entire column at once (regardless of whether the column has summarized fields, totals, or both).

You can resize a field by clicking the field to select it and then using the handles to change the width or height of the field. You can also right-click the field and select Size and Position from the shortcut menu to open the dialog box shown in Figure 16.15, which gives you precise control over the field's size.

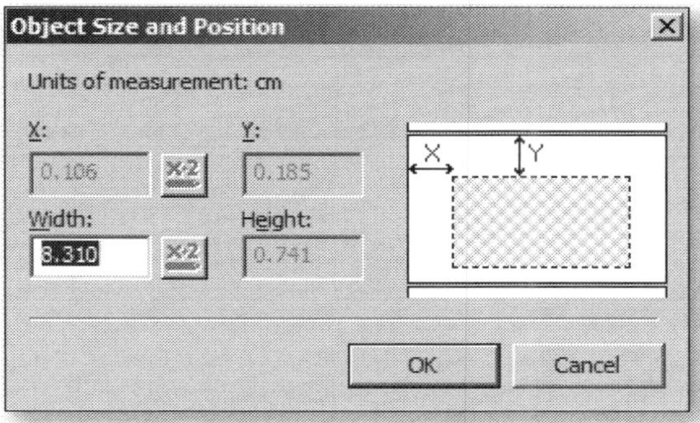

Figure 16.15 The Object Size and Position dialog box

When you have finished resizing your fields, columns, or rows, click anywhere outside of the OLAP grid to deselect the object with which you were working.

Applying a Pre-formatted Style

Because the majority of time spent working with OLAP grids is for formatting them, Crystal Reports includes a number of pre-formatted styles in which the formatting attributes have been set for you. This eliminates the need to individually change each of the formatting settings and also provides a way to apply uniform formatting to OLAP grids that may appear on different reports.

To apply a pre-formatted style to your OLAP grid object, insert or edit an existing OLAP grid and, in the Format OLAP grid dialog box, click the Style tab to open the dialog box shown in Figure 16.16 below.

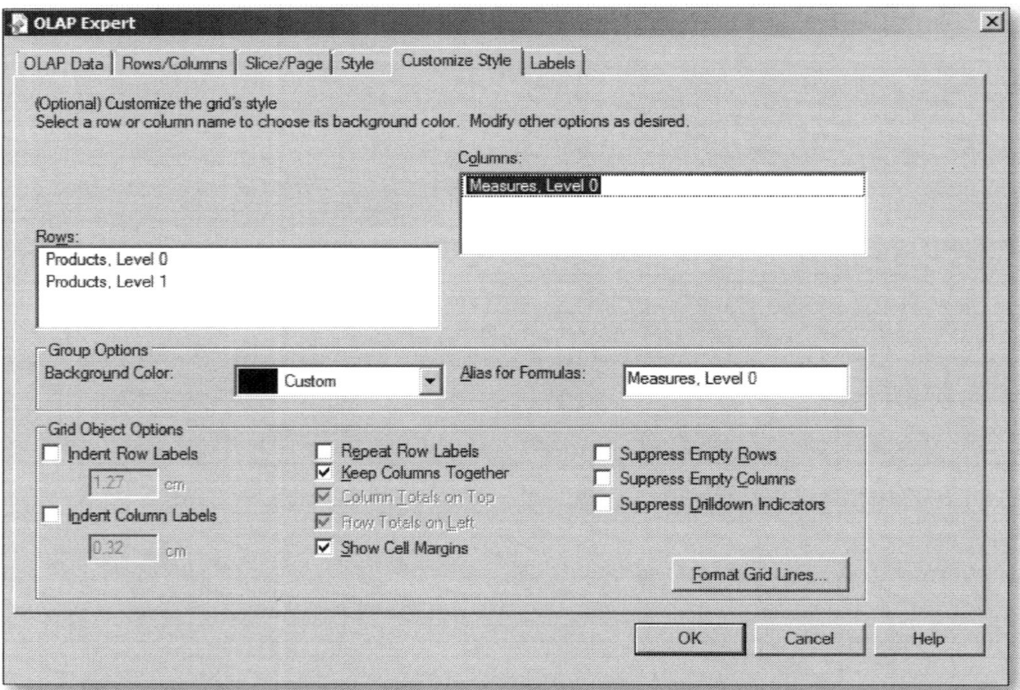

Figure 16.16 You can apply a pre-formatted style to your OLAP grid

To see a preview of the style's formatting, click the style. The pane on the right side of the dialog box will display a preview of a sample OLAP grid with the formatting attributes applied. When you have found the style you want to apply, click OK to return to your report and see the style applied to your OLAP grid.

Customizing a Style

If you don't particularly care for some of the formatting attributes of a particular style, you can customize the style using the Customize Style tab, also found in the OLAP grid Expert and shown in Figure 16.17.

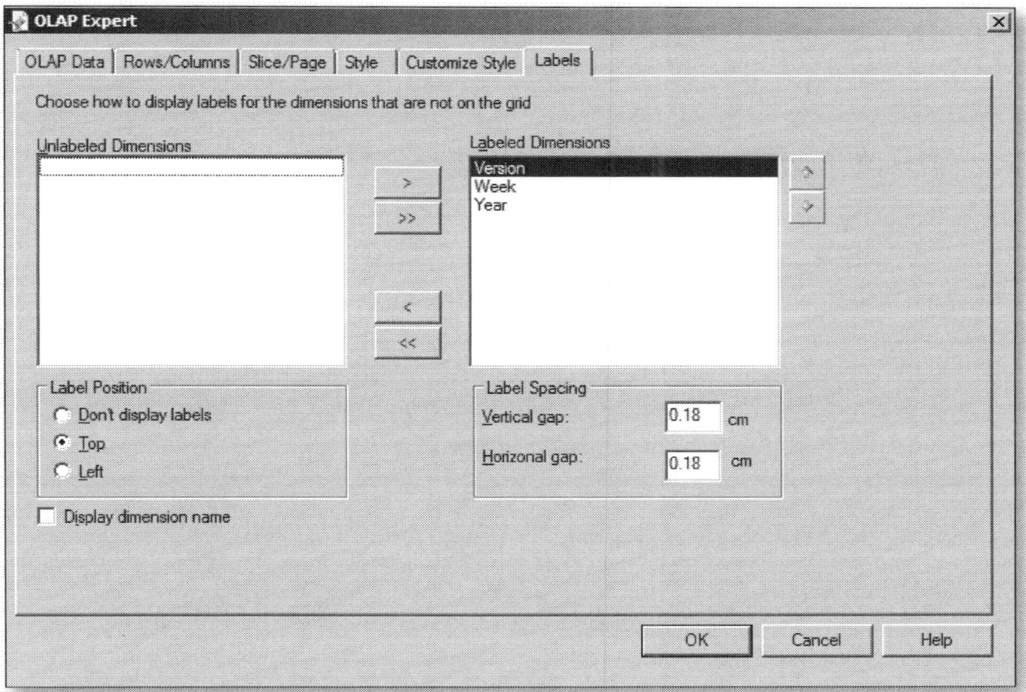

Figure 16.17 You can further customize a pre-formatted style using the OLAP grid Expert

To change the formatting of a particular row or column, highlight the row or column from the lists provided, and then use the drop-down box immediately below to change the background color and set other options, including the showing of labels, aliases for formula fields, and so on.

On the bottom half of the dialog box are the options for formatting the grid that appears around your OLAP grid by default. Table 16.1 summarizes the gridline options available on the Customize Style tab.

Table 16.1 OLAP grid Style Options

Option	Description
Show Cell Margins	Displays the internal cell margins in your OLAP grid
Indent Row Labels	For each row, indents the labels that appear on the left side of the OLAP grid
Repeat Row Labels	Repeats row labels on any new pages
Keep Columns Together	Attempts to keep all columns together on the same page
Row Totals on Top	Moves the row totals from their default location at the bottom of the OLAP grid to the top
Column Totals on Left	Moves the column totals from their default location on the right side of the OLAP grid to the left
Suppress Empty Rows	Suppresses any empty rows in the OLAP grid
Suppress Empty Columns	Suppresses any empty columns in the OLAP grid
Suppress Row Grand Totals	Suppresses the grand totals that would appear by default at the bottom of the OLAP grid
Suppress Column Grand Totals	Suppresses the grand totals that would appear by default on the right side of the OLAP grid

You can also control the grid lines that appear in your OLAP grid by clicking the Format Grid Lines button to open the dialog box shown in Figure 16.18, on the next page.

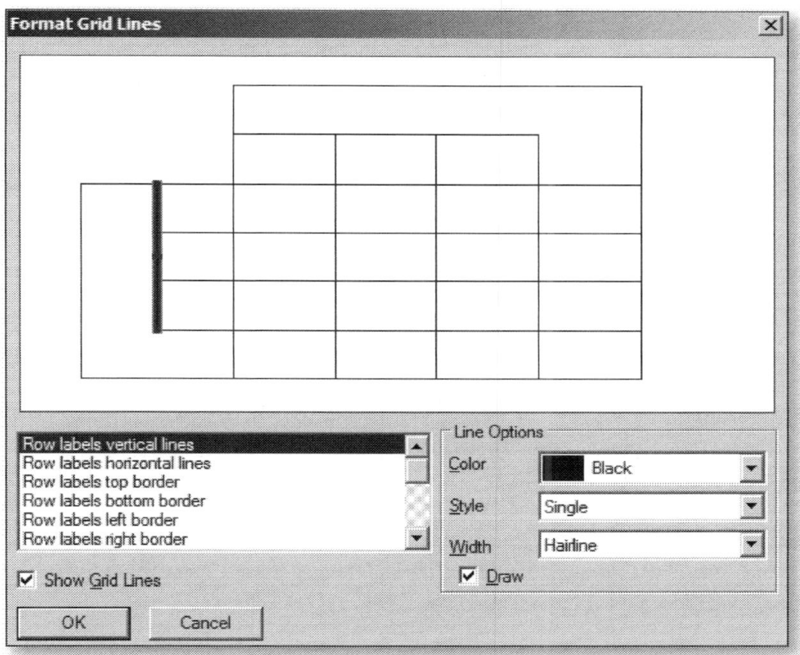

Figure 16.18 You can format individual grid lines as well

Using this dialog box, select the part of your OLAP grid or OLAP grid that you want to format and choose the line color, style, and width. The following grid lines are available:

- Row Labels Vertical Lines
- Row Labels Horizontal Lines
- Row Labels Top Border
- Row Labels Bottom Border
- Row Labels Left Border
- Row Labels Right Border
- Column Labels Vertical Lines
- Column Labels Horizontal Lines
- Column Labels Top Border
- Column Labels Bottom Border
- Column Labels Left Border
- Column Labels Right Border
- Cells Vertical Lines
- Cells Horizontal Lines
- Cells Bottom Border
- Cells Right Border

When you have finished setting the grid lines for your OLAP grid, click OK to return to your report design. The changes you have made should be reflected in your OLAP grid.

ANALYZING OLAP GRID DATA

In addition to summarizing data, the data in OLAP grids can also be manipulated to help analyze the data presented and find information quickly.

Changing OLAP grid Orientation

Sometimes when creating OLAP grids, we don't have a good indication of what the resulting summarized data is going to look like. To make OLAP grids a bit easier to read and to bring at least one of its features into line with Microsoft Excel, you can pivot OLAP grids by right-clicking the OLAP grid and selecting Pivot OLAP grid from the shortcut menu.

This action will cause the rows and columns in the OLAP grid to be switched, so a report with different countries across the top in columns would then have the countries showing as rows.

This feature is available only at design time. When users view the report through any of the export formats or through one of Crystal Report's Web delivery methods, this feature is not available. You would have to pivot the OLAP grid to the desired configuration before distributing or publishing your report.

REPORTING FROM SAP DATA SOURCES

Crystal Reports has long leveraged a tight integration with SAP data sources, starting with the original integration kit.

WORKING WITH SAP ERP DATA

When connecting Crystal Reports to SAP ERP data, you have the ability to report from a number of interfaces through the SAP Integration Kit. These interfaces include:

- Classic InfoSets
- SAP Queries
- ABAP Functions
- Tables and Clusters (via OpenSQL)

Within Crystal Reports itself, these interfaces appear within the Data Explorer, as shown in Figure 16.19 and when you connect to them, they behave as if they were any other relational tables/views that you would use as a data source for your Crystal Reports.

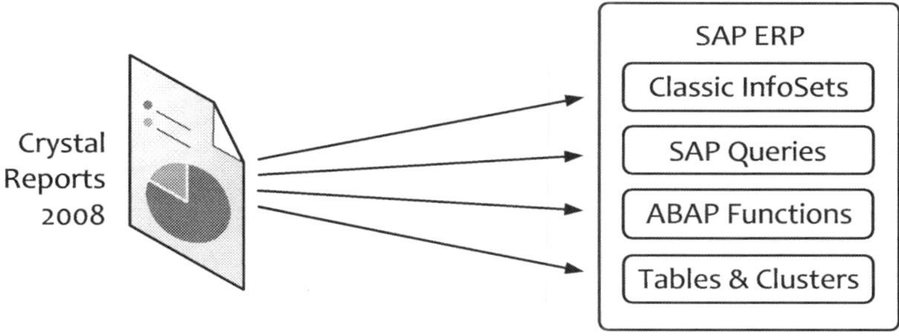

Figure 16.19 Connecting directly to SAP ERP data

Classic InfoSets & SAP Queries

When you need to provide complex data sets to end-users or report developers, you can leverage the functionality within SAP to build a query or InfoSet that contains the data required. This is similar to using a stored procedure with relational database — the actual logic for the data set will be executed when the report is run. From within Crystal Reports, queries and InfoSets appear as tables within Crystal Reports and can be used in the same way.

ABAP Functions

The concept behind reporting off an ABAP function is that we may need to lend some processing power to Crystal Reports to get the data set we need for reporting. Using an ABAP function, you have the ability to create a dataset programmatically and incorporate complex logic that may not be possible within Crystal Reports. From the Crystal Reports designer, you can report off ABAP Functions, providing they have a defined return type for each of the output parameters, and that they don't have a whole table as an input parameter.

Tables and Clusters

You can also report directly off the underlying transparent tables, pool and cluster tables, as well as any views you may have created. These structures will appear as tables within Crystal Reports and you can use the table fields to build your report.

Since the basic workflow of creating a Crystal Reports is exactly the same as a relational database, you can leverage all of your report design skills once you have connected to your SAP ERP data. To connect to SAP ERP data, follow these steps:

1. Ensure that you have the SAP Integration Kit installed and configured on the computer where Crystal Reports is installed.

> **For more information on installing and configuring the SAP Integration Kit, visit http://help.sap.com and navigate to the BusinessObjects documentation section.**

2. From within Crystal Reports, select File > New > Blank Report to open the Database Expert.
3. Navigate to Create New Connection and double click either "SAP InfoSets" or "SAP Table, Cluster and Function".
4. Enter your login details for the SAP environment you would like to connect to.
5. Once you have made a successful connection, navigate to the data structure that you would like to use as the source of your report, then use the arrows to move it to the list of Selected Tables and click OK.

You can now design your report as you normally would, dragging on fields from the Field Explorer to your report canvas, inserting groups, formulas, etc.

WORKING WITH SAP BW DATA

SAP BW is the data warehousing platform from SAP, which provides a number of OLAP-based structures or cubes for reporting purposes. Many SAP customers use SAP BW as their standard datawarehousing platform, as there is a large amount of "business content" (i.e. pre-defined structures, cubes and queries) that can be leveraged to create a data warehouse layer with SAP data.

When connecting Crystal Reports to SAP BW data, you have the ability to report on a number of different SAP BW structures and queries, including:

- BEx Queries
- DSO (Data Source Objects)
- InfoProviders
- MultiProviders
- InfoCubes

Even through you can create reports from a number of different SAP BW data sources, as shown below in Figure 16.20, a best practice for creating reports off SAP BW data is to first create BEx queries to provide the data for your report.

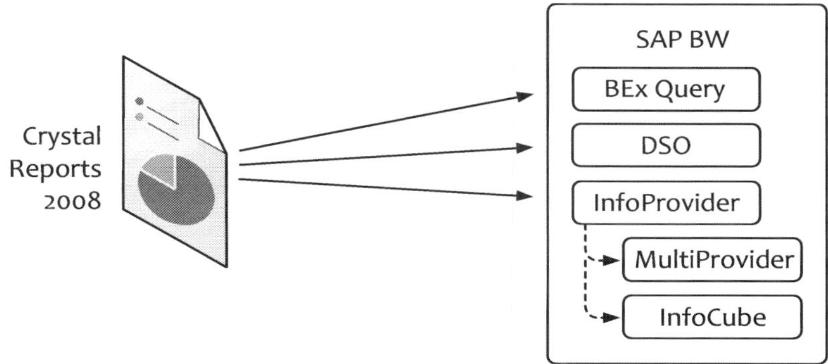

Figure 16.20 Connecting directly to SAP BW data

The BEx toolset offers a powerful query tool and interface for creating queries, leveraging all of the power of the SAP BW platform. By creating a BEx query, you also limit the data that is returned to Crystal Reports, making for more efficient reports and leveraging the processing power of your SAP BW server.

To create a report off a BEx query, follow these steps:

1. Ensure that you have the SAP Integration Kit installed and configured on the computer where Crystal Reports is installed.
2. From within Crystal Reports, select File > New > Blank Report to open the Database Expert.
3. Navigate to Create New Connection and double-click "SAP BW".
4. Enter your login details for the SAP environment that you would like to connect to.
5. Once you have made a successful connection, navigate to the BEx query that you would like to use as the source of your report, then use the arrows to move it to the list of Selected Tables and click OK.

   With your data source selected, you can now design your report as you normally would.

ACCESSING ERP DATA SOURCES USING INTEGRATION KITS

ERP systems generally run on top of a relational database, but they have their own business rules and data structures. Crystal Reports has a number of data drivers available for ERP systems, including:

- Baan™
- PeopleSoft®
- PeopleSoft Enterprise One (formerly JD Edwards)

Some of these drivers and integration kits may be sold separately from Crystal Reports 2008.

To configure these drivers and access information, you will definitely need the help of your application administrator, because the setup is not for the faint of heart. A number of technical white papers are available at the SAP Help site (http://help.sap.com) which provide the technical background necessary to get started. Often there will be specific requirements for client software and configuration or specific OEM versions or add-ins to Crystal Reports required to use these drivers.

WORKING WITH OTHER DATA SOURCES

There are also a number of other data sources that can be used as the source for your Crystal Report — in the following section we will be looking at a few of these.

BUSINESSOBJECTS UNIVERSES

If you are using BusinessObjects Edge or BusinessObjects Enterprise, you can now use Crystal Reports 2008 to report from BusinessObjects Universes as well. A Universe is a meta layer provided to mask the complexity of the underlying database, allowing a developer to create complex joins and manipulate data behind the scenes. From a user perspective, a Universe provides a well-organized view of the underlying data, including proper table and field names, view-time security, and more.

To report from a BusinessObjects Universe, you will need to be using Crystal Reports in conjunction with one of the BusinessObjects server platforms, including BusinessObjects Edge and BusinessObjects Enterprise.

For more information on creating reports from a universe, check out Chapter 17.

NON-TRADITIONAL DATA SOURCES

Crystal Reports also includes a number of drivers for non-traditional data sources, including Act!®, Microsoft Exchange, and Microsoft Logs. Most of these data sources have their own setup and configuration requirements and do not fit into the standard data source categories that can be accessed through a native or ODBC driver. Some of the available formats follow:

- Act! 3.0
- Microsoft Exchange
- Microsoft Outlook
- Lotus Domino
- Web/IIS Log Files

All of these data formats are available in the Database Expert, either in the main or the More Data Sources section, as shown in Figure 16.21 on the following page, and most require some additional parameters (file name, location, server name, and so on).

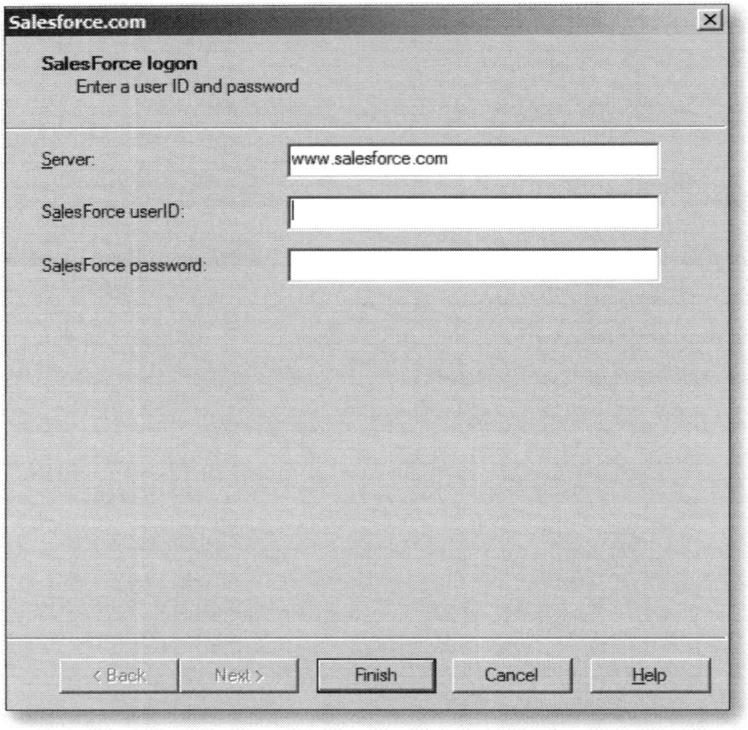

Figure 16.21 Connecting directly to Salesforce.com data

For a complete list of the information required by these data sources, see the Crystal Reports Help file by clicking Help > Crystal Reports Help from within the Report Designer.

USING THE VISUAL LINKING EXPERT

Relational databases are usually split into a number of different tables. These tables can be rejoined together to create complex queries. In Crystal Reports, these joins are created by using the Visual Linking Expert to visually draw a line between two key fields. If you select more than one table in the Database Expert, a separate Links tab will appear in the Database Expert. This dialog box, shown in Figure 16.22, is used to specify the relationship between the tables and views you have selected for your report.

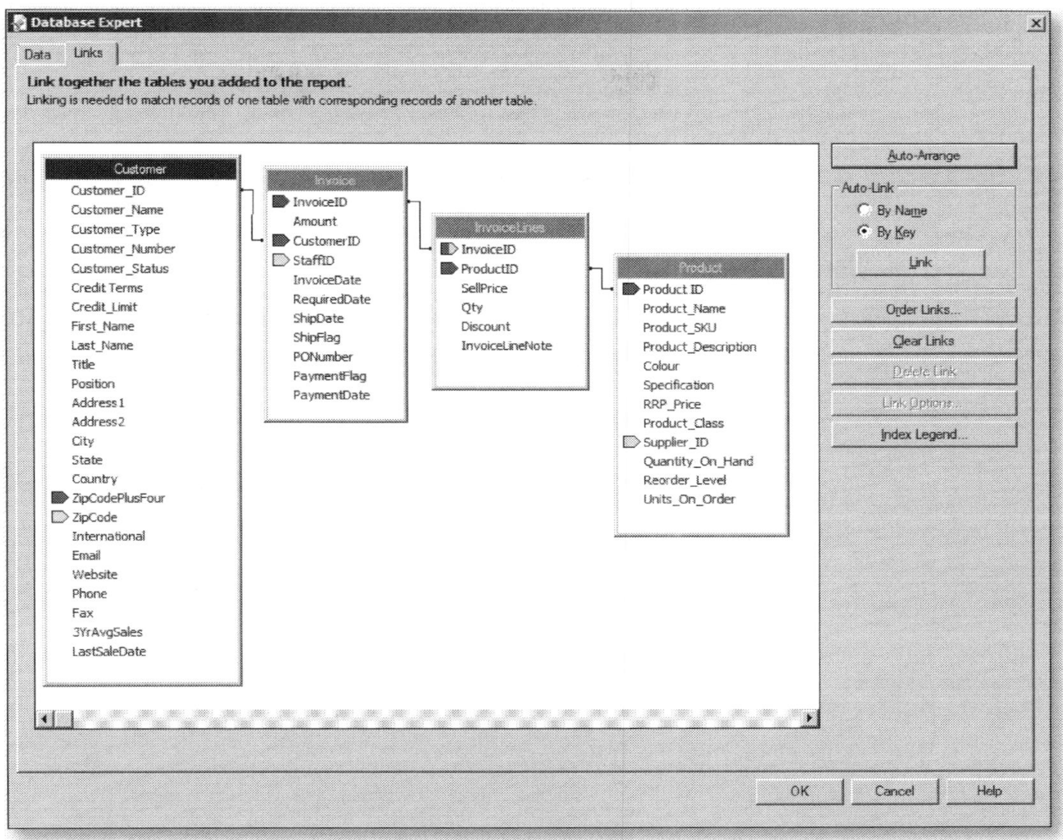

Figure 16.22 Visual Linking Expert

You can make this dialog box larger by dragging the bottom-right corner.

The links that are shown in the Links dialog box correspond to the SQL joins that exist between the different tables in your data source.

Use the Visual Linking Expert to indicate the relationship between your database tables or database files

Crystal Reports will first go through the table, views, and so on that you have selected in your report and will attempt to auto-link these structures based on the keys and indices it finds as well as the field name, type, and length.

If you are using a PC-type database such as Access or dBase, Crystal Reports will display the different indexes that are present in the table. You can click the Index Legend button to open a key to all of the indexes (for example, red = first index, blue = second index, and so on).

Unless you are using a very well-structured database, smart-linking usually gets it wrong, so you can remove the links between tables or views by clicking the line between the two tables and pressing the Delete key on your keyboard. You can also use the Clear Links button to clear all of the links at once.

Once you have determined the correct links, you can add them to the dialog box by dragging one field on top of another to draw a line between the two tables. To change the type of join between tables, right-click the link you have drawn and click Link Options.

A dialog box opens which allows you to select the join type (inner join, outer join, and so on) as well as the type of link. Once you have finished specifying the join type, click OK to return to the Links tab. One final clean-up option you can use (if you are working with a large number of tables and links) is the auto-arrange button, which arranges them neatly on the Links page.

SUMMARY

So whether your data is held in a file format, database, ERP system or even through a web service, Crystal Reports 2008 can quickly access the data to create presentation-quality reports. This chapter was all about how to access your corporate data — in the following chapter we are going to be looking at how to use these reports in conjunction with Crystal Reports Server 2008.

Chapter 17
Reporting from Universes

In This Chapter

- What is a Universe?
- Understanding Objects
- Creating a Report from a Universe Query
- Editing Universe Queries
- Filtering Queries
- Working with Prompts
- Advanced Query Techniques

INTRODUCTION

If you are using Crystal Reports to design reports for use with BusinessObjects Edge or BusinessObjects Enterprise, you have the ability to base your reports on a BusinessObjects Universe.

To create a Crystal Report from a Universe, you first will need to create one or more queries to retrieve the data you want to see in your reports. Queries can be based on relational or OLAP data sources and leverage all of the business logic that is present in the Universe.

The first step in creating a query is selecting the universe you want to base your query on. From that point, a report developer will select the dimensions and measures that they want to see in their report. From their selection of objects, a SQL query will be generated by the Web Intelligence server and submitted to the database.

The database will then return the data retrieved by the query to Crystal Reports and the report developer can then format the report as required.

WHAT IS A UNIVERSE?

A universe is a semantic layer that is unique to BusinessObjects. Universes are created by experienced IT professionals and are designed to take the complexity out of report development — all of the database tables, joins, calculations, business logic, etc. can be encapsulated within the universe.

In addition, instead of representing the database as "database objects", a universe will represent the tables, fields, etc. as "business objects" using terminology that is familiar to business users, as opposed to cryptic table and field names used in the actual data source. This means that when creating reports, users will select objects that are familiar to them.

UNDERSTANDING OBJECTS

When working with a universe, there are a number of objects that you can select in your query and include in your report, as shown in Figure 17.1 below.

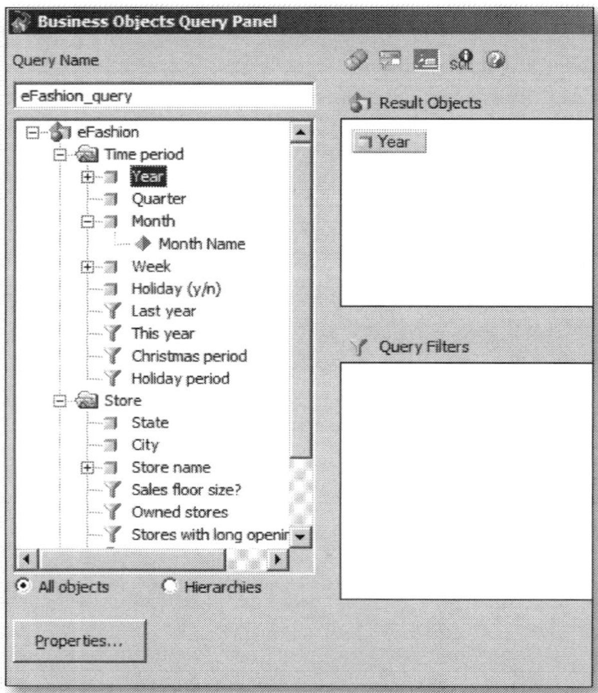

Figure 17.1 Universe Objects

Objects are elements in the universe that map to data in the relational database. Each object is defined with commonly used business terms, such as Country, State, City, Year, Revenue, Customer, Customer Address, etc. These objects are detailed in the sections below.

CLASSES

A Class is a local grouping of the related Objects. It behaves like folders containing objects. Each class can also contain one or more subclasses. Classes help the user easily find the objects that represent the information that the user wants to use in a query. An example of a class would be a Store class which contains the following objects: Country, State, City and Store Name.

DIMENSION

A Dimension retrieves data that provides the basis for analysis in a report. Dimensions are typically character-type data, such as Customer Name, Store Name, Year, Quarter, Month, Product ID, Product Description, etc.

DETAIL

Detail provides descriptive information about the dimension. A detail object is always attached to a dimension object for which it provides additional information about the dimension. An example would be the Month Name for the dimension object Month.

PRE-DEFINED QUERY FILTER

Pre-defined Query Filters are used to restrict the data returned by the object used in the query against the database. An example would be to restrict the data to only return the last financial year's data.

MEASURE

A Calculated value that can be used in the report. An example would be the Sales Revenue figures, Qty of products sold, Discount and profit margins.

CREATING A REPORT FROM A UNIVERSE

Now that you know a little bit more about the objects you can select in your queries, we need to look at how to create a query that we can base a Crystal Report on. To start, open the Crytal Reports designer and select File > New Report. From the "Available Data Sources" list, you will see Universes at the bottom, as shown below in Figure 17.2.

Figure 17.2 Universe option in Available Data Sources

When you double-click this data source, you will be prompted for your BusinessObjects server name, as well as a user name and password and authentication method, as shown below in Figure 17.3.

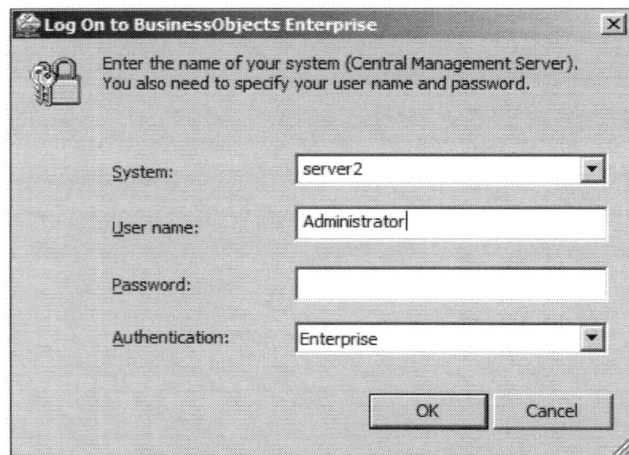

Log On to BusinessObjects Enterprise

Enter the name of your system (Central Management Server).
You also need to specify your user name and password.

System: server2

User name: Administrator

Password:

Authentication: Enterprise

OK Cancel

Figure 17.3 BusinessObjects Login

Once you have successfully logged into BusinessObjects, you will be presented with a list of universes that are available to you. To select a universe, click on the universe name and click Open as shown at the bottom of the dialog in Figure 17.4.

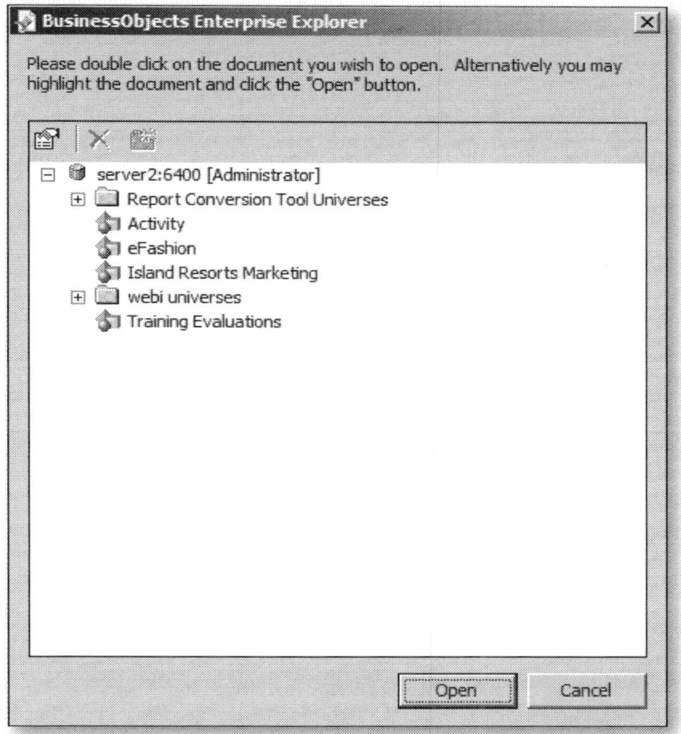

Figure 17.4 Universe selection

This will open the query panel shown in Figure 17.5 below. To select objects, you can drag-and-drop them from the universe into the Result Objects or Query Filters pane or you can double-click them to add them to the Result Objects. You can remove objects from the query panel by clicking on them, then pressing the Delete key on your keyboard.

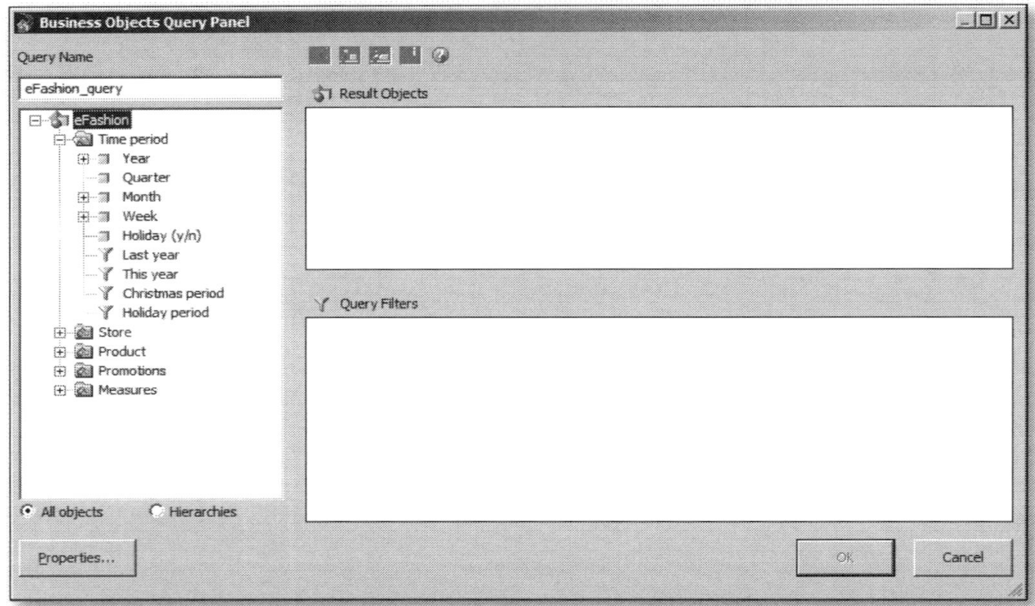

Figure 17.5 Query Panel

Often when you are working with a very large universe, it can be difficult to find the particular object you are looking for. You can jump to an object by clicking on the universe name, then start typing the name of the object.

When you have finished selecting the objects for your query, you can click OK which will return you to the Database Expert, as shown in Figure 17.6 on the next page. To use your query in the report, click to select it and then use the arrows to move it to the right-hand list of Selected Tables.

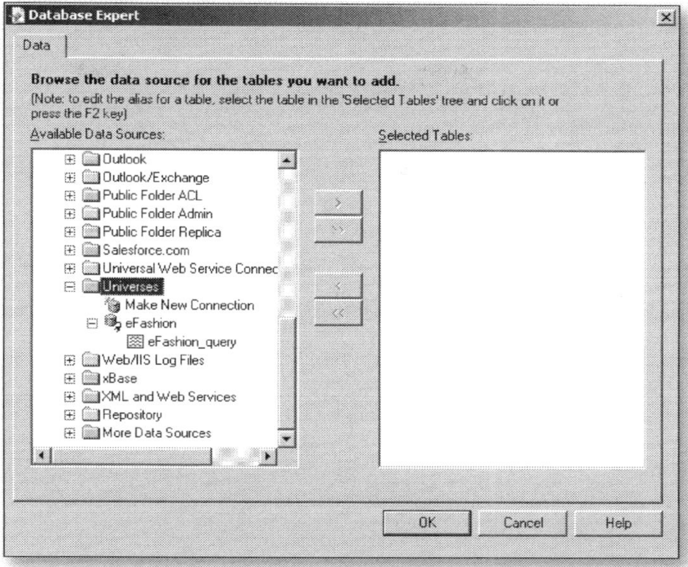

Figure 17.6 Database Expert

Once you have selected your query, you can then click OK to return to your report design. Your query fields will now appear in the Field Explorer, just as if you had selected a database table. In our example, we selected the dimensions Store Name, Year and Quarter and a Sales revenue measure — these objects now appear as fields in Crystal Reports, as shown below in Figure 17.7.

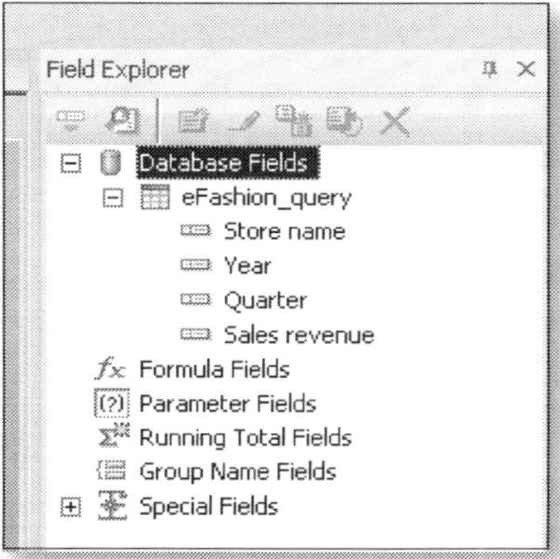

Figure 17.7 Field Explorer

EDITING UNIVERSE QUERIES

Once you have created a query, you may want to go back and edit the query later to add additional dimensions, measures or filter the query results. From within Crystal Reports, you can select Report > Query Panel to open the dialog shown below in Figure 17.8.

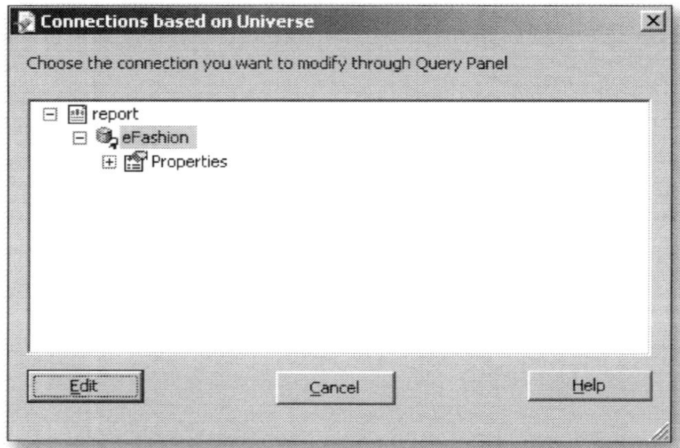

Figure 17.8 Universe query list

To edit a query, click on the name of the query and then click the Edit button at the bottom of the dialog. In addition to editing queries, this dialog also allows you to see the properties associated with the query, including the Universe name, user, etc. To see these properties, expand the Properties node to display the information shown in Figure 17.9.

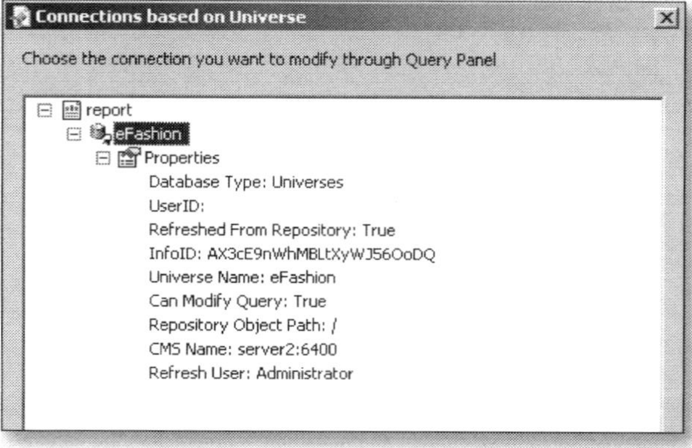

Figure 17.9 Universe query properties

In addition to changing the query itself, you can also point your query to a new universe using the Set Location function within Crystal Reports. To invoke this, select Database > Set Location which will open the dialog shown below in Figure 17.10.

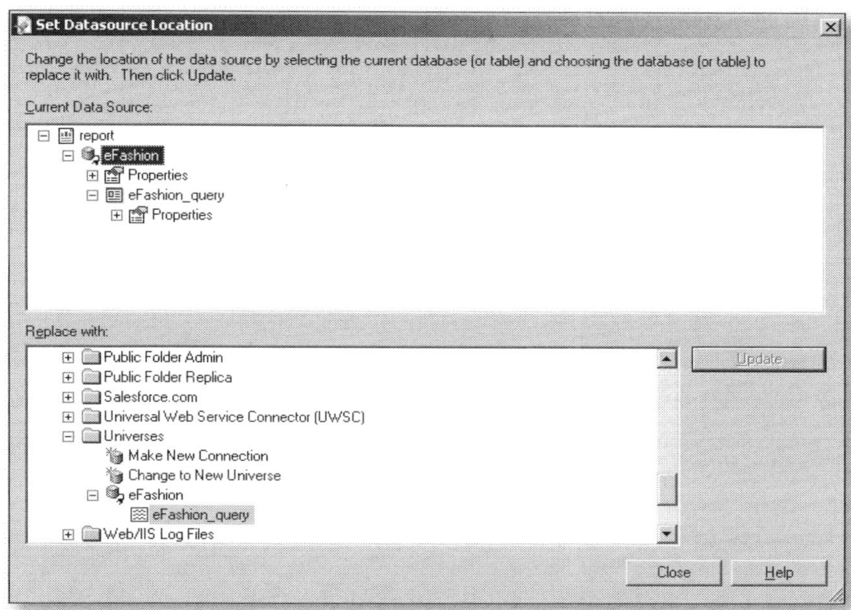

Figure 17.10 Set Datasource Location options

You will notice that there is an additional option under Universes to "Change to New Universe". If you double-click this node, it will allow you to select another universe to point your query to.

FILTERING QUERY DATA

When creating a Crystal Report off a universe, chances are you don't want to return all of the records in your database. To restrict your query data, we can use query filters to restrict the data that is returned to your Web Intelligence document.

In addition to providing you only the data you need for your reporting requirements, query filters also make your reports more efficient by reducing the number of records that are returned to Web Intelligence.

Behind the scenes, when you add a query filter to your query, the SQL that is generated is limited by a WHERE clause that limits the data which is returned.

If you were to have a look at the SQL that is generated by the report, there is now a WHERE clause added to the SQL statement that will limit the data which is returned to the query.

To make life easier, you may have some pre-defined condition filters defined in your universe. A universe designer may create pre-defined condition filters for commonly used scenarios where you need to filter data.

Pre-defined condition filters are shown in the Data pane of the Query Panel with a funnel icon. To add a pre-defined condition filter to your query, all you need to do is drag the pre-defined condition from the Data pane into the Query Filters area of your query.

CREATING A BASIC QUERY FILTER

A basic query filter has three components — first, is the object that you want to use to filter the data — this can be a dimension or a measure field. Secondly, you will need to select an operator. An operator is used to perform some comparison against the third component, an operand that you select.

For example, if you wanted to filter a report for all of your customers from the State of California, your filter would look something like this:

Object	Operator	Operand
State	Equal To	'California'

To create a basic query filter from the Query Panel, follow these steps:

1. Select a dimension or measure field by dragging the field from the data pane on the left hand side of the dialog to the Query Filters area of the Query Panel.
2. Once you have dragged a dimension or measure field into the query filter area, select an operator from the list below:
    - Equal To
    - Not Equal To
    - Different From
    - Greater Than
    - Greater Than Or Equal To
    - Less Than
    - Less Than Or Equal To
    - Between
    - Not Between
    - In List
    - Not In List
    - Matches Pattern
    - Different From Pattern
    - Both
    - Except
3. Select an operand from the list below:
    - Constant
    - Value(s) from List
    - Prompt
4. Once you have selected your operand, you can then continue to work with your query, or you can click "Run Query" to run the query and see the results.

Alternately, if you want to create a query filter using an object you have selected in Result Objects, you can click on the object, then click on the "Quick Filter" icon in the toolbar.

REMOVING A QUERY FILTER

To remove a Query Filter, you can click on the filter in the Query Filters pane and press the delete key. Or alternately, click to select the object and then use the "X" in the Query Filters toolbar to delete the objects. You can also use the multiple "X" icon to remove ALL of the query filters you may have created.

From the Query Filters toolbar, you also have the option to perform a query-on-a-query and do database ranking. These are advanced query concepts that are covered later.

Web Intelligence features a number of operators which you can use to create query filters, including complex operators that can be used to narrow your data to your specific requirements.

WORKING WITH OPERANDS

In addition to being able to set your operator, you can also set an operand, which in this dialog is labelled as "Operator", as shown in Figure 17.11.

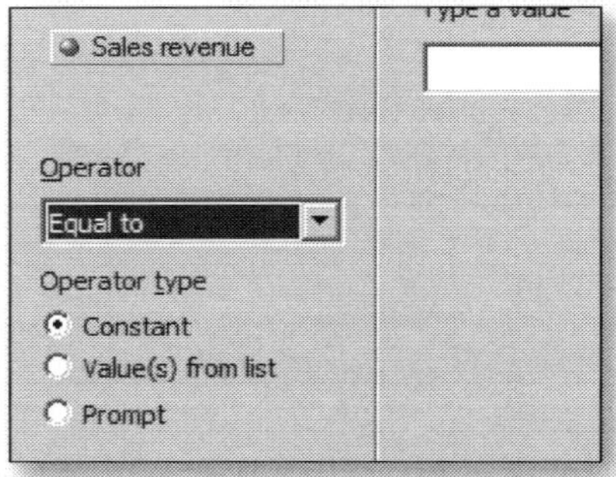

Figure 17.11 A list of available operands

The following operands are available for use with Query Filters:

Operand	Description
Constant	A constant value that you type in
Value(s) from List	A value (or values) that you select from a list of values
Prompt	A prompt that will prompt the user to enter some value

By combining objects, operators, and operands together, you can create powerful filters to narrow your data to just the records you want to see. For example, if you select the "Equal to" operator and then select "Value(s) from List", you will be presented with a list of distinct values from that dimension which you can select from, using the dialog shown below in Figure 17.12.

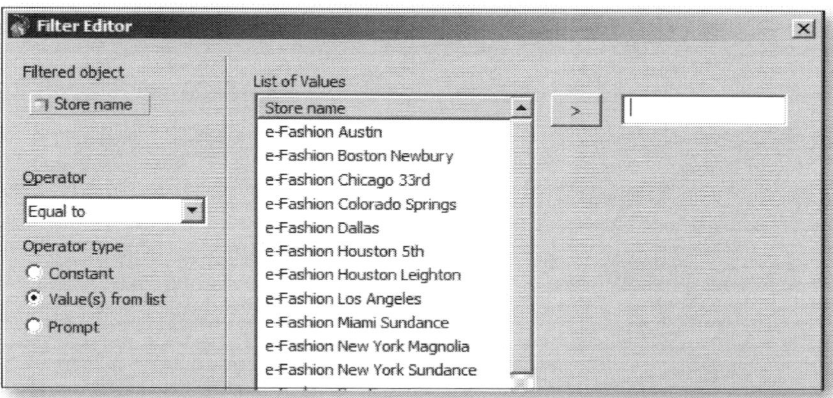

Figure 17.12 Single Values from List

To select a value, highlight it in the list of values, then click the right arrow icon (>) to move it to the selected box to the right. If you were to change your operator to "In List", this dialog would change slightly and allow you to select multiple values from your list of values, as shown below in Figure 17.13.

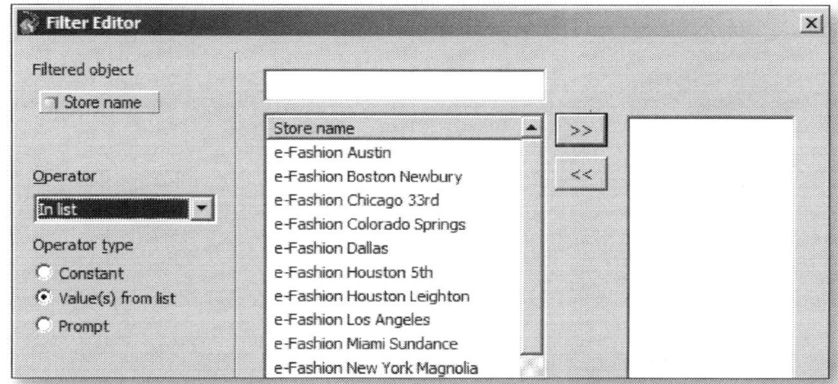

Figure 17.13 Multiple Values from List

COMBINING QUERY FILTERS

When creating query filters, it is possible to combine multiple query filters together. The advantage of combining query filters is that you can narrow your query further and further to just the information you want to see in your report. An example of a combined query filter is shown in Figure 17.14.

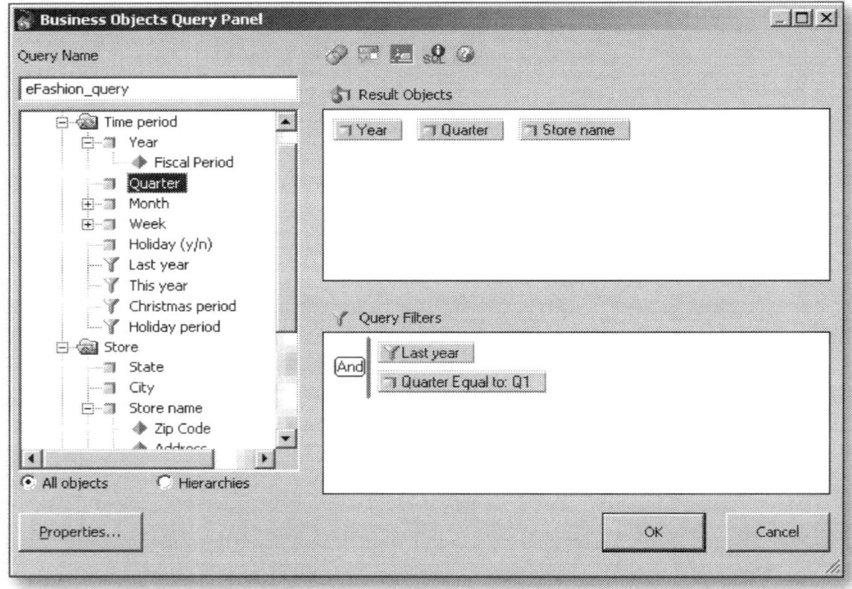

Figure 17.14 An example of a combined query

In this example, two query filters have been combined — the first is a pre-defined condition filter to return only non-English speakers. The second is a prompt filter that will prompt the user to enter a suburb. These two filters will be added to the SQL statement as a WHERE clause when the query is run.

To combine two or more query filters, follow these steps:

1. Within the Query Panel, create a query filter as you normally would by adding a dimension or measure object to the Filters panel, then select an operand and value.
2. Create a second query filter by dragging a second dimension or measure object to the Filters panel and set the operand and value for the second query filter.
3. You will notice by default the combined query filter is joined by an "And" statement. You can change this to "Or" (and vice-versa) by double-clicking the "And" operand.

When combining query filters, you may occasionally need to nest your query filters to tie two or more conditions together. The technique we use for this is called "nesting" and it allows you to set the precedence of query filters that will be evaluated by the query.

For example, you may want to select all of the customers who are in California — AND — have an order value over $10,000 — OR — have been invoiced since 01 January, as shown in Figure 17.15.

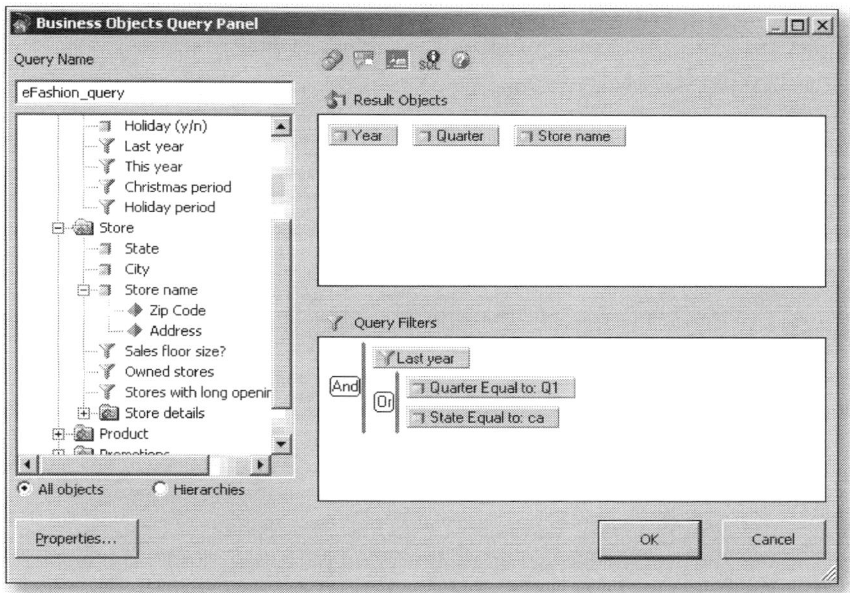

Figure 17.15 Another combined query

To create a nested query filter, follow these steps:

1. Within the Query panel, create a combined query filter as you normally would, adding multiple conditions.
2. To nest your query filters, drag and drop a query filter up and to the right (this technique takes a little practice to get right!).
3. To un-nest your query filters, drag and drop your query filter down and to the left to remove the nesting.

WORKING WITH PROMPTS

An easy way to create flexible reports is to use prompts in your query filters. By adding a prompt to your query, the end user will be prompted to enter a value whenever the report is refreshed. This will allow you to create one report which can suit many different users and uses.

For example, a prompt could be created to ask the user to enter a particular suburb they would like to see in the report. This prompt value is then passed in the SQL "where" clause to filter the data that is returned from the database.

To create a prompt, you will need to be in the Query Panel and have some idea of the dimension or measure you would like to use for the prompt. Once you know which object you want to base your prompt on, you can create one by following these steps:

1. From within the Query Panel, drag dimension or measure field from the Data pane into the filters pane at the bottom of the dialog.
2. Use the drop-down list to select an operator from the list.
3. For the operand, select the option for "Prompt". You'll notice at this point that Web Intelligence automatically creates a prompt based on the dimension or measure that you have selected. You can change the name of this prompt by changing the text shown in Figure 17.16.

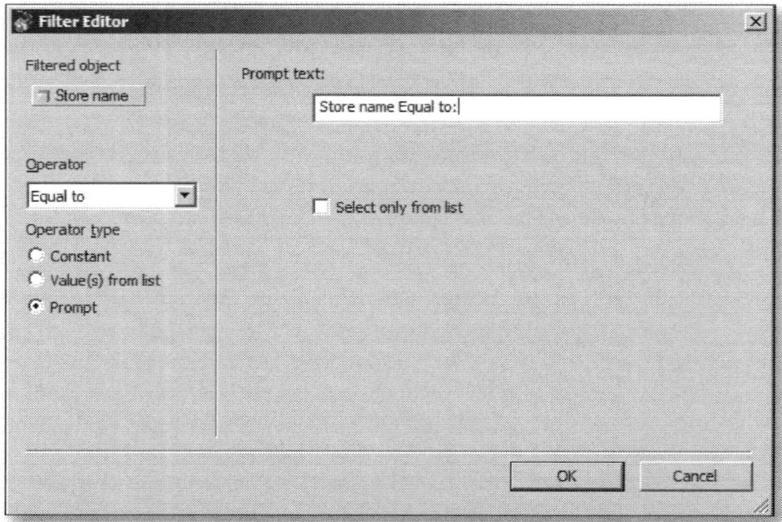

Figure 17.16 A default prompt name

In this example, since we selected "Store name" for the dimension, the name of the prompt by default is "Enter Store name", but we could change that to anything we like.

4. When you have finished creating your prompt, click "Refresh Query" to be prompted for a value. Your report results should now be formatted to only show the suburb you entered.

To display a prompt on your report, there is a special free-standing cell called "Prompt Summary" that will allow you to see what the end user has entered. To add this information to your report, drag a "Prompt Summary" cell from the Templates area in the Report panel.

When the Crystal Report is run, a standard Crystal Reports parameter dialog will be presented, as shown below in Figure 17.17.

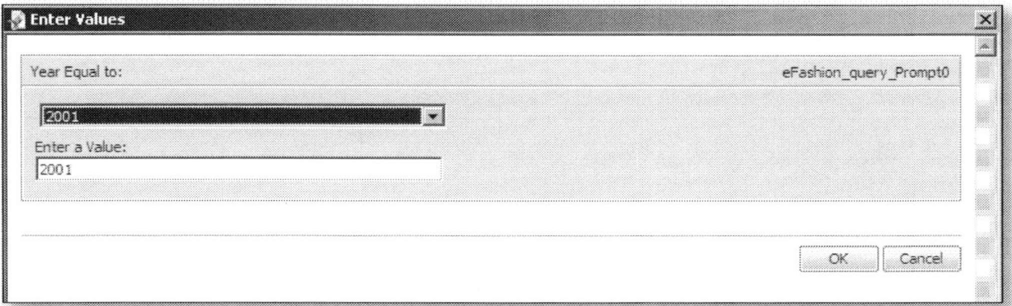

Figure 17.17 Standard prompt dialog

This prompt will also appear as a Crystal Reports parameter field and be visible in the Field Explorer, as shown over the page in Figure 17.18.

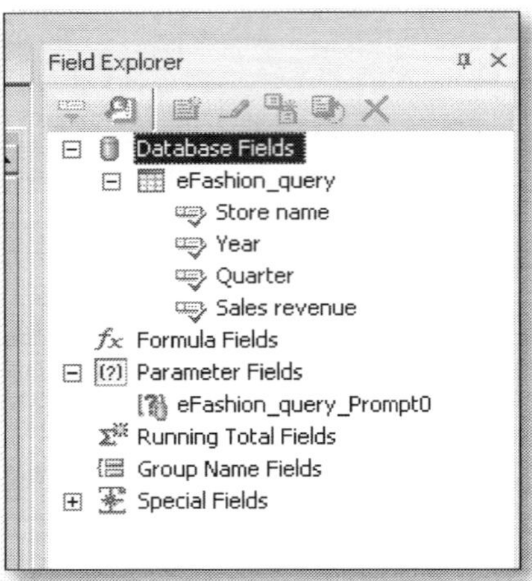

Figure 17.18 Prompt in Crystal Reports

As this parameter field is actually a Prompt that was created in the Query Panel, there are some properties of the parameter field that can be changed and some that cannot. To edit the underlying prompt, you would need to edit the query by selecting Database > Query Panel and make any modifications in the Query Panel itself.

With that said, there are some properties that you can change from within Crystal Reports by right-clicking the Prompt/Parameter Name and selecting Edit from the right-click menu. This will open the Edit Parameter dialog shown below in Figure 17.19.

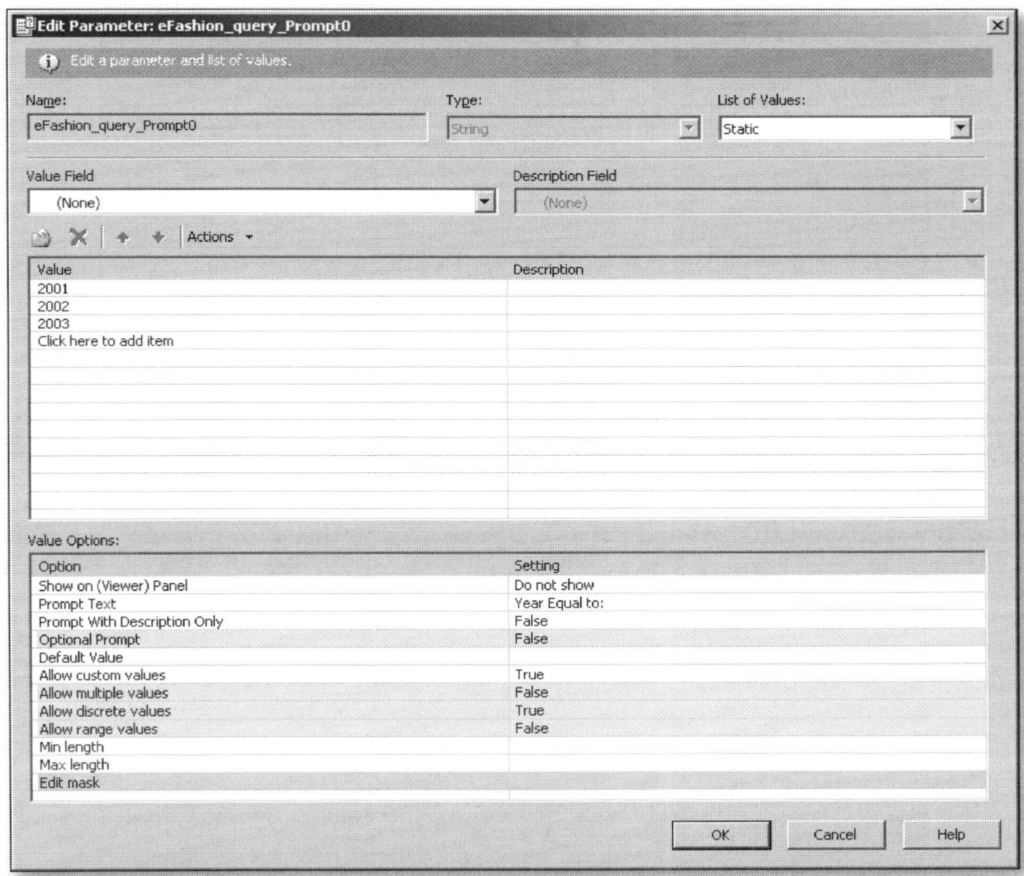

Figure 17.19 Prompt/Parameter properties

The properties that you can change will be enabled, but some of the properties in this dialog will be grayed out, indicating that you are unable to change these properties or must actually change them from the Prompt definition in the Query Panel.

ADVANCED QUERY TECHNIQUES

When using a Universe as the source of data for your Crystal Reports, you have the ability to use advanced query techniques like unions and subqueries to retrieve the results you need.

The Query Panel features a toolbar with a few icons that may be of use to you — the icon with the green and yellow circles (a Venn diagram) will allow you to create Union, Intersection and Minus queries from the query panel, which are called "combined queries". This involves creating two or more queries in which the data from these queries is unioned (similar to a SQL union statement) or the intersection of the two queries is returned, or where the negative result is returned from a minus query.

If you want to create a combined query, each query you create must have the same number of objects and these objects must be of the same database type. When the report is run and objects presented, only the objects in the first query will be displayed, which can be a bit confusing at times. You can combine up to 8 queries to get the desired results, but if you are creating more than 2 or 3 queries, you may want to look at doing some of the work at the database level just to make your life a bit easier.

In the area of the Query Panel, you'll also find an icon that will allow you to create subqueries, like the one shown below in Figure 17.20.

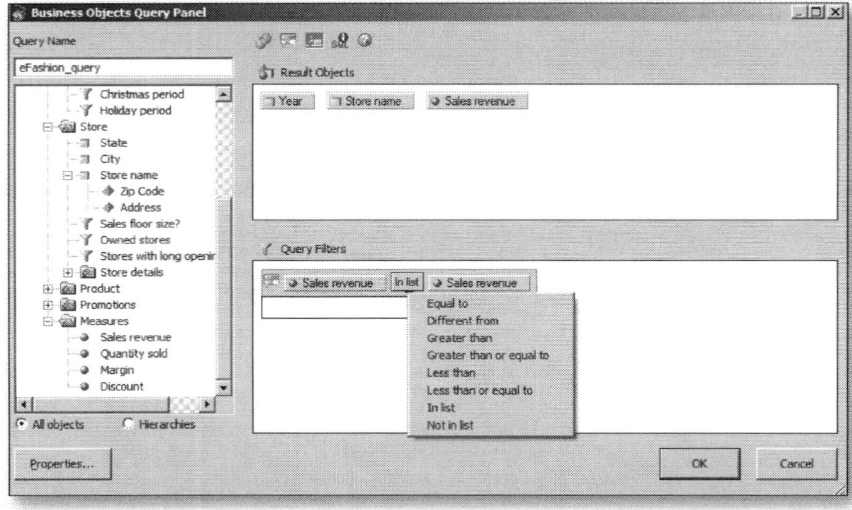

Figure 17.20 An example of a Subquery

Subqueries are a useful technique where you want to retrieve a data set based on the results of another. For example, you may want to find out how much revenue you have generated from a single category of products, like Board Games, and then return a list of all products that have generated more revenue than Board Games.

In order to do this type of query, you would need to use a subquery to first find out how much revenue that Board Games had actually generated. The reason we use a subquery for this type of calculation is that you could run the report over any time period and the revenue for Board Games may change. By using a subquery, you can ensure that your query will always be dynamic and will return the correct results.

For an example of a Crystal Report with a subquery like the one described above, check out the Chapter 17 folder in the book download files.

SUMMARY

If you are using Crystal Reports as part of a BusinessObjects Edge or BusinessObjects Enterprise deployment, you will definitely want to take a look at using Universes to take some of the complexity out of your reports. By adding a flexible metadata layer where the database is presented as "business objects" (as opposed to "database objects") you abstract a lot of the technical work behind the scenes that is required to actually create the report.

Our look at the server-based technology offered by BusinessObjects doesn't stop there. In the next chapter we will be taking a deep-dive into Crystal Reports Server, which gives you a flexible, secure, web-based framework that you can use to deliver your reports to a wider audience.

Chapter 18
Working with Crystal Reports Server

In This Chapter

- Working with the Report Designer
- Report Design Environment
- Customizing the Design Environment

INTRODUCTION

Crystal Reports is a fantastic tool for creating reports, but eventually you will want to share those reports with other users. In previous chapters, we looked at the different export formats that Crystal Reports supports, so you could export the report from Crystal Reports and send it to other users, but there is a much easier way to distribute reports.

Crystal Reports Server 2008 (CRS) is a server-based product that allows you to publish and share reports with a wider audience through a web-based interface. Crystal Reports Server provides the ability to securely publish the reports you have designed, which can then be refreshed, scheduled, distributed via e-mail, FTP, file server and more.

Crystal Reports Server is based on the same server-based technology and code-base as BusinessObjects Edge and BusinessObjects Enterprise. In fact, the end-user and administration interfaces are exactly the same. The differences between these platforms are listed in the diagram shown in Figure 18.1, over the page.

Crystal Reports Server is a great choice for small-to-medium enterprises looking to distribute reports and dashboards, as well as departments within larger organizations that don't currently use BusinessObjects as their reporting platform. However, if you want to leverage the BusinessObjects universe layer, or if you are looking to do ad-hoc reporting, OLAP analysis or data exploration, you may want to look at using either BusinessObjects Edge or BusinessObjects Enterprise. The good news is that both of these platforms support the same features found in Crystal Reports Server (and much, much more!).

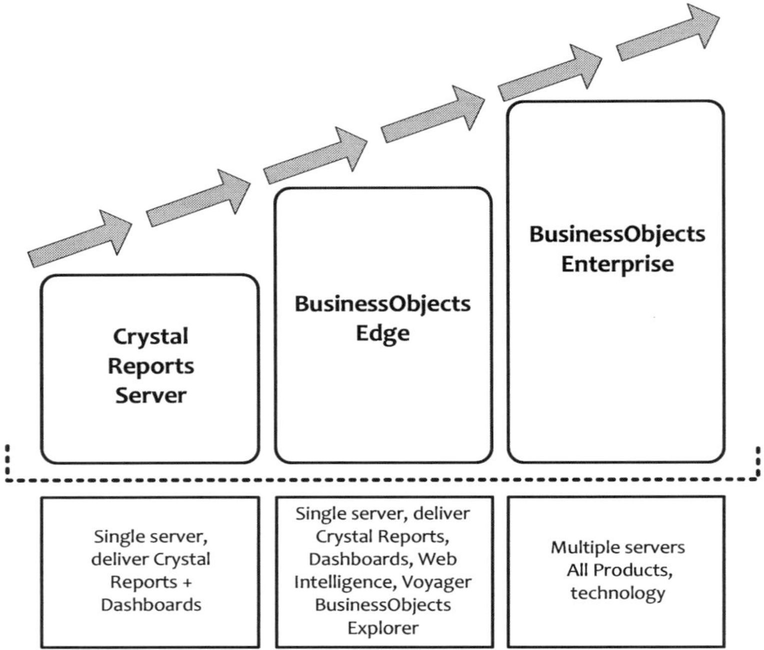

Figure 18.1 BusinessObjects server technologies

Some of the features in Crystal Reports Server are dependent on your licensing—Crystal Reports Server is sold as both "Named User Licenses" (NUL) and "Concurrent Access Licenses" (CAL), and at the time of writing this book, the ability to publish dashboards, for example, is only included with NUL licenses. For more information on the licensing for Crystal Reports Server, visit www.crystalreports.com

---

CRS FEATURES & FUNCTIONALITY

Crystal Reports Server provides the ability for you to publish and share your reports with multiple users on a secure, web-based platform. Some of the features and functionality include:

- Easy installation and configuration
- Single-server architecture where all components are installed on a single Windows or Linux server
- Zero administration, web-based interface for accessing reporting content and server management

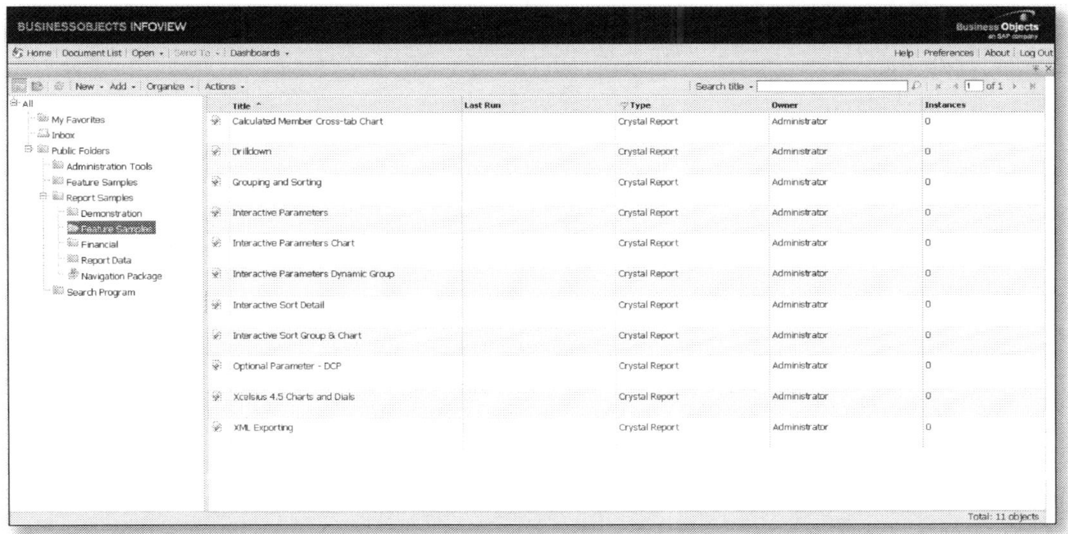

Figure 18.2 End-User interface for Crystal Reports Server

- Ability to organize content into folders and apply security based on users/groups
- Refresh reports on demand or schedule them to be refreshed on a regular basis
- Distribute reports to users via e-mail, FTP, file share and more
- "Burst" reports to multiple users with publications, with each user getting their own "slice" of the data in their report
- Publish and view dashboards created with Xcelsius

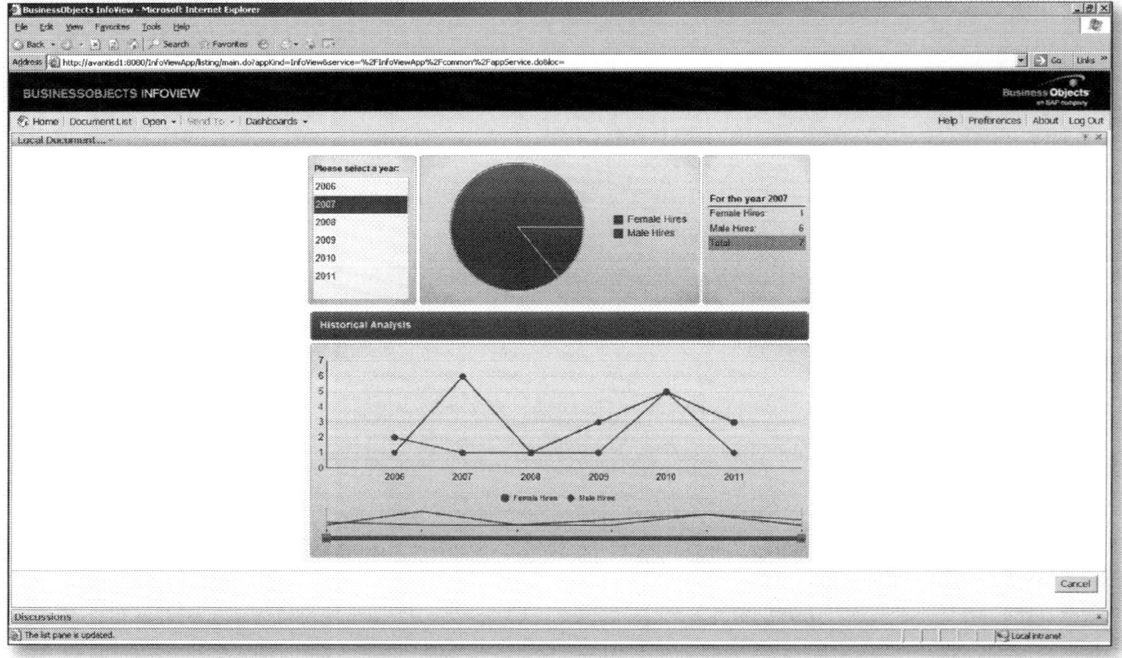

Figure 18.3 A dashboard published to Crystal Reports Server

- Access reporting content from within Microsoft Office using the SAP BusinessObjects Live Office add-in
- Integrate into your Microsoft SharePoint Server environment
- Develop custom .NET or Java applications to take advantage of the processing power, features and functionality that Crystal Reports Server provides

CRS ARCHITECTURE

Crystal Reports Server is available for both Windows and Linux, with a single-server architecture, where all of the components are installed on a single server. This architecture is by design, as Crystal Reports Server was designed for the small-to-medium enterprise (SME) market and for departmental use in larger organizations. If you do need scalability for a large number of users, you may want to consider moving up to BusinessObjects Enterprise, which allows you to use multiple servers in your architecture.

When you install Crystal Reports Server, all of the server components will be installed to this server. The key server components are shown in Figure 18.4.

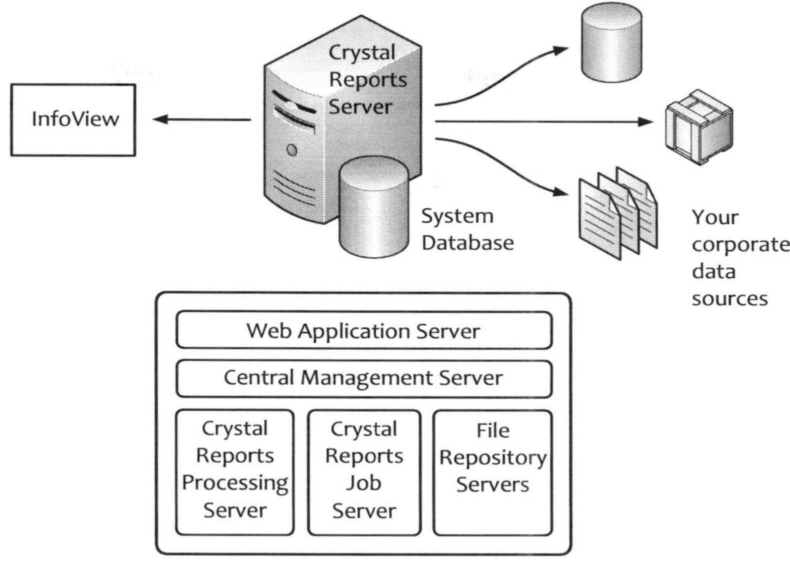

Figure 18.4 CRS key server components

In the following sections, we will be looking at all of these components in detail. It's important to remember that there are other components of the Crystal Reports Server architecture, but we are only looking at the most important ones at this point.

This chapter was not designed to be the exhaustive reference on Crystal Reports Server, but should provide enough information for you to get started. For the complete Crystal Reports Server reference library, visit http://help.sap.com and navigate to the BusinessObjects documentation, where you will find all of the documentation for Crystal Reports Server.

CENTRAL MANAGEMENT SERVER

The Central Management Server (CMS) is the "brains" of the Crystal Reports Server architecture, providing a link to the system database (or repository) and handling requests, including login requests, requests to view reports, scheduling requests, etc. The CMS also routes requests to the appropriate servers and manages the security and workflow within Crystal Reports Server.

CRYSTAL REPORTS PROCESSING SERVER

The Crystal Reports Processing Server handles report processing when a report is viewed "On Demand" (i.e. run immediately when the user requests the report). When a user requests a report to be viewed on demand, the CMS accepts the request and then passes it on to the Crystal Reports Processing server to be fulfilled. The server then gets a copy of the report template from the Input File Repository Server (covered later) and proceeds to run the report, connect to the database, retrieve the data, etc. and returns the report pages to the user who is viewing the report.

CRYSTAL REPORTS JOB SERVER

The Crystal Reports Job Server handles report processing when a report is scheduled. When a user schedules a report, this request is handled by the CMS. When it comes time to schedule the report, the CMS routes the request to the Crystal Reports Job Server, which in turns gets a copy of the report template from the Input File Repository Server and then proceeds to run the report, connect to the database, retrieve the data, etc. The Job Server creates an "instance" of the report with saved data and saves this file in the Output File Repository. This instance can then be viewed whenever the user next logs in to InfoView.

INPUT FILE REPOSITORY SERVER

The Input File Repository Server manages the directory where all of the report templates are stored. These are all of your reports that you have created for use with Crystal Reports Server and are accessed whenever a user runs a report on demand, schedules a report, etc.

OUTPUT FILE REPOSITORY SERVER

The Output File Repository Server manages the directories where all of your report instances are stored. A report instance is created when a report is scheduled and run. These instances are actually a copy of your report template, but with saved data.

INSTALLING CRS

Installing Crystal Reports Server will take you approximately 1 – 4 hours, depending on the server you are installing on, amount of RAM, etc. You will need to obtain a copy of the software and your license keys before starting the installation. If you purchased the software online, the software can usually be downloaded and your license keys will be provided via e-mail or at the time of purchase. Keep these license keys in a safe place, as you may need them again.

CHECKING A SUPPORTED PLATFORM

Before you do the install, you should check that you are installing Crystal Reports Server on a "supported platform". BusinessObjects provides a list of tested and supported platforms on their web site, or alternately you can Google for "Crystal Reports Server 2008 Supported Platforms" to find the PDF. This document lists all of the operating systems, versions, databases, drivers, browsers, etc. that SAP has tested and will support if you run into problems. This document is updated for each service pack, so make sure you check out the right version of the document for the software you are installing. At the time of writing, the current service pack for Crystal Reports Server 2008 was Service Pack 3.

The list below is a condensed list of the minimum requirements from the SAP web site. Please check the supported platforms to verify the minimum requirements for the version you are installing.

- Processor: 2.0 GHz Pentium 4-class processor
- Memory: 2 GB RAM
- Disk space: 5.5 GB hard disk space
- Operating systems: Microsoft Windows Server 2003 SP1 and SP2, Windows Server 2008, Red Hat Advanced/Enterprise Server versions 4 or 5, SuSe Linux Enterprise Server versions 9.0 SP3 or 10
- Browsers: Firefox 3.0, Safari 3.0, IE 6.0 SP2, IE 7.0
- SharePoint Portal Integration Kit (32 bit editions only): Microsoft SharePoint Portal Server 2007, Microsoft Windows Sharepoint Services (WSS 3.0) 2007
- Java Portal Integration Kit: Oracle Portal Server 10g R3, WebLogic Portal Server 9.2x, WebLogic Portal Server 10.0.x, WebSphere Portal 6.0.x.x

INSTALLING CRS ON WINDOWS

Before you install Crystal Reports Server, check one last time that you have met the requirements listed in the Supported Platforms document. To do the installation, you should be logged on to the server as a local Administrator and have the installation media on a local drive or CD to speed things up. You should also have your license keys handy and paper and a pen to note down the details you enter during the installation.

Once you are logged in, you can kick off the installation by running the setup.exe found in the installation files. The installation will ask you to accept the licensing agreement, as well as enter your license key.

When you are installing a new installation of Crystal Reports Server, you will be prompted for which database you would like to use for your repository database with a dialog like the one shown in Figure 18.5. Remember, this is the database where all of the internal information about CRS is stored. By default, Crystal Reports Server 2008 ships with a MySQL database that is installed and configured by default. This is a good choice for most installations, as it will be installed locally and the database itself doesn't grow very large. There is also zero administration you will need to do on the database and it runs in the background without impacting the performance of the server.

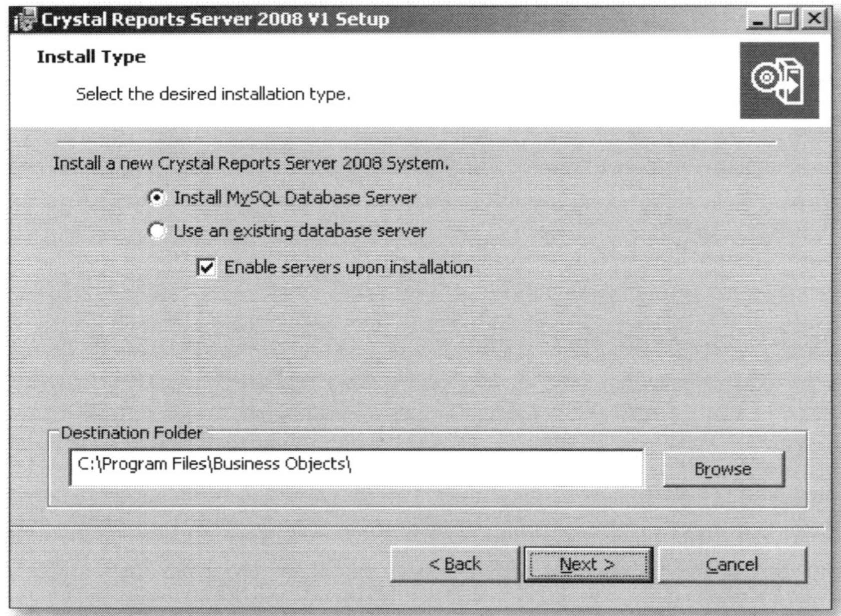

Figure 18.5 Repository database options

If you would like to use another database, you can choose to use an existing database, from one of the database platforms listed below:

- MySQL
- Sybase
- DB2
- Oracle
- SQL Server

You don't have to configure these data sources right away — in fact, my preference is to do the default installation on MySQL, then move the data source later. It is a quick and painless process and by using MySQL first, you can check the installation, perform some baseline testing, etc. It also means you don't have to wait for a blank database to be created, or get user credentials, etc. for the different platforms.

For more information on the requirements for pointing to each of these data sources, check out the "Crystal Reports 2008 Quick Installation Guide for Windows" available at http://help.sap.com

In this instance, we are going to assume that you are using the MySQL database that is included with the Crystal Reports Server install. You will be prompted using the dialog shown in Figure 18.6.

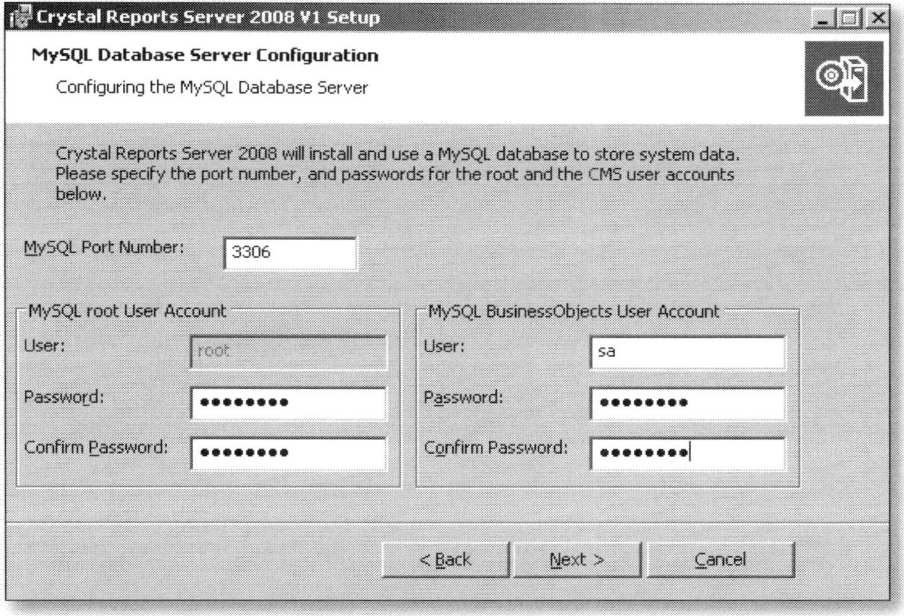

Figure 18.6 MySQL database server options

When you select the MySQL option, you will be presented with the dialog shown in Figure 18.7 below, where you can enter a port for the MySQL server (3306 is the default) as well as passwords for the "root" and "sa" accounts. Make a note of what these are, because you may need them later. You may also want to check the box for "enable remote root access", as you can access this MySQL database remotely (from your desktop, for example).

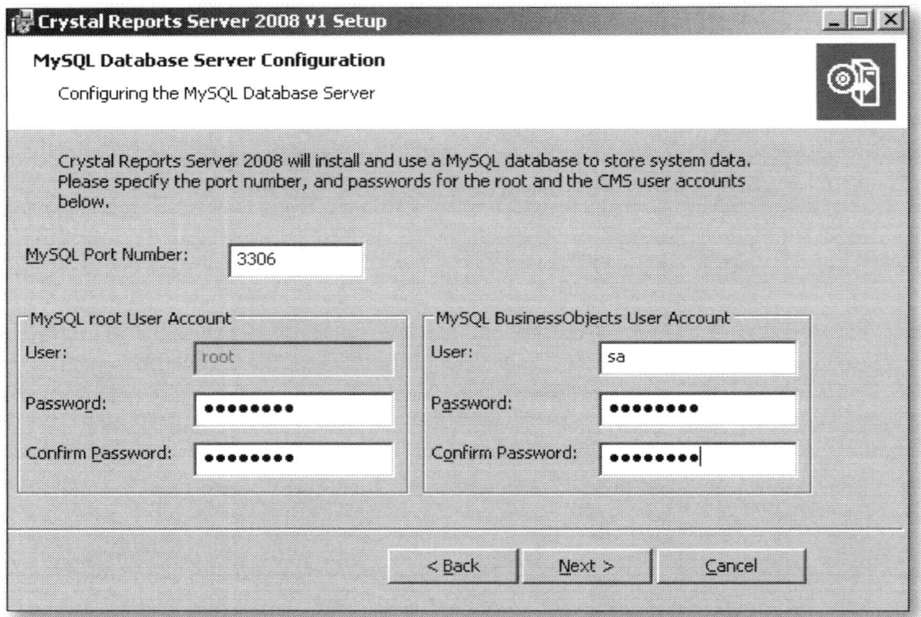

Figure 18.7 MySQL Database ports and passwords

In the next step of the setup, you will be prompted on whether to use an existing Web Application server or whether to install the version of Tomcat included with the installation. Again, it is always a good choice to select the default, perform some baseline testing, then move to another web application server. In this example, we are going to assume that you are installing the Tomcat option, as shown in Figure 18.8 on the next page.

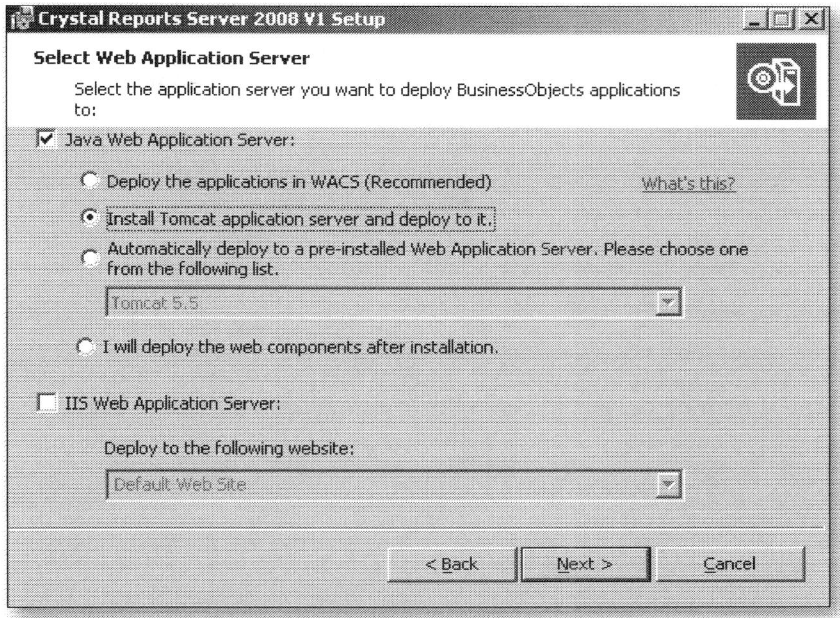

Figure 18.8 MySQL Database ports and passwords

From that point, the setup will present some default port numbers — leave these numbers as is, as most of the documentation will refer to the default port numbers, so it makes your life a little bit easier. With all of the options selected, the final step is the actual installation itself which may take some time (and is the perfect opportunity to go get a cup of coffee). When the installation is finished, you will be presented with a "Finish" dialog and your software has been installed.

WORKING WITH INFOVIEW

Users access Crystal Reports Server content through InfoView, a web-based application that provides the ability to view reports on demand, schedule reports, organise content into folders, apply security, etc.

To access InfoView, you can either click on the InfoView icon in the Crystal Reports Server program group, or alternately use the following URL shortcut:

http://servername:8080/InfoViewApp

When you access InfoView, you will be prompted for a user name and password from the Login screen shown below in Figure 18.9.

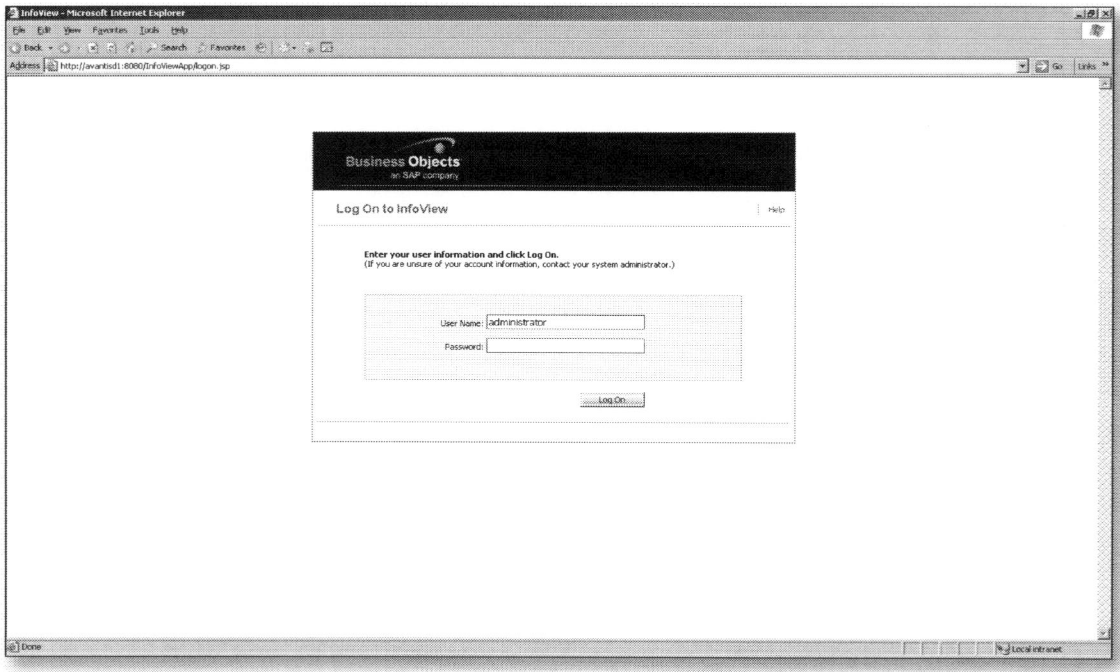

Figure 18.9 InfoView Login Screen

If you are logging into InfoView for the first time, the default user name is "Administrator", using whatever password you entered during the installation process. If you ticked the box to configure this password later, the password for the Administrator account is blank.

Once you are logged into InfoView, you will see the home page with a number of useful links. To see your report content, click on Document List, which will bring up the view shown below in Figure 18.10.

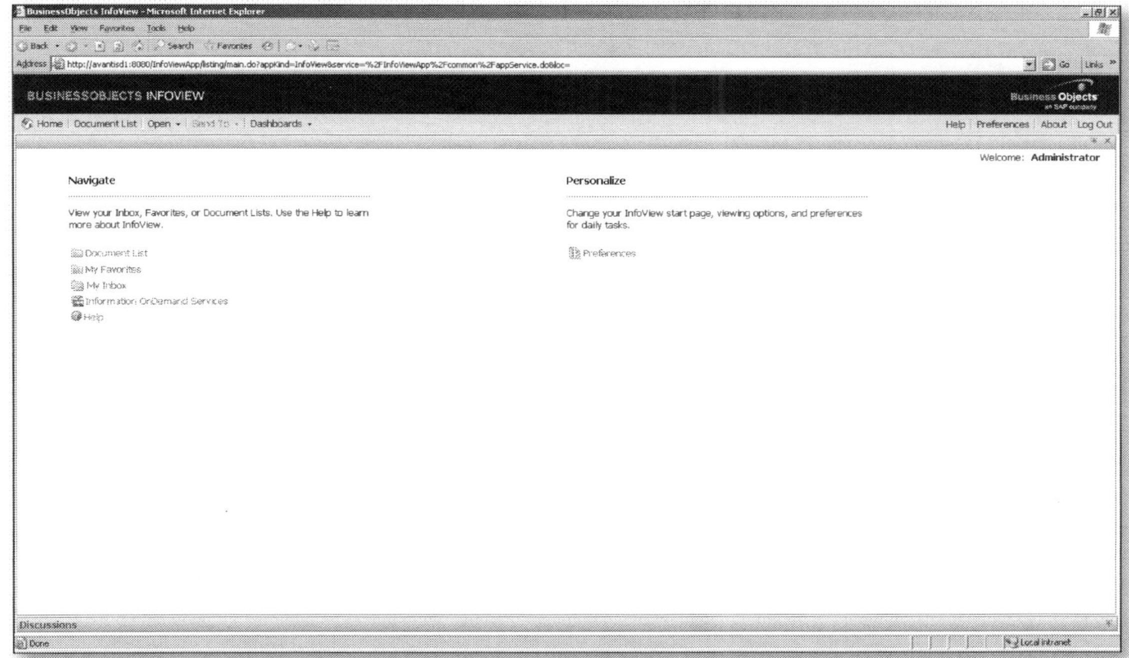

Figure 18.10 Document list

INFOVIEW INTERFACE

The InfoView interface consists of 3 main areas — the toolbars, document list and content list, as shown below in Figure 18.11.

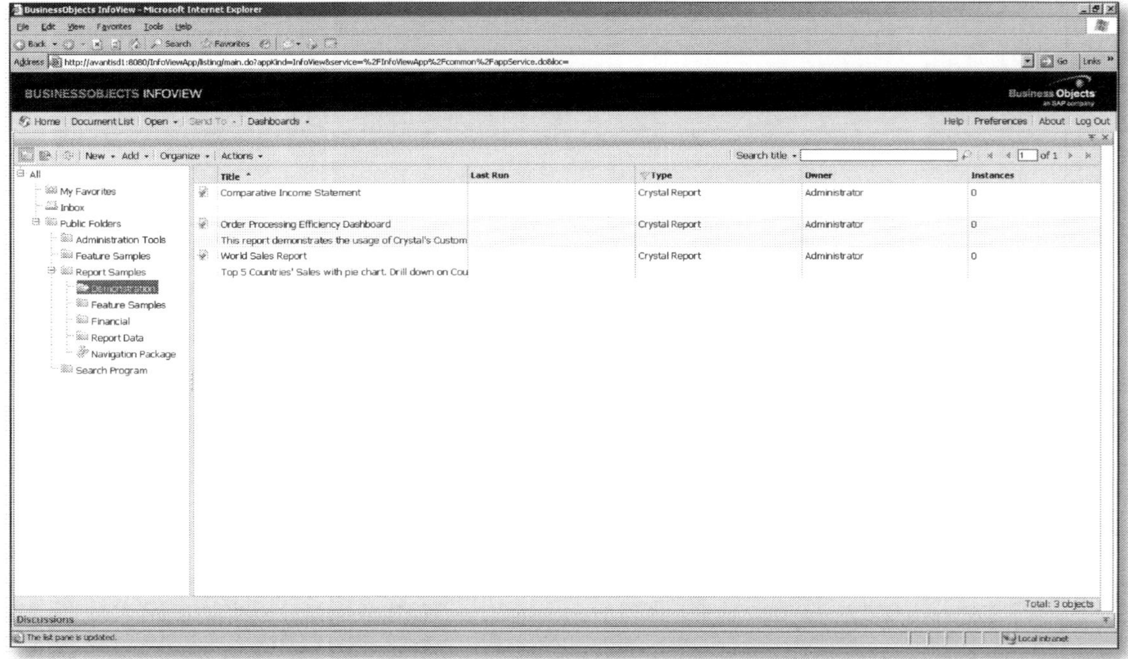

Figure 18.11 Infoview interface

The document list is used to navigate through your report content and has three nodes:

My Favourites — Is a personal folder for each user, which only they can view. Similar to "My Documents" on your own computer.

Inbox — The BusinessObjects Inbox, where administrators or users can schedule reports to be delivered to the Inbox and viewed when they login to InfoView.

Public Folders — Where all of the publically available content published to CRS is available. The folders you see underneath this folder are determined by your rights.

As you navigate through these nodes, you will notice that the content list changes to show the content in the different areas. In the example shown in Figure 18.12, we have navigated through to the Report Samples folder and the content list is now showing all of the reports available in that folder.

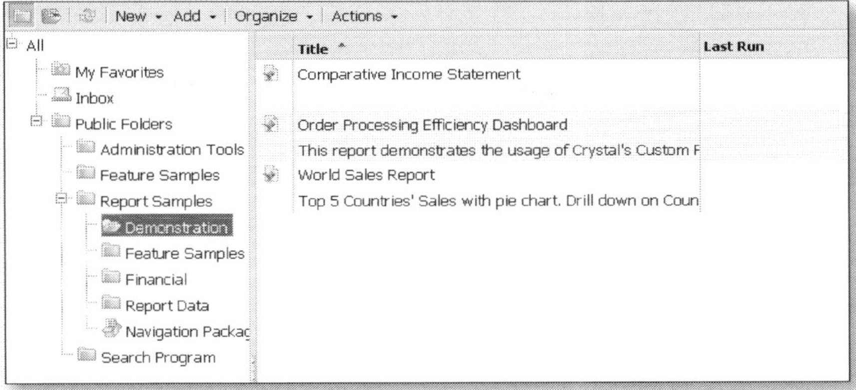

Figure 18.12 Content List

Finally, the toolbars at the top of the InfoView user interface show all of the different actions that are available when working with folders, reports and other content accessed through InfoView.

FINDING CONTENT

The first thing an end-user will want to do when working with InfoView is to find their reports. There are two primary ways of finding content within InfoView — the first is to navigate through the folder structure that appears by clicking the "plus" icon beside each folder to expand the folder, then clicking on a subfolder. This will bring up a list of all of the reports and other content that reside in that folder.

Alternately, you can search for content using the search toolbar located at the top of the page, and shown below in Figure 18.13.

Figure 18.13 Search toolbar

To perform a search through all of your report content, make sure that you click on Public Folders first. This is essential to ensure that you are searching all of the publicly available report content. Once you have clicked on Public Folders, enter your search term and click OK. An additional node will appear in the folder list, showing you your search results. This node is shown below in Figure 18.14.

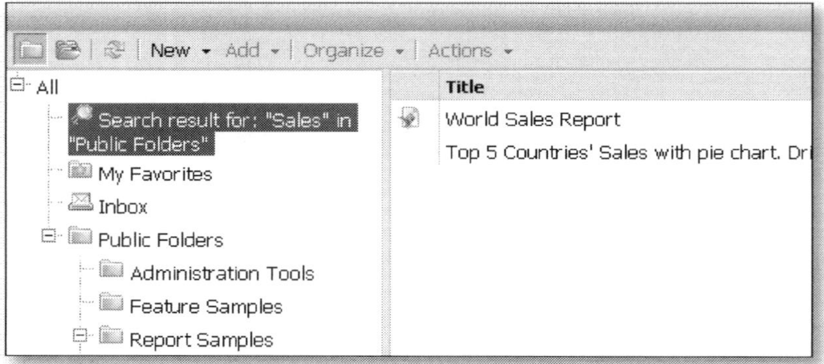

Figure 18.14 Search results node

By default, the search toolbar will allow you to search by the title of the report, but you can also change your search options to search by keywords, description or even search through the report content itself. To change the search options, click on the right-click menu beside the search box to open the dialog shown below in Figure 18.15. You can also do an advanced search, where you can search by owner, time and date, etc.

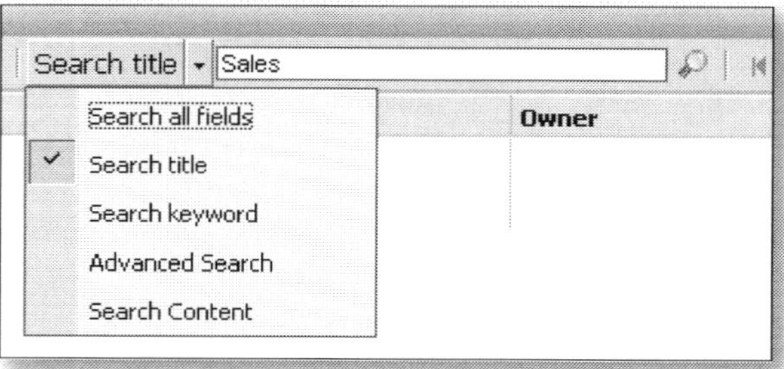

Figure 18.15 Search options

If you want to search within the report content itself, the Search service needs to index the report content on a regular basis. For more information on how to configure the search service, check out the Crystal Reports Server 2008 Administrator's Guide available at http://help.sap.com

PUBLISHING REPORTS TO CRS

Now that you have had a look at how users will access your reports, it's time to see how you can publish your reports to Crystal Reports Server. First of all, there are multiple ways you can publish your reports, but we are going to focus on the two most popular methods — through the Crystal Reports designer itself, and through the Central Management Console (CMC) or InfoView.

To publish a report through the Crystal Reports designer, open your report using File > Open from the location where you have saved it on your desktop, local drive, etc. Next, select File > Save As to open the dialog shown below in Figure 18.16.

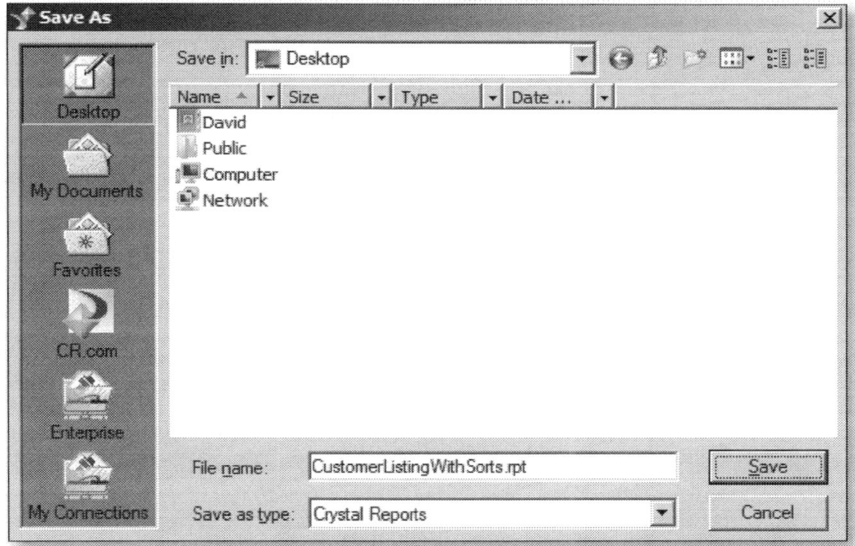

Figure 18.16 Save options

Click on the icon for "Enterprise" which will open a logon dialog box and prompt you to enter your server name, as well as your Crystal Reports Server user name/password, etc. as shown below in Figure 18.17.

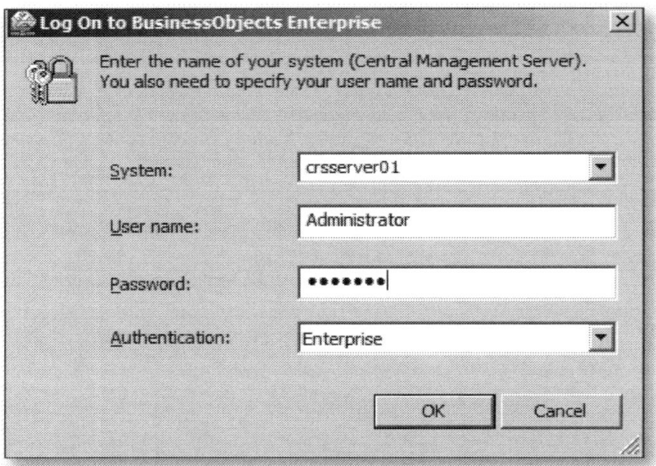

Figure 18.17 CRS Login options

Once you have logged into Crystal Reports Server, you can then navigate through the folder structure to select where you would like to save the report on the server. Once you have navigated to where you have saved the report, click OK to save the report and return to your report design.

The second method we are going to look at for publishing your reports is through the CMC or InfoView. With both of these web-based applications, you have the ability to publish a Crystal Report to the server using only your web browser.

To publish a report through the CMC, follow these steps:

1. Open your web browser and navigate to the CMC (http://localhost:8080/CmcApp) or alternately open the CMC from the shortcut in the Crystal Reports Server program group.
2. Then login to the CMC, which should present the default home page, as shown below in Figure 18.18.

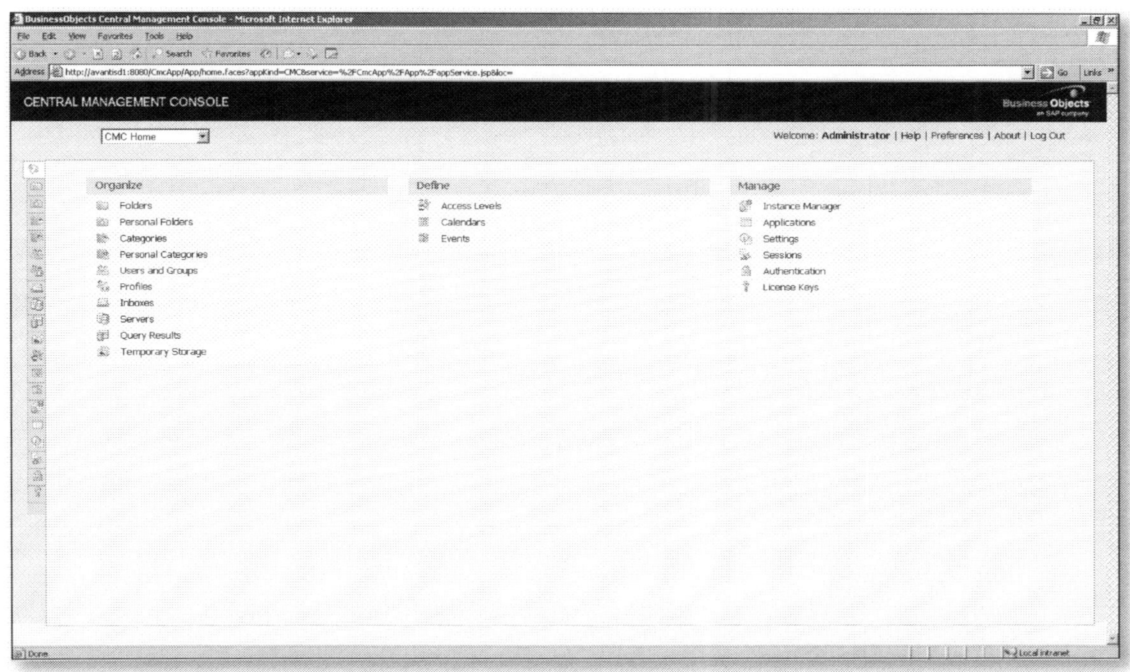

Figure 18.18 CMC Home Page

3. Click on Folders to navigate to the folders page, then navigate through the All Folders node to find the folder where you want to publish your Crystal Report.
4. Next, right-click on the folder and select Add > Crystal Reports. This will open the page shown below in Figure 18.19.

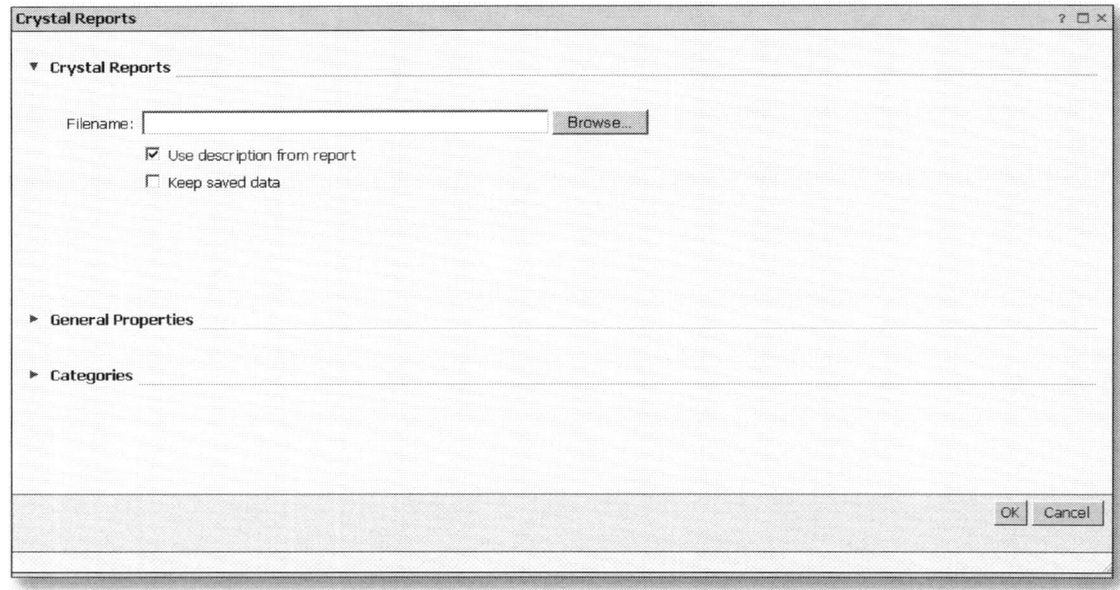

Figure 18.19 Report publishing options

5. Click on the Browse button to browse your location computer for your Crystal Report (.RPT) file and then click OK.
6. To upload your report, click OK on the report publishing page.

You can now view and schedule your report — this technique can also be used to publish reports through InfoView, providing the user has the rights to publish reports.

TROUBLESHOOTING PUBLISHED REPORTS

Often when you first publish reports to Crystal Reports Server, you will not be able to run them as there are a few "gotchas" you may run into. In the sections below, we have outlined some of the most common problems and how to fix them.

DATABASE CREDENTIALS

Often, when you publish a report based on a secure datasource (for example, a database that requires a username/password) the user will be prompted for a login every time the report is run or scheduled. This can be annoying and wastes valuable time — you can set the database user name and password once for each report by following these steps:

1. Login to the CMC.
2. Navigate to the folder where you have published the report, right-click on it and select Database Configuration from the right-click menu. This will open the dialog shown below in Figure 18.20.

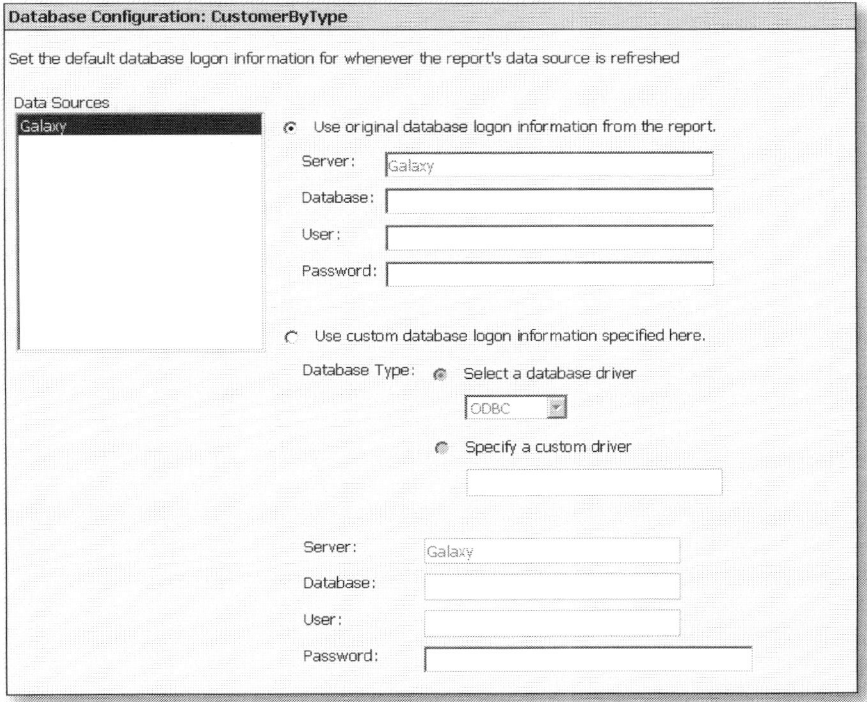

Figure 18.20 Database configuration options

3. Enter a user name and password for your database/data source, then scroll down to the bottom of the page to select the option of "Use same database logon as when report is run" then click OK to return to your report listing.

Whenever a user runs the report, they will not be prompted for the database username/password and the credentials you have entered here will be used instead.

One of the techniques that may make your life a bit easier is to create a "generic" database username and password that you can use when designing and publishing reports. Preferably this username would have a password that would not change, so you wouldn't have to update the password details or deal with angry users when their reports won't run due to an expired password. It may be a fight with your DBA or network administrator to get this type of account set up, but it is well worth it to cut down on the administration time you will spend.

SERVER DATA SOURCES

Whenever you publish a report created using an ODBC or native driver to Crystal Reports Server, you will need to make sure that you have the exact same ODBC driver or database client setup on the server.

For example, the sample reports that are included with this book were created off an Access ODBC driver with a Data Source Name of "Galaxy". If you want these reports to run on the server, you will need to create the exact same data source name, using the same ODBC driver setup. Likewise, the database must be stored in a location where the server can get access to it.

On the native driver front, if you are reporting off an Oracle database you will need to install and configure the Oracle client on the server, and perform some configuration so the client is configured to access your Oracle data (i.e. create an entry in TNSNAMES.ORA or copy the file from your local PC).

TROUBLESHOOTING WITH CR DESIGNER

For trouble-shooting purposes, you may want to install a copy of the Crystal Reports 2008 designer on the server itself. This can often be useful when trying to track down database and connection errors, as you can open the report in the designer. From there, you can refresh the report and view any error messages that are returned, directly from the server environment. Again, server admins often don't like to install client applications on a server, but in this case it is a handy troubleshooting tool.

ORGANIZING REPORTS WITH FOLDERS

With your reports published, it's time to look at how to organize your reports and other content. In CRS, content is stored and secured using folders (similar to how Windows stores content in folders on your computer).

When you install CRS, the installation will create a number of sample folders for you , as shown below in Figure 18.21.

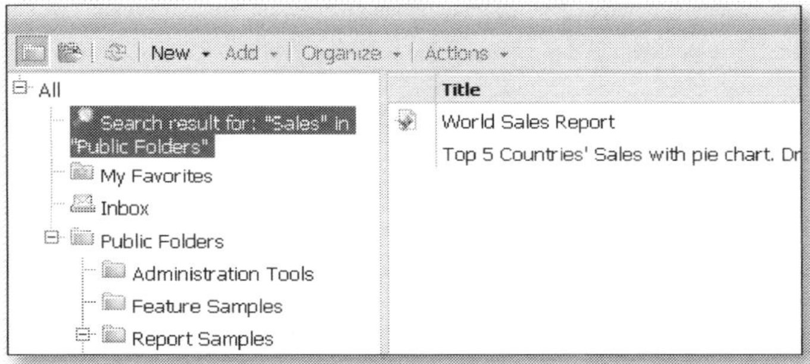

Figure 18.21 Default folders

You will want to create your own folder structure to logically organize your reports. To create a new public folder, follow these steps:

1. Login to InfoView.
2. Click on Document List.
3. Navigate to Public Folders, then right-click on the folder and select Add>Folder from the right-click menu shown in Figure 18.22.

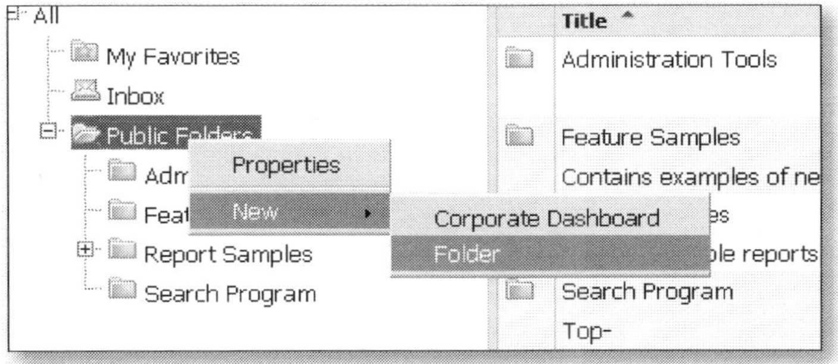

Figure 18.22 Right-click menu

4. Enter a name for your folder, then click OK to create your folder.

If you would like to move content from one folder to another, you can copy/paste/move documents using the technique below:

1. Login to InfoView.
2. Navigate to the Report Samples > Demonstration folder and click on the folder to display the content list.
3. Click on the check box beside the World Sales.
4. Right-click on the report title, then select Organize > Copy.
5. Navigate to the Feature Samples folder and right-click on the folder.
6. From the right-click menu, select Organize > Paste.

The World Sales Report should now be copied to this folder. You can also use this same technique to cut and paste reports form one folder to another.

A little later in the chapter, you'll learn how to apply security to your folders to restrict access to certain users and groups.

VIEWING/SCHEDULING CRS REPORTS

When viewing reports through InfoView, there are two different methods — the first is to "View on Demand" and the second way is to "Schedule" the report then view the instance. When you view a report on demand, the report is run on the server and presented in the browser window while you wait. This is the default behaviour when you double-click the report in InfoView, or when you right-click and select View from the right-click menu, as shown below in Figure 18.23.

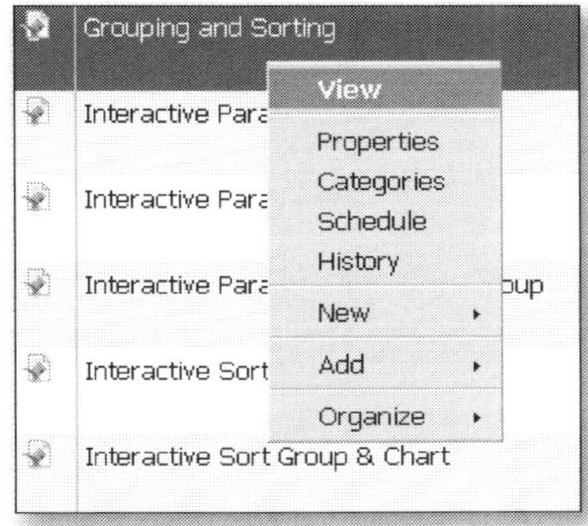

Figure 18.23 View option within InfoView

This will open your report in the Crystal Reports DHTML viewer, as shown over the page in Figure 18.24. This viewer allows you to refresh the report, respond to parameters, navigate through the report pages, export, print and more, all through your web browser with no other software to install.

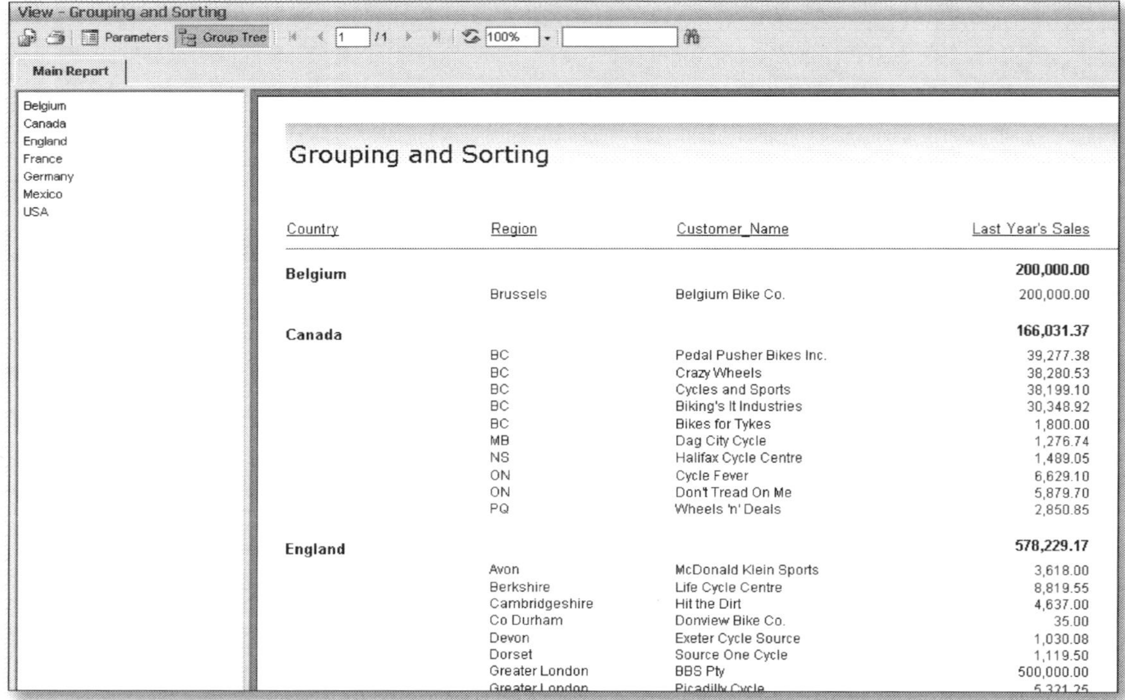

Figure 18.24 Crystal Reports DHTML Viewer

For printing, you can choose to install a small ActiveX print control for printing directly to your local or network printers, or alternately you can print through PDF but this will require Adobe Acrobat Reader to be installed on your PC.

Viewing a report "on demand" is good for reports that don't take very long to run, or reports that may require a number of parameters to be entered before it can be run. However, for very large reports that take a long time to run (20 minutes or more), your web browser window or session may time out waiting for the report to finish. In this instance, it would be better to schedule the report and then view the report instance that is generated.

The easiest way to think of scheduling is that there is a "monkey in the box" that is actually running the report for you. When you schedule a report, you can select a time, format, destination, etc. When the schedule time is reached, the monkey will then run the report for you and generate a report instance. A report instance is a copy of the report that is saved with the data. If you scheduled a report to run every day, at the end of a week you would have 7 report instances. These instances are kept on the server and can be viewed at any time.

To create a schedule for a report, follow these steps:

1. Navigate to InfoView, either through the URL (http://servername:8080/InfoViewApp) or by clicking on the InfoView icon from the Crystal Reports Server program group.
2. Click on Document List.
3. Navigate through the folders to find the report that you wish to schedule.
4. Right-click on the report to select Schedule from the right-click menu.
5. By default, CRS will schedule the report to run "Right Now" using the default settings and format. Click the "Schedule" button to schedule your report.

Once you have scheduled your report, you will be taken to the instance History list. This list shows all of the instances that have been scheduled and run for this report. The history list will present a list of instances with the following statuses:

Pending — Meaning the report is yet to run

Running — The report is currently running

Paused — The report schedule is paused

Recurring — The report has been scheduled to run on a recurring basis

Failed — The report failed to run or returned an error message

Success — The report has been run successfully and is ready to be viewed

You can click on the Refresh icon in the toolbar to see your report change statuses as it moves from Pending to Running to Success. Once the report is finished, you can click the instance date/time to view the report itself.

You can get back to this History list at any time by right-clicking on your report in InfoView and selecting History from the right-click menu.

CREATING USERS/GROUPS

To access content from CRS and InfoView, a user must login using a unique user name and password. CRS supports many different types of authentication, including Windows AD, LDAP and more. The configuration of these authentication methods is beyond the scope of this book, but don't despair! The default authentication method for CRS is called "Enterprise" authentication where you maintain users and groups yourself within the CMC.

SETTING UP A NEW USER

To create a new Enterprise user account, follow these steps:

1. Login to the Central Management Console (CMS) as Administrator.
2. Click on the icon for "Users and Groups".
3. From the toolbar, click on the "New User" icon to open the dialog shown below in Figure 18.25.

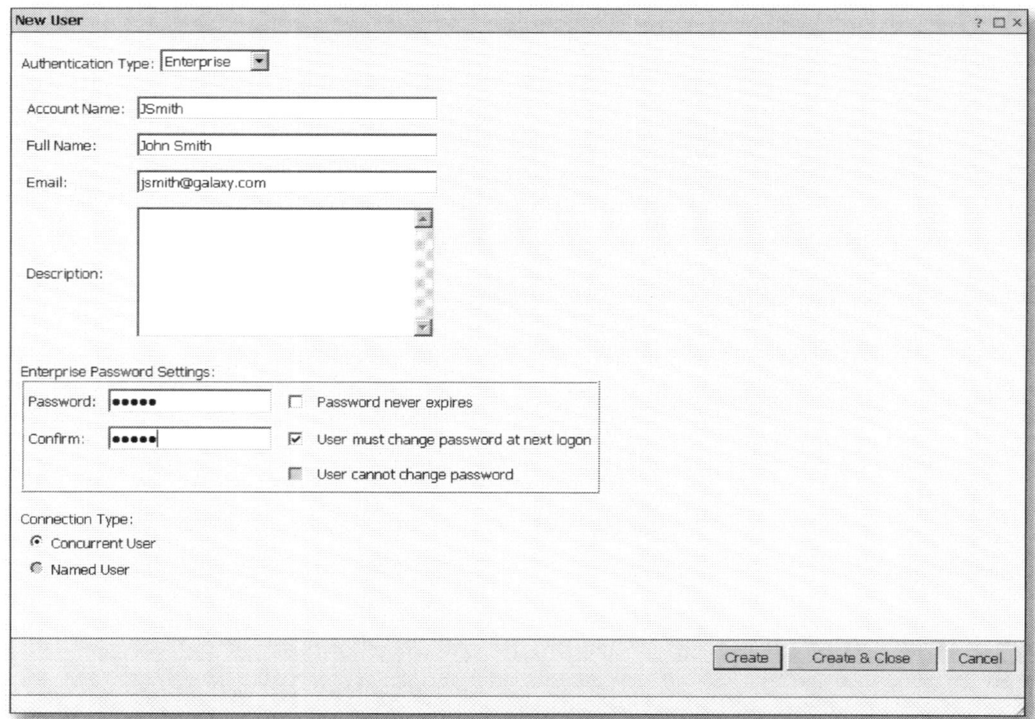

Figure 18.25 New user dialog

4. Enter an account name, as well as a description and password.
5. Click "Save and close" to return to the list of users.

By default the user will be asked to change their password on their first login. They can change their password any time by clicking on the "Preferences" link in InfoView.

You may be prompted for a strong password with a combination of upper/lower case, at least 6 characters, etc. You can change these defaults in the CMC by navigating to Authentication > Enterprise and then right-click to select Properties.

SETTING UP A NEW GROUP

A group is a collection of users and a user can be a member of multiple groups. Groups are used when applying security — for example, you may want to restrict access so only the "Marketing" group can see the "Marketing Reports" folder.

To create a group, follow these steps:

1. Login to the Central Management console.
2. From the home page, click on Users and Groups.
3. Click on the New Group icon or select Action > New Group to open the dialog shown below in Figure 18.26.

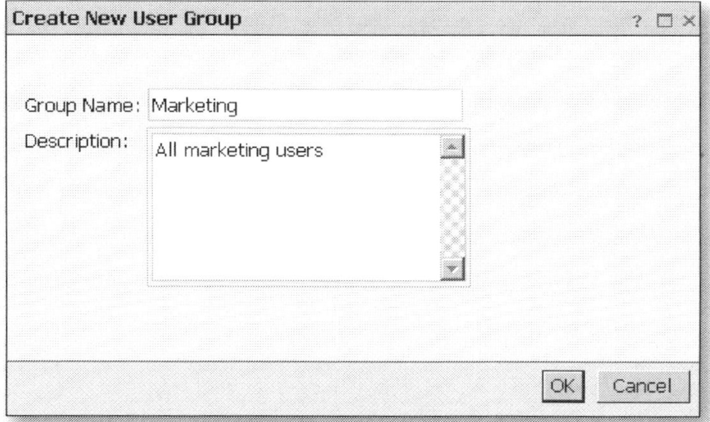

Figure 18.26 New Group Options

4. Enter a name for the group, then click "Save and close" to return to the users/group listing.

   Now that we have your group created, you'll need to add some users to the group.

   To add a user, follow these steps:

1. Login to the Central Management Console.
2. From the home page, click on Users and Groups.
3. Click on the user list and navigate to the user you wish to add and right-click on the user.
4. From the right-click menu, select "Join Group" to open the dialog shown below in Figure 18.27.

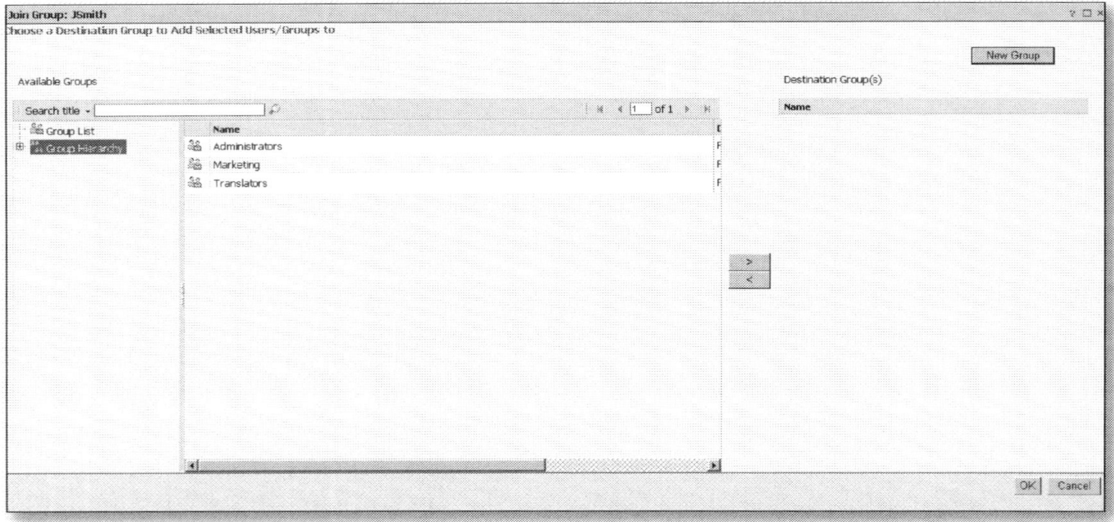

Figure 18.27 Selecting groups to join

5. Click to highlight the groups you wish to join from the list on the left, then use the arrow key to move them to the right.
6. When you have finished, click OK to return to the list of users/groups.

Once you have learned a little more about the security options in CRS, we are going to look at how to use groups to apply application and folder security.

But first, we need to look at the basic security concepts in CRS.

UNDERSTANDING CRS SECURITY

Crystal Reports Server (CRS) provides a scalable, robust platform for delivering business intelligence applications and content to end users. Security plays an important role within CRS, ensuring that users have appropriate access to these applications and content. Over the years, the CRS security model has evolved to provide a system that is both easy to use, with pre-defined access levels and flexible enough to provide more granular access to applications, content and metadata.

Using the security model within CRS, you can secure content based on users and groups. From the last chapter, you know that you can either maintain a list of users and groups within CRS or leverage external directories such as Windows Active Directory, Windows users or LDAP.

The CRS security model works on the basis of granting rights to these users and groups, controlling which applications they have access to, what features they can use and what content and data they can access. This object-centric approach means that you can also delegate administration, giving users the ability to administer aspects of the security model.

In the following sections, we will be looking at the CRS security model in-depth, starting with a view from 10,000 feet and then drilling down into the rules that govern the security model and how inheritance works. Towards the end of the section we will look at how to actually apply security to your CRS environment and some of the best practices for implementing a security model for your organization.

SECURITY MODEL OVERVIEW

Within Crystal Reports Server, there are three separate areas where security can be applied — on applications, objects and metadata, as shown in Figure 18.28.

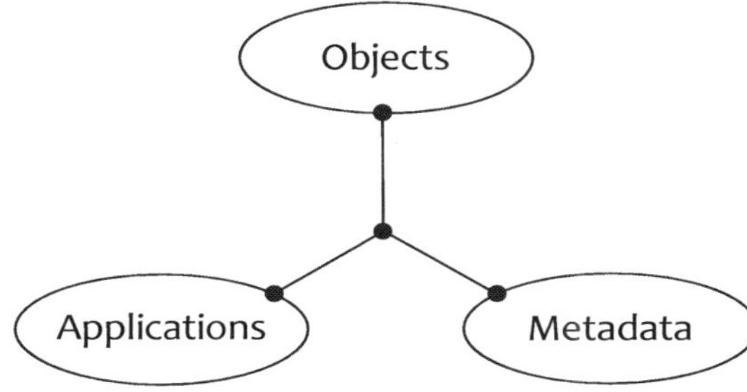

Figure 18.28 The three security areas

Security is applied to these three areas on the basis of rights. A right is a permission granted to a user or group to perform a specific action. For instance, if you are applying security to a folder, there is a right "View Objects" that can be applied to allow a user or group to view the folder and its contents.

Likewise, on the application rights side you may grant users or groups the right of "Create Desktop Intelligence Documents" to give them access to Desktop Intelligence and allow them to create documents.

Rights allow granular control of security within CRS, but if you were try to set individual rights on every object, you would spend a lot of your time doing nothing but security administration! Luckily, there is also the concept of pre-defined access levels, in which rights are grouped together and given a name.

For example, all of the rights you may want to assign to a user who can schedule and view reports are grouped together in a pre-defined access level called "Schedule" that you can apply to objects, as shown in Figure 18.29.

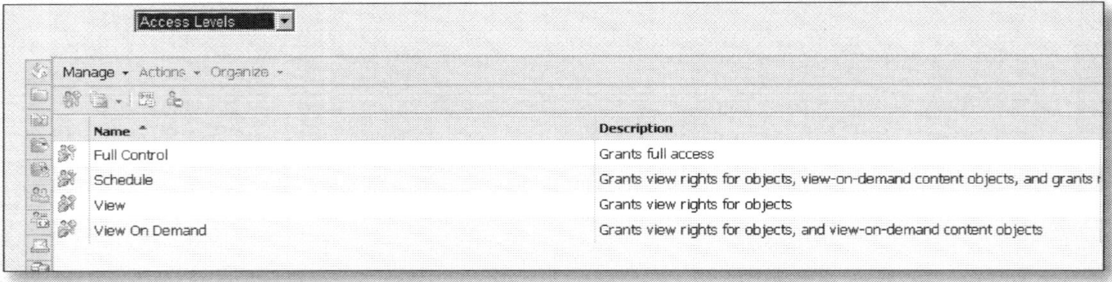

Figure 18.29 An example of the pre-defined access rights within CRS

So in most cases, you have the choice between applying one of these pre-defined access levels or picking the individual "Advanced" rights you want to grant or deny for a particular user or group.

Application Rights

Crystal Reports Server is comprised of a number of different applications, each with a specific use. For example, administrators may use the Central Management Console to administer CRS, while end-users will access their reports and other content through InfoView. Report developers may use Desktop Intelligence or Web Intelligence to create reports and documents for users and so forth.

Each one of these applications has rights that can be assigned to a particular user or group. These rights determine what access a user (or group of users) will have to that application, including which features they can use, and what they can do with the application.

With application rights, there are not any pre-defined access levels, as each of these applications has a different set of rights that are specific to that application. One of the best practices we have for administering applications is creating groups that relate directly to application access.

For example, you may want to create a group called "Universe Designers" and then grant that group access to the Designer. To grant a user access to the Designer, all you need to do is add them to the appropriate group.

This also follows one of our best practices that we will at look near the end of the chapter — keeping security at the group level, as opposed to assigning rights for individual users.

Object Rights

For managing objects, CRS features a security model that includes Global, Folder and Object level rights, as shown in Figure 18.30.

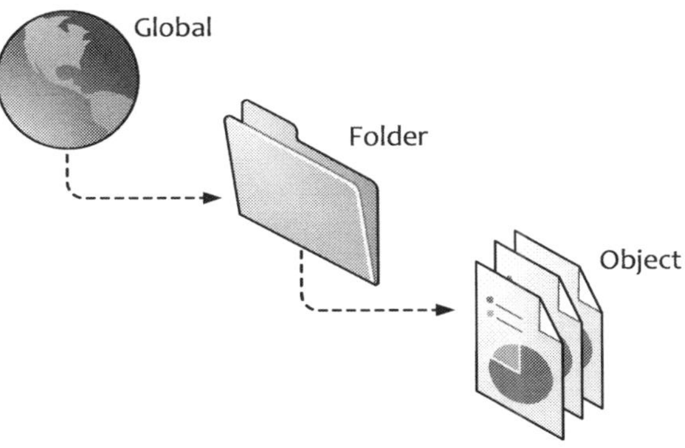

Figure 18.30 The hierarchy of object rights

All content published to Crystal Reports Server is stored in a folder, and object rights rely heavily on this folder structure for applying and maintaining security across a large number of objects.

This folder security concept is similar to how the Windows operating system stores files. If we look at a typical folder structure for Crystal Reports Server, you may find folders created for different departments (Finance, Sales, Marketing, etc.) or for different types of reports (Accounting Reports, Sales Reports, etc.).

A typical folder structure showing a combination of these techniques is shown in Figure 18.31, where a number of top-level folders have been created within InfoView for various departments and then sub-folders have been created below to store different types of reports.

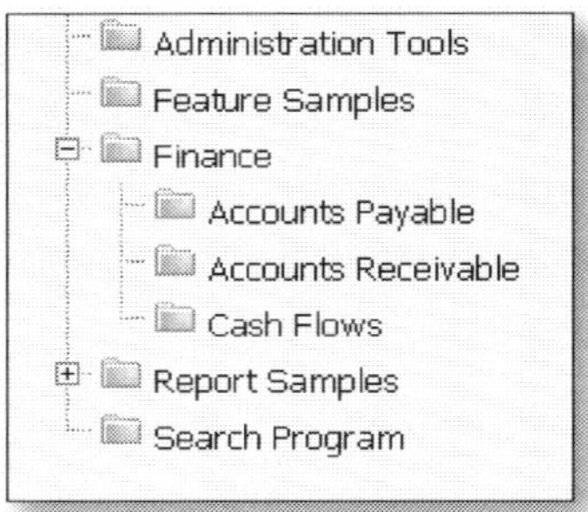

Figure 18.31 A typical folder structure for storing reports and content

How the folders are organized is up to you and your requirements, but it is important to understand that all content that is stored within Crystal Reports Server (reports, documents, dashboards, etc.) is stored in a folder and the security is driven by this folder/object model. It is also important to note that the folders within CRS are not actual file system folders — these folders are defined in the repository database as objects themselves.

In the following sections we will be looking at an overview of the object rights hierarchy, starting at the top with global rights, then folder and individual object rights.

Remember that this is just the overview — in the sections that follow we will drill down into rights, access-levels and how inheritance works, as well as the rules that govern the security model.

Global Rights

Global level rights are set at the very top of the security model hierarchy and these rights can be inherited by the top-level folders and folders and objects underneath. These rights can be set for a particular group or user and can be set by assigning pre-defined security roles or by picking individual rights to assign. When a new folder is created, these global level rights are applied to that folder by default.

When you first install CRS, there are two groups that are granted global rights. The "Everyone" group is a default group that is created when CRS is installed and this group contains all of the users within CRS. This group will be granted "Schedule" rights at the global level. The is a second default group, called "Administrators" that contains the Administrator account (and other users who need administrative rights) and this group is granted "Full Control".

"Schedule" and "Full Control" are pre-defined access levels that are explained a little later in the chapter.

One of the best practices for CRS is to grant the Everyone group "No Access" at the global level — this setting will be inherited by any folders and objects underneath this level. You can then go set the security for folders and objects, granting access to the appropriate groups and users.

Folder Rights

Folder level security in CRS works much like the folder security within Microsoft Windows. Within Windows you have the ability to specify which groups or users have access to a particular folder and its contents by applying security to that folder.

Within Crystal Reports Server, rights are applied to users and groups at the folder level and can be inherited by any sub-folders or objects that are present in the folder.

For example, you may have a "Finance" folder and you want to grant access only to the Finance group. Any reports or other objects stored in that folder will inherit the rights by default, so anyone who is granted access to the Finance folder will be able to see the folder and its contents.

Object Rights

Crystal Reports Server also gives you the ability to create object-level rights to create a very granular security model. Object-level rights are applied to each object and again can be applied to either users or groups.

For example, you have granted the Finance group access to the Finance folder, but you want to limit access to the "Executive Salary Report" to only one user. You can use the object level rights to set the security for this specific object.

In terms of best practices for your security model, you should always keep security at the Folder and Group level. This allows the most flexibility and over time will mean less administration effort.

Business Views

CRS supports a second metadata layer in Business Views. Business Views were developed by Crystal Decisions as a way to provide a metadata layer for Crystal Reports. Security within Business Views work on the basis of defining a filter and then applying that filter to a user or group. Filters within Business Views are created using the Crystal Reports formula syntax and when a report is run or viewed, this filter is used to filter the contents of the report, showing the user only the data that they are authorized to view.

Business Views are created using the Business View manager which you can use to apply security and filters at various levels within the view. The diagram shown in Figure 18.32 shows the basic structure of a business view and the areas where security can be applied are shaded gray.

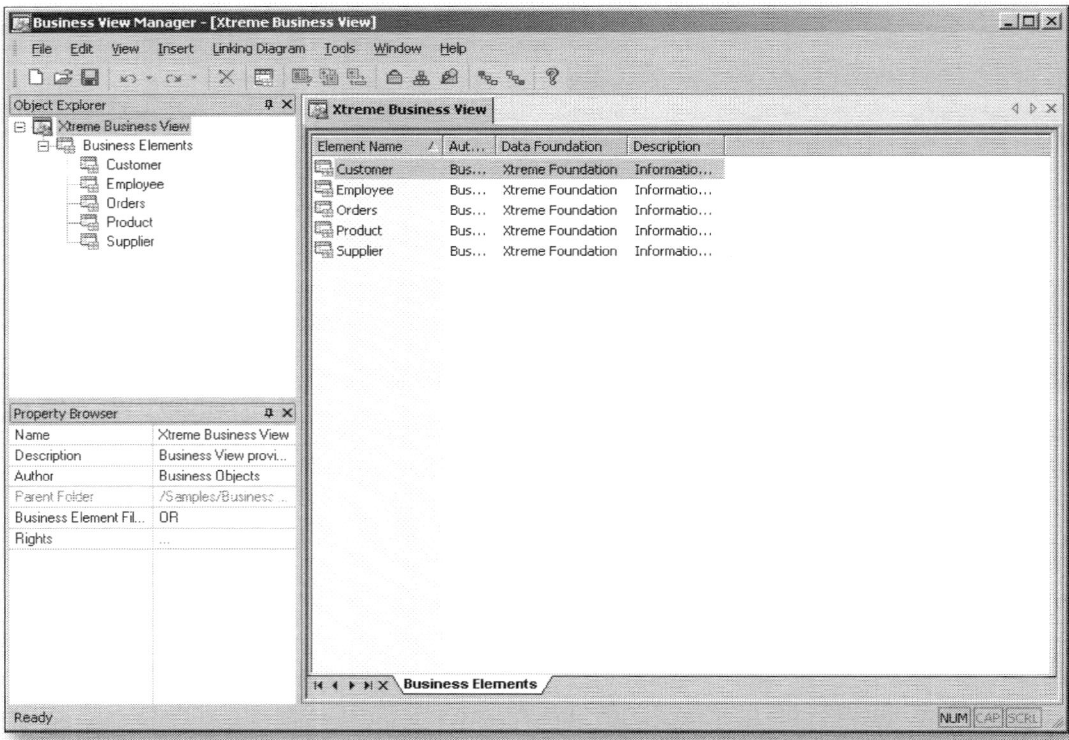

Figure 18.32 Business View Manager

For example, you could create a data source connection and then grant or deny the right to use that particular connection. Or you could create a filter at the data foundation or business element level to add row-level security to the view.

Business Views were developed specifically for use with Crystal Reports and Crystal Reports Explorer, so they are not as widely used as Universes. In terms of which metadata you should use for your projects, universes have wider support across the platform, while Business Views have some definite advantages when it comes to view-time security with Crystal Reports. Most organizations will use universes for their reporting and analysing reports and add Business Views when there is a strong requirement for view-time security for Crystal Reports.

WORKING WITH RIGHTS

The security model within Crystal Reports Server is based on the concept of a right, which is a permission granted to a user or group to perform a specific action. In addition to individual rights, there are also groups of rights called “access levels” that can help simplify the administration of your security model.

Now that you know a bit about how the security model works, we can drill-down further into rights and access levels, as well as how inheritance works and the rules that govern the security model.

So whether you are administering a security model that caters for five users or five hundred, it all starts here.

Predefined Access Levels

Within CRS, there are five pre-defined access levels that are provided to make administering the security model easier. These access levels are:

- No Access
- View
- Schedule
- View on Demand
- Full Control

These access levels were created to cater for typical user roles that you may have within your organization. You can apply these access levels to folders and objects using the Rights tab within the Central Management Console, as shown below in Figure 18.33.

Title	Description
Full Control	Grants full access
Schedule	Grants view rights for objects, view-on-demand conter
View	Grants view rights for objects
View On Demand	Grants view rights for objects, and view-on-demand co

Figure 18.33 Predefined Access Levels

The first access level is "No Access" and it is used to deny a particular user or group access to a folder or object.

The next access level is "View", which was created to cater for users who may need to view report instances, but don't need to schedule reports themselves or view reports on demand. This access level is often used in "locked down" installations where administrators do all of the scheduling for end-users.

For example, the administrators may schedule all of the reports to run at 5am each morning and users are limited to viewing the instances that have been generated. This keeps users from scheduling reports in an ad-hoc fashion and keeps the database load low during the day, as users are just viewing report instances that have already been run.

"Schedule" is the next access level, which includes all of the rights that are part of View, but also includes the ability to schedule reports. This access level provides the ability for end users to schedule and run reports, but NOT run reports on demand. Often this access level will be given to users and a separate job server will be set up to process scheduled reports. This helps administrators minimize the impact that report processing will have on the overall system performance.

The next access level is "View on Demand" which provides all of the abilities of "Schedule" but also adds the ability for users to view reports on demand, where the report is run immediately when they request it. This access level is common among users who have diverse needs and may need to view existing report instances, schedule a report to run at a specific time and/or view the report on demand.

The final access level is "Full Control" which provides all of the abilities of "View on Demand" but also gives the user the ability to copy and delete objects and so forth.

The specific rights associated with each access level are broken down in Table 18.1.

Table 18.1 Predefined Access Levels

Access Level	Included Rights
No Access	No rights included
View	View objects
	View document instances
Schedule	View objects
	View document instances
	Define server groups to process jobs
	Copy objects to another folder
	Schedule to destinations
	View document instances
	Print the report's data
	Export the report's data
	Edit objects that the user owns
	Delete instances that the user owns
	Pause and resume document instances that the user owns
View on Demand	View objects
	Schedule the document to run
	Define server groups to process jobs
	Copy objects to another folder
	Schedule to destinations
	View document instances

Access Level	Included Rights
View on Demand (cont'd)	Print the report's data
	Refresh the report's data
	Export the report's data
	Edit objects that the user owns
	Delete instances that the user owns
	Pause and resume document instances that the user owns
Full Control	Add objects to folder
	View objects
	Edit objects
	Modify the rights users have to objects
	Schedule the document to run
	Delete objects
	Delete server jobs to process jobs
	Delete instances
	Copy objects to another folder
	Schedule to destinations
	View document instances
	Pause and resume document instances
	Print the report's data
	Export the report's data

It is important to note that these pre-defined access levels were originally created to help manage security on report objects, which can be viewed, scheduled, viewed on demand, etc. These access levels can also be applied to other objects where the rights don't necessarily correspond to the object. For example, if you are applying security to a dashboard, for example, you may use the access levels of "No Access" and "View" to deny and grant access to the dashboard. But it wouldn't make sense to apply "Schedule" access as you can't schedule a dashboard object.

Advanced Rights

In addition to the pre-set access levels, you can also set individual rights by setting the Advanced rights for a particular folder or object, as shown in Figure 18.34.

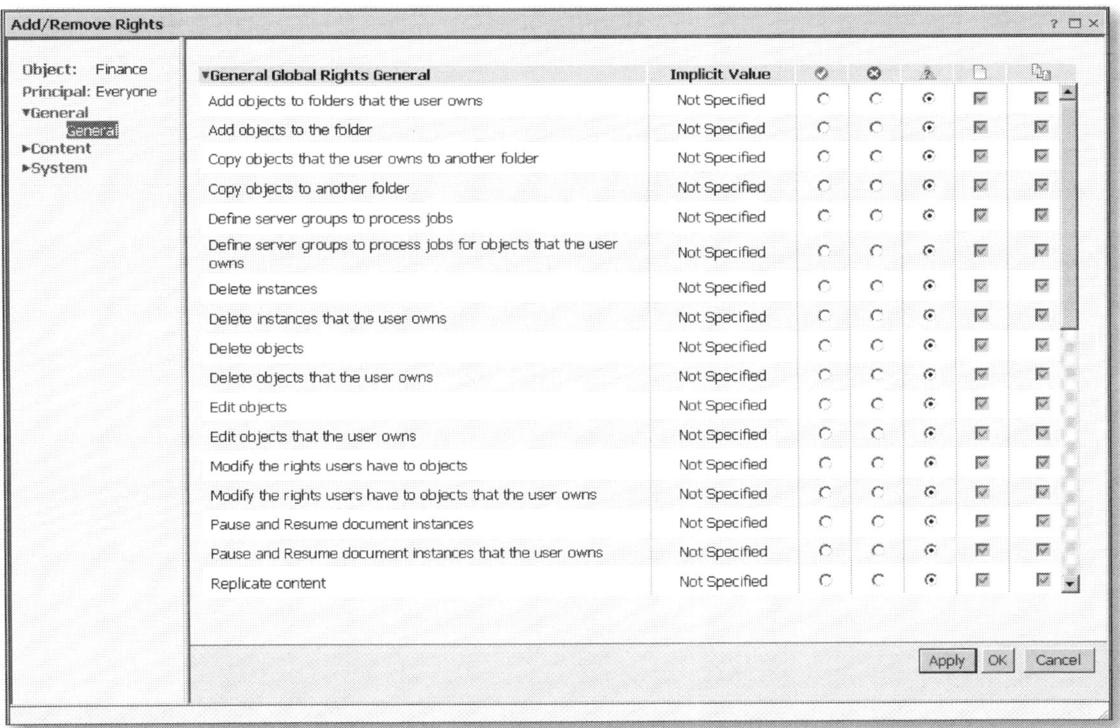

Figure 18.34 Advanced rights

Advanced rights are organized into logical groups based on the application, content type or format they are related to. When selecting advanced rights, there are four states in which a right can exist.

Table 18.2 Advanced Rights

Right	Description
Not Specified	Indicates that the right has not yet been specified
Inherited	Indicates that the right is set and has been inherited from a parent folder or object
Explicitly Granted	Indicates that the right has been explicitly granted
Explicitly Denied	Indicates that the right has been explicitly denied

It is important to note that when determining net or effective rights on an object, that explicitly granted rights take precedence over any rights that were inherited.

In addition, explicitly denied rights always override any granted rights. For example, if a user has inherited "View" access to a particular folder and then the administrator chooses to "Explicitly Deny" that right, that denial takes precedence.

We'll be looking at some of the other security model rules a little later in the chapter, but first we need to look at how inheritance works in respect to rights.

Understanding Rights Inheritance

Earlier, when we first started looking at the CRS security model, you saw that object level rights can be set at three levels — the global level, on folders and on objects. By default, CRS is setup to allow the rights to flow down through these different levels, as shown in Figure 18.35 on the following page.

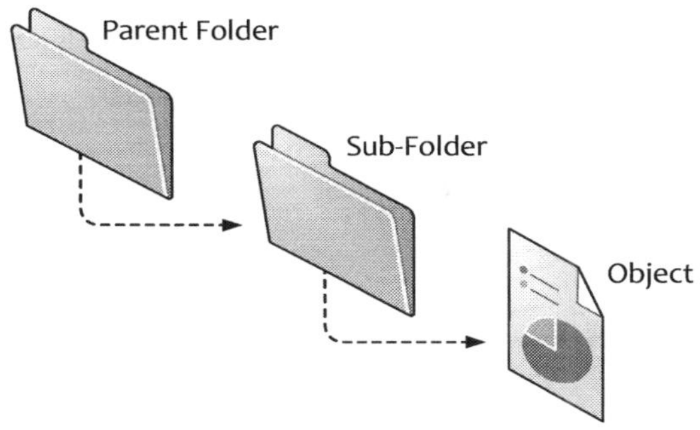

Figure 18.35 Rights hierarchy

This concept is called inheritance, where any rights set on a parent object are passed on to the child object. Considering that within CRS, you can have folders and sub-folders, this inheritance can pass through multiple levels of folders and content, as shown in Figure 18.36.

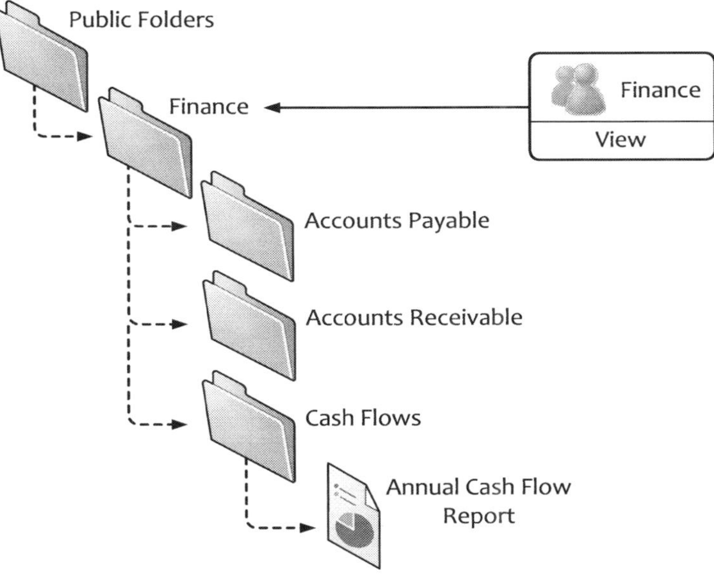

Figure 18.36 Inheritance in action

To see this in action, imagine you have created a group called Finance and given them "View on Demand" access at the Global level. Then you have created a top-level "Finance Reports" folder and underneath it created subfolders for Accounts Payable, Accounts Receivable and Cash Flow reports and then folders underneath, and so forth.

All of these folders and objects would inherit "View on Demand" access for the Finance group, as this access level was set at the global level. So the first rule of inheritance is that objects can inherit access levels and rights from their parent objects.

The second rule of inheritance is that you can override inheritance by setting the access level or rights at a lower level. For example, you may decide that for the Cash Flow folder (and any folders or reports underneath it), you only want the Finance group to be able to schedule and view reports. So you can apply the "Schedule" access level to only this folder, as shown in Figure 18.37.

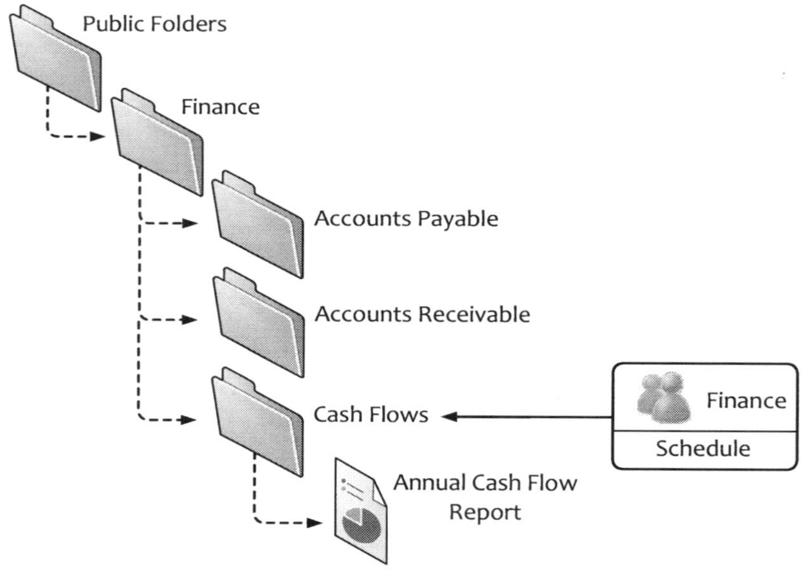

Figure 18.37 An example of inheritance and setting rights at a lower level

So our second rule of inheritance is that inheritance can be overridden by setting the access level or rights at a lower level.

When working through your security model, you may discover situations where you need to explicitly grant or deny rights on specific folders or objects. For example, you may want to restrict access to the "Annual Cash Flow" report to a single user who is the Finance Manager, as opposed to everyone in the Finance group.

You can use the Advanced rights to secure that one report so that the Finance group can't view the object and only the Finance Manager can view the object. This brings the third rule of inheritance into play — explicit rights always override any other rights (including any inherited rights).

Finally, our last rule of inheritance is that inheritance is turned on by default, but you can turn off inheritance for specific folders and objects through the CMC.

Within the CMC, folders and objects have a "Rights" tab and if you select the "Advanced" access level, you will see a check box where you can turn off inheritance for that particular folder or object.

For example, to turn off inheritance for the Accounts Payable folder shown above you would follow these steps:

1. Login to the Central Management Console.
2. Click on Folders.
3. Click on the Finance folder.
4. Click on the Rights tab.
5. On the rights tab, use the Access Level drop down list to select Advanced.
6. In the Advanced rights dialog, click the check box beside "Accounts Payable" will inherit rights from its parent folders.
7. Click OK to apply your changes and return to the rights tab.

Now any rights that are applied at the Finance folder level will NOT be inherited by the Accounts Payable folder, as shown in Figure 18.38.

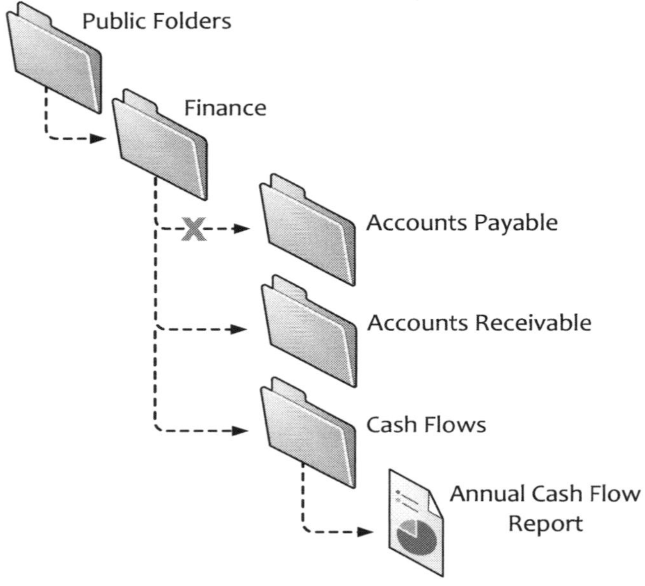

Figure 18.38 An example of where inheritance has been broken

The best practice for inheritance is to break it only when absolutely necessary. Otherwise, you may end up creating a security model that is unnecessarily complex and difficult to manage. If you think you need to break inheritance to meet a specific requirement, carefully consider if there is any other way you can accomplish the same result without breaking inheritance in the security model.

ADDITIONAL SECURITY MODEL RULES

Now that you understand a bit more about the security model, we can look at some of the additional rules that govern this security model. To start, a group can contain users or sub-groups, and these users or sub-groups inherit the rights set at the group level.

For example, you may have a group called Sales that has two sub-groups called "Domestic" and "International". These two groups would inherit the rights set for the Sales group. So our first security model rule is groups or user inherit rights from their parent group.

Our second rule has to do with conflicts between two or more groups. Within the CRS security model, a user can be a member of more than one group, and these groups can have separate security rights on different folders and objects.

For example imagine you had a user, Alice, who was a member of both the Sales and Marketing groups. Both of these groups have been granted access to a top-level folder named "General Reports". The Sales group has "View on Demand" access, while the Marketing group has "No Access", as shown in Figure 18.39.

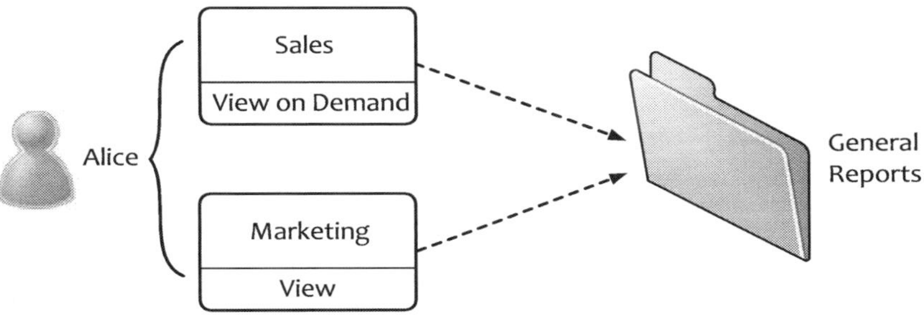

Figure 18.39 Conflicting rights between two groups

So for this folder, Alice would have "View on Demand" access, as our second rule is that when a user is a member of two (or more) groups and there is a rights conflict, it is the least restrictive right that is applied.

CONFIGURING SECURITY

So now that you understand how the security model works, it's time to have a look at how to actually go about implementing security within Crystal Reports Server.

Within CRS there are a number of different administration tools that can be used to administer various aspects of the security model and in the following sections, you will find the step-by-step instructions on how to work with these tools and apply the security concepts we have looked at.

SECURITY TOOLS

The main tool used to manage security within Crystal Reports Server is the Central Management Console (or CMC). The CMC is a web-based application that you can use to configure application and object-level security, in addition to performing other configuration and administration tasks.

The CMC can be launched from Start menu by navigating to Programs and then Crystal Reports Server 2008 > Crystal Reports Server > Central Management Server. Alternately, if you performed the default installation on either the .NET or Java platform using the included Tomcat application server, you can launch the CMC directly from the following URL's:

.NET — http://servername/CmcApp

Java — http://servername:8080/CmcApp

In the following sections we will be looking at how to use the CMC to apply application, folder and object level security.

CONFIGURING APPLICATION LEVEL SECURITY

Application level security provides the ability to grant or deny rights related to the applications that make up Crystal Reports Server, which include:

- Central Management Console
- Content Search
- Discussions
- Encyclopedia
- InfoView
- Report Conversion Tool

Each of these applications has its own unique set of rights that can be granted or denied to a group or user. Since each application is different, we can't use the pre-defined access levels as they really don't apply to the different applications. So to set the security for an application, we need to grant or deny specific rights.

While each application is different, most of these applications have some rights in common. These common rights are listed below, where "XXX" is the application in question.

- Log on to XXX and view this object in the CMC
- Edit this object
- Modify the rights users have to this object
- Securely modify rights users have to objects

So to grant the Sales group access to InfoView, we would apply the right of "Log on to InfoView and view this object in the CMC" using the steps outlined below (assuming that you already created a group called "Sales"):

1. Login to the Central Management Console.
2. Click on Crystal Reports Server Applications. This will open a list of applications that can be secured using the CMC.
3. Click on InfoView and then click on the Rights tab.
4. If the Sales group does not appear in the list of groups and users, click on the Add/ Remove button. Then highlight the group from the list of Available Groups and use the arrows to add it to the list on the right-hand side, then click OK.
5. Back in the rights tab, locate the Sales group and click on the "Advanced" link that appears beside the group. This will open the dialog shown below in Figure 18.40.

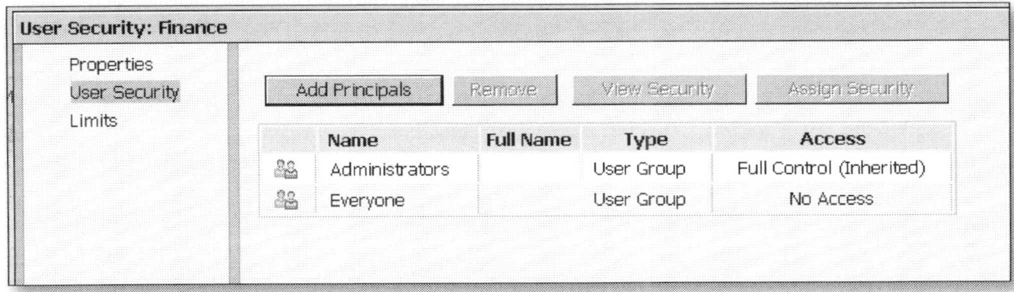

Figure 18.40 InfoView Advanced Rights

6. Click on the radio button underneath Explicitly Granted to grant the right to “Log on to InfoView and view this object in the CMC”.
7. Click OK to return to the rights tab.

The “Sales” group has now been granted access to login to InfoView and, regardless of what type of right you want to assign, which application, etc., the procedure is similar. In the following sections we are going to look at how to apply security to different applications, as well as the available rights for each.

Central Management Console

The Central Management Console (CMC) provides the tools to administer and configure Crystal Reports Server and by default, the rights shown below in Table 18.3 are granted to the Everyone and Administrators group.

Table 18.3 Central Management Console Common Rights

Everyone group	Administrators group	Right
Explicitly Granted	Explicitly Granted	Log on to the CMC and view this object in the CMC
	Explicitly Granted	Edit this object
	Explicitly Granted	Modify the rights users have to this object
	Explicitly Granted	Securely modify rights users have to objects

InfoView

Table 18.4 InfoView Common Rights

Everyone group	Administrators group	Right
Explicitly Granted	Explicitly Granted	Log on to InfoView and view this object in the CMC
	Explicitly Granted	Edit this object
	Explicitly Granted	Modify the rights users have to this object
	Explicitly Granted	Securely modify rights users have to objects

Table 18.5 InfoView Advanced Rights

Change user's preferences	Use the Preferences option within InfoView to change the appearance of InfoView, report viewers, etc.
Organize	Use the Organize button within InfoView to move or copy objects
Search for simple text	Search for simple text using the search text box at the top of the InfoView user interface
Do an advanced search	Search using the Advanced Search dialog, where a user can search by location, keyword, title, etc.
Filter object listing by object type	Allow the user to utilize a drop-down list to filter what content is displayed within InfoView (i.e. only show Crystal Reports, WebI reports, etc.)
View the favorites folder	Allow the user to view their favorites folder, where they can store personal documents, reports, etc.

View the Inbox	Allow the user to view their Business Object Inbox, where scheduled reports pushed to their Inbox can be viewed
Create categories	Allow the user to create categories to tag objects
Assign categories	Allow the user to assign categories to objects
Send documents	Allow the user to use the "Send to" drop down button within InfoView to send objects to e-mail, FTP, file location, Inbox, etc.
Create dashboards	Allow the user to create personal dashboards using the single dashboard builder
Create folders	Allow the user to create public folders from within InfoView

CONFIGURING FOLDER/OBJECT SECURITY

When administering Crystal Reports Server, the most common administration task is configuring folder/object security. But with that said, using inheritance and some best practices can actually make this job a bit easier.

To start, you always want to apply security at the folder level to a group of users, using an access level — this way any objects that are in the folder will inherit the rights. Likewise, any users who are members of the group will inherit the rights from the group.

So if we were to look at a common example, you might take the Finance Folder and then give the Finance Group the access level of "Schedule" which would allow them to schedule reports in that folder and then view the instances.

To add security to a folder, follow these steps:

1. From within the Central Management Console, navigate to Folders.
2. Right-click on a folder and from the right-click menu, select User Security as shown in Figure 18.41.

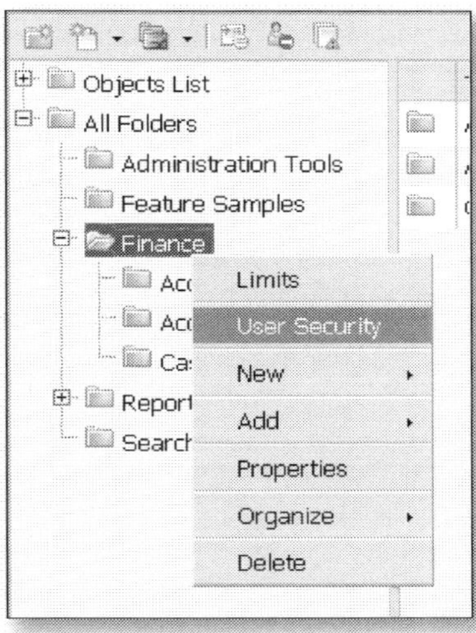

Figure 18.41 Adding User Security to a folder

3. This will bring up a list of security principals (i.e. groups and users) who already have some access levels or rights set on the folder. Click the "Add Principals" button shown at the top of the page and in Figure 18.42.

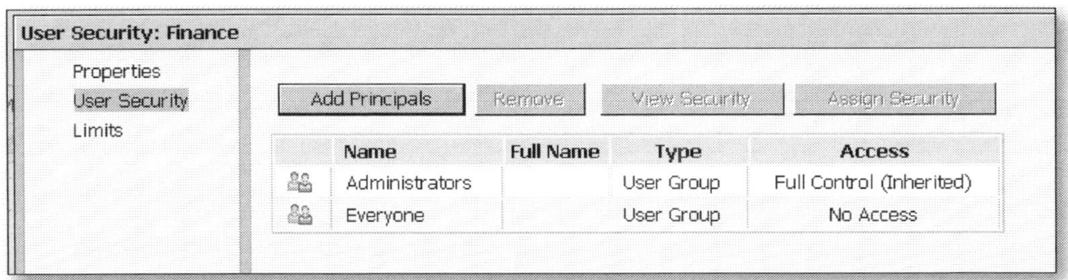

Figure 18.42 Working with Security Settings

4. To add a security principal, you will need to select it from the list or alternately you can search for it, using the dialog shown in Figure 18.43 over the page.

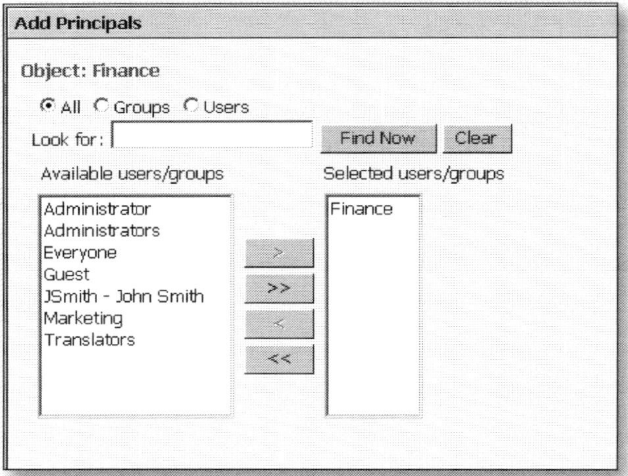

Figure 18.43 Adding security principals

Change the radio buttons to display groups, users or both — to search for a group or user, enter a search term in the box marked "Look for" and then click the "Find Now" button.

5. Once you have selected your security principals, click to highlight the principal you added and then click Assign Security to open the dialog shown in Figure 18.44.

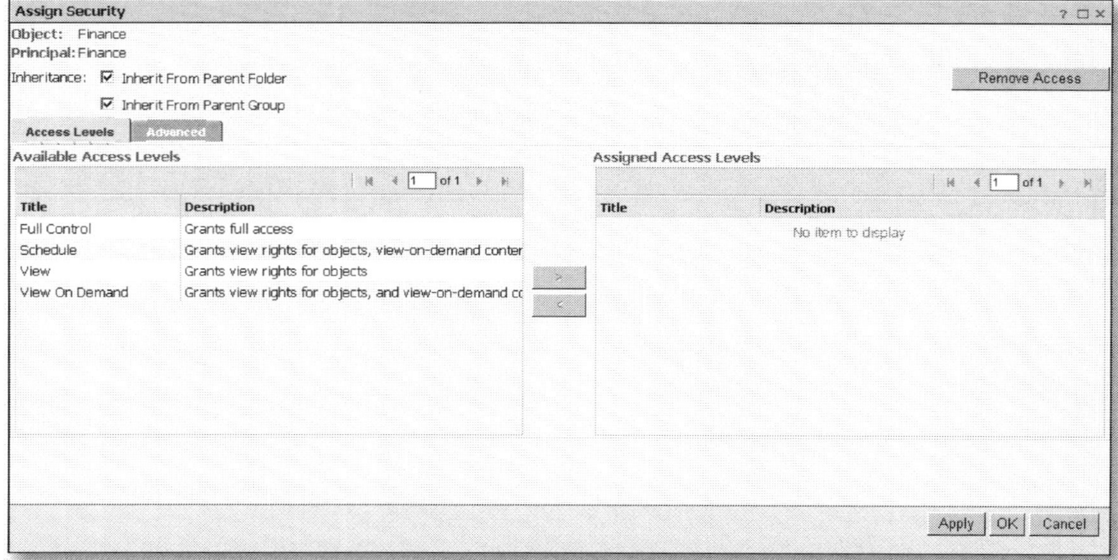

Figure 18.44 Assigning security

6. Click to select an available access level, then click the right-arrow to select the access level. Once you are finished, click OK to return to the list of security principals.

You should now see your security applied to the principal, as well as the net rights which that particular principal (user or group) has to the folder.

BEST PRACTICES FOR SECURITY

When looking at creating a BusinessObjects security model for your organization, you will want to keep in mind that whatever model you decide to create should be flexible enough to cater for future requirements. In addition, your security model should not be so complex that it becomes cumbersome to implement and maintain over time. With these goals in mind, the following best practices should be used when implementing a security model:

1. Plan your security model ahead of time.

Go through all of the different departments or areas that plan to use Crystal Reports Server and document the potential users that will want access to CRS content.

2. Secure objects rights at the Folder and Group level.

Just because you can apply security using individual users doesn't mean you should. Imagine that you decide to take every report, document, dashboard, etc. stored in Crystal Reports Server and apply security based on individual users.

So for Alice, a user in the Marketing team, you would go through each individual marketing report and grant Alice access to that report. That may work in the short term, but what about when Alice leaves the company? How are you going to identify all of the objects that Alice has access to? And how are you going to quickly change all of the rights to give access to Alice's replacement? You would need to go to each individual object, remove the rights Alice has to the object and put in the rights for her replacement.

By keeping the rights at a folder and user level, you can ensure that the security model can easily be maintained. In this situation, you could create a Marketing folder to contain all of the reports and grant folder access to the Marketing group, of which Alice is a member. Should she leave the company, you simply remove her from the group and add her replacement to the same group.

There may still be instances where you may want to grant object level security to users, but it should be used sparingly.

CRS ADMINISTRATION BASICS

As an administrator, there are not too many things you need to worry about when working with Crystal Reports Server. Once the initial installation and configuration are complete, it is set up to run with little or no administration required. This is fortunate, as you probably have enough reports to design to keep you busy.

For administration, there are two main applications you will need to become familiar with. On the server itself, there is a Windows based application called the Central Configuration Manager (CCM) which you can use to restart all of the Crystal Reports Server services at once. You can find this tool in the Crystal Reports Server 2008 program menu which is shown below in Figure 18.45.

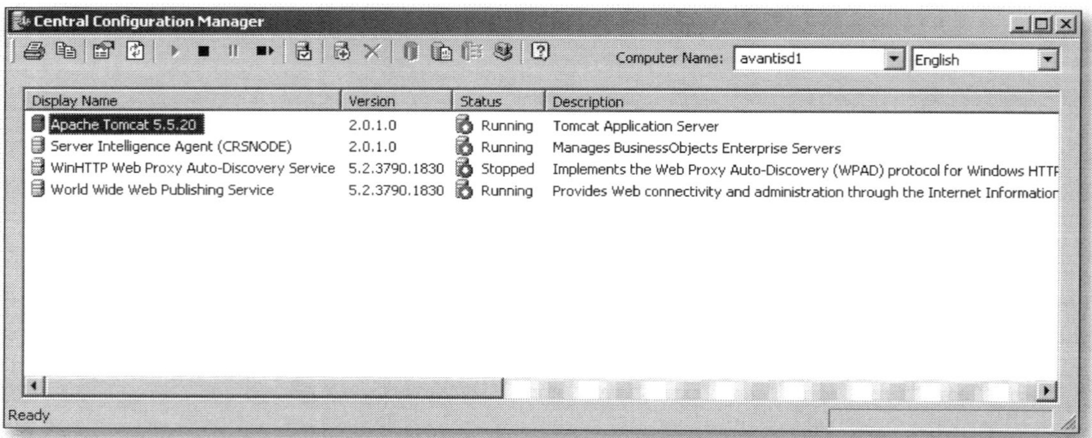

Figure 18.45 Central Configuration Manager

Since all of the CRS services are controlled by a single Server Intelligence Agent (SIA) they will be restarted whenever you restart the SIA service using the CCM controls. The SIA service is the only "true" Windows service that is installed and configured in the Windows service list and all of the other services run as child processes. In addition to the SIA, you can also stop/start/restart the Apache Tomcat web server if required.

If you do experience errors with CRS, another good Windows-based utility is the Event Viewer which will provide more detail on the error, and some indication of where you should concentrate your troubleshooting efforts.

Another tool that you are probably already familiar with is the Central Management Console (CMC) which is a web-based administration tool that you can access from your web browser. Within the CMC, you can control the different CRS servers from the Servers area, which is shown below in Figure 18.46.

Figure 18.46 Central Management Console Servers Area

From the servers area, you can stop/start/restart the servers, as well as force their termination. Forcing termination is the equivalent of issuing the "kill" command on Windows to force a particular thread ID to stop. The difference between stopping the server and forcing termination is that if you stop a server, it will perform whatever code has been associated with the stop command, allowing the server to stop gracefully.

Another key concept in the servers area is the concept of "enabling" and "disabling" servers. When a CRS server is disabled, it will no longer accept jobs or work, but will continue to process whatever it is working on. If you ever need to shut down CRS during production hours, disabling the services will allow CRS to close down gracefully and ensure no running jobs were lost. To make this type of shut down effective, you would need to disable the servers then monitor to make sure any jobs that were running had finished before stopping or restarting the server.

The controls used to stop/start/restart/force terminate as well as enable/disable a server are located in the Service Control Toolbar, which is shown below in Figure 18.47.

Figure 18.47 Service Control Toolbar

Keep in mind that CRS is very robust, so you can usually reboot the machine with no consequences, other than having to re-run any jobs that failed as a result of the shut down. You may have a few users who may also lose their sessions, so try to keep restarts out of hours and on weekends.

Another bit of administration that you may be required to perform is setting up the Destinations on the different Job servers. One of the features of CRS is the ability to schedule reports and send them via e-mail as an attachment or give access to them through a hyperlink. This functionality works out of the box, but there is a little bit of one-time configuration that you will need to do at the start.

To configure e-mail as a destination, navigate to the Servers area of the CMC and right-click on the Crystal Reports Job server. Then on the right-click menu, select the Destinations option, as shown below in Figure 18.48.

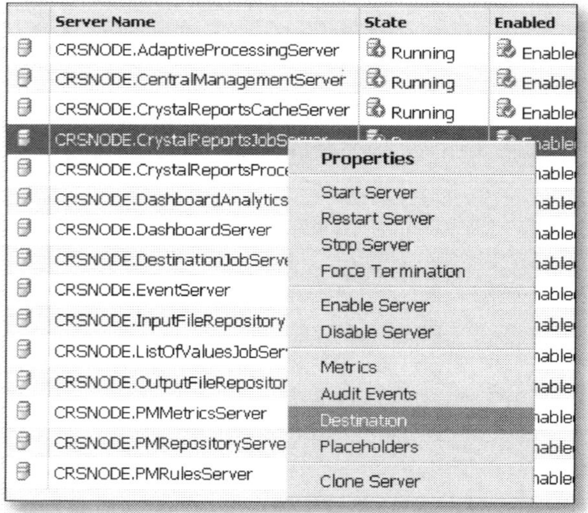

Figure 18.48 Destination Configuration

By default, when you install CRS, the only destination that is enabled is the CRS Inbox (not to be confused with your e-mail inbox). To enable e-mail, you will need to select the Email option from the drop-down list and then click the Add button to add it to the list of available destinations. From that point, you will need to enter the details of your e-mail server and the account that you plan to use to send e-mails from the server. Some administrators will set up a separate e-mail account for this purposes (i.e. CrystalServer. DoNotReply@company.com) or use an existing e-mail address for a person or group.

You will need to check with your e-mail administrator to see what details will be required — some organizations allow open relaying of messages to users inside their organization, while others require an authenticated user to be able to send e-mails. Whatever details your e-mail administrator provides, you will need to put them into the dialog shown in Figure 18.49.

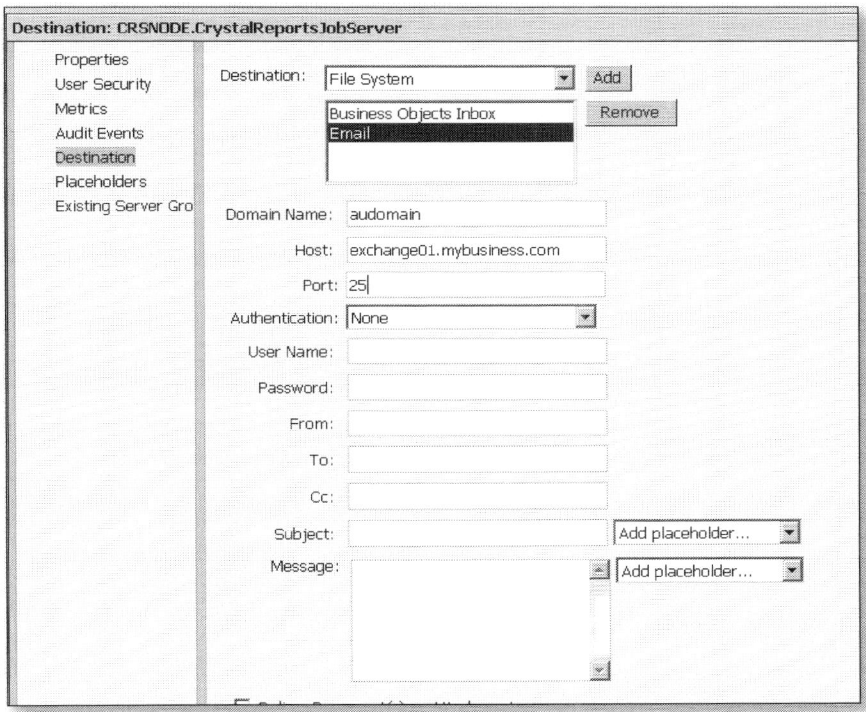

Figure 18.49 E-mail Destination Options

Since it is only the Job-type servers that process and e-mail reports, you will need to do these same configuration changes on each Job server. If you wish to test out this functionality, you can login to InfoView, find a Crystal Report and then right-click on the report to select "Schedule". From the schedule dialog, there will be a Destination setting where you can enter an e-mail address, subject, message, etc. as well as an attachment or link to the report. Simply schedule the report to be e-mailed to yourself and check the History to ensure that the report is shown as a "Success". If it has failed, check the error message as it will often give you the clue you need on which of your e-mail destination options to tweak.

SUMMARY

So, for distributing reports, you can't beat Crystal Reports Server — with a powerful framework for viewing and distributing reports, it eliminates the need for the report developer to refresh and distribute reports. Featuring a web-based front-end, it is easy for users to login, view their reports, schedule them, interact and export them, as well as distribute via e-mail. When comparing Crystal Reports Server to some of the other solutions on the market, including standalone report viewers and desktop-based tools, Crystal Reports Server provides a scalable, robust solution for report and dashboard delivery that won't break the bank.

And one last note on this chapter — we have covered the basics of CRS administration, to help you get up and running quickly. But there is so much more the platform can do, and literally thousands of pages of documentation and guides to help you make the most of your investment. You can check out the Crystal Reports Server documentation on http://help.sap.com to find out everything you ever wanted to know about the platform.

Appendix A
Developer Resources

This appendix lists some of the online resources you may find useful — these links were current at the time of the publishing of this title. The links below have also been included on a resource page that can be found at → **http://bit.ly/fUUOZ9**

COMMUNITY SITES

Crystal Developers Journal

http://www.crystaldevelopersjournal.com

Featuring "how-to" tutorials, tips, tricks and techniques for making the most of your investment in SAP Crystal Reports, SAP Crystal Reports Server and SAP Crystal Dashboard (Xcelsius).

BusinessObjects Board (BOB)

http://www.forumtopics.com/busobj

BOB is one of the oldest and most respected BusinessObjects communities on the web, featuring forums on all of the BusinessObjects products, including Crystal Reports as well as updates on local, national and international user group meetings. If you are looking for an answer, you will probably find it on BOB!

SAP Developer Network – Crystal Reports

http://www.sdn.sap.com/irj/boc/crystalreports

The SAP Development Network (SDN) is a community-based site sponsored by SAP that provides samples, downloads, blogs, forums and more.

GENERAL REPORT DESIGN RESOURCES

Product Documentation (Click on BusinessObjects > All Products)

http://help.sap.com/

Sample Reports + Databases

https://websmp230.sap-ag.de/sap(bD1lbiZjPTAwMQ==)/bc/bsp/spn/bobj_download/main.htm

Crystal Reports Viewer 2008

http://www.businessobjects.com./product/catalog/crystalreports_viewer

Download Data Direct ODBC Drivers

https://www.sdn.sap.com/irj/scn/weblogs?blog=/pub/wlg/11484

Support for SAP Crystal Reports

http://www.sdn.sap.com/irj/boc/gettingstarted

Search SAP Notes (Knowledge Base)

https://www.sdn.sap.com/irj/sdn/businessobjects-notes

Supported Platforms Documentation

https://www.sdn.sap.com/irj/sdn/go/portal/prtroot/docs/library/uuid/c0aeebe8-f748-2b10-b9b2-ac0f525c8a25

Technical Support for Crystal Reports

https://www.sdn.sap.com/irj/sdn/businessobjects-support

SERVICE PACKS/UPDATES

Critical Updates

http://resources.businessobjects.com/support/additional%5Fdownloads/

Hot Fixes

https://websmp130.sap-ag.de/sap(bD1lbiZjPTAwMQ==)/bc/bsp/spn/bobj_download/main.htm" >Hot Fixes

Service Packs

https://websmp130.sap-ag.de/sap(bD1lbiZjPTAwMQ==)/bc/bsp/spn/bobj_download/main.htm

Crystal Reports Server

http://www.businessobjects.com/products/reporting/crystalreports/server/default.asp

.NET DEVELOPER RESOURCES

Sample Code

http://www.sdn.sap.com/irj/boc/samples

Runtime Packages

http://resources.businessobjects.com/support/additional_downloads/runtime.asp

NET SDK documentation

https://boc.sdn.sap.com/developer/library

Visual Studio Integration Manager

https://boc.sdn.sap.com/vsim

.NET Developer Forums

https://www.sdn.sap.com/irj/sdn/forum?forumID=313

Licensing for Developers

http://www.businessobjects.com/product/catalog/crystalreports/licensing.asp

JAVA DEVELOPER RESOURCES

Sample Code

http://www.sdn.sap.com/irj/boc/samples

Crystal Reports for Eclipse

http://www.sap.com/solutions/sapbusinessobjects/sme/reporting/eclipse/index.epx

JAVA SDK documentation

https://boc.sdn.sap.com/developer/library

Java Developer Forums

https://www.sdn.sap.com/irj/sdn/forum?forumID=315

USER GROUPS/NEWSLETTERS

BusinessObjects Users Groups

http://www.busobjects.com/usergroup/join_group.php

SAP BusinessObjects Newsletter

https://www.sdn.sap.com/irj/scn/updateprofile

Appendix B
Integrating Crystal Xcelsius

One of the features that was introduced with Crystal Reports 2008 was the ability to embed Xcelsius visualizations into your Crystal Reports. This provides the ability to extend Crystal Report's visualizations with all of the different components, charts, dials and more that Xcelsius offers.

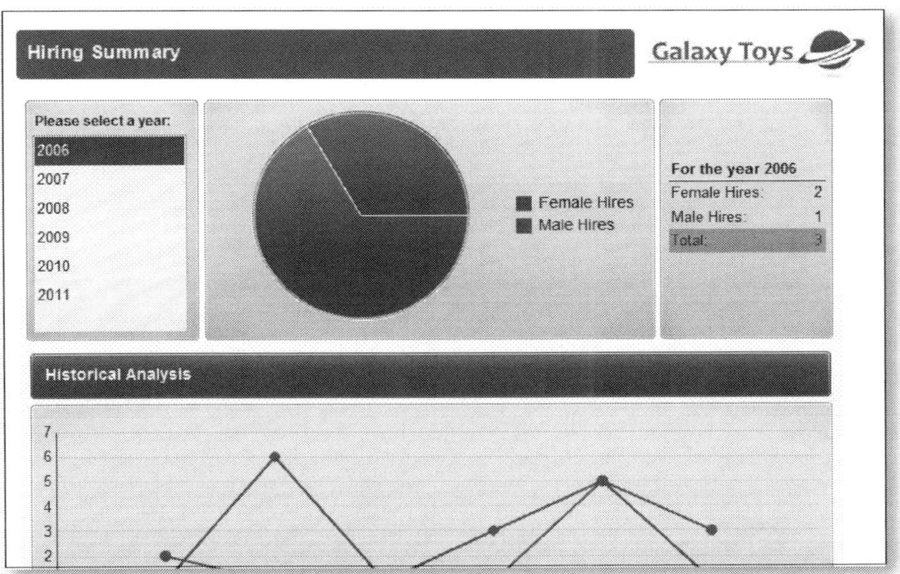

Figure B.1 An example of a Crystal Reports with Xcelsius content

In order to get the two products working together, there are three steps that you need to perform.

Step 1: Create your Xcelsius Visualization

Step 2: Add the Crystal Reports Data Consumer

Step 3: Embed Flash Content in your Crystal Report

In the following sections, we are going to work through these steps so you can see how this integration works first-hand.

BASIC INTEGRATION TECHNIQUE

To integrate Xcelsius visualizations into your reports, you will initially need to create your Xcelsius visualization. In this first example, you are going to be creating a visualization that shows sales history for the past 6 months, as shown in Figure B.2.

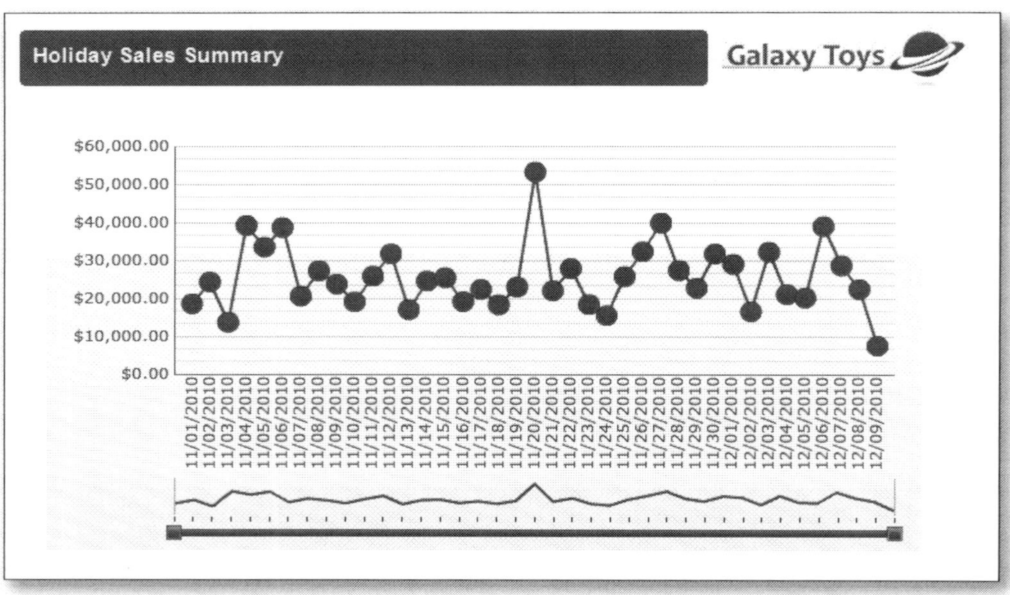

Figure B.2 An example of a Crystal Reports with Xcelsius content

STEP 1: CREATE YOUR XCELSIUS VISUALIZATION

To create your Xcelsius visualization, follow these steps:

1. Open the Xcelsius 2008 designer and select File > New > New with Spreadsheet and open the SALESHISTORY.XLS file provided with the book downloads.
2. Under File > Document Properties, change the size of your visualization to 700 pixels by 350 pixels.
3. Next, select View > Components and locate the Line Chart component and drag it on to your dashboard canvas.
4. Right-click on the Line Chart you added to the canvas and select "Properties" from the right-click menu.

5. At the bottom of the General properties page, click the radio button for data "By Series" and click the "plus" icon to add a series.
6. Rename the series from "Series1" to "Sales History".
7. For the "Y" values, click the range selector and select B5 to B160.
8. For the Category labels (the "X" axis,) click the range selector and select A5 to A160. With this setting complete, the properties should look like the ones shown in Figure B.3.

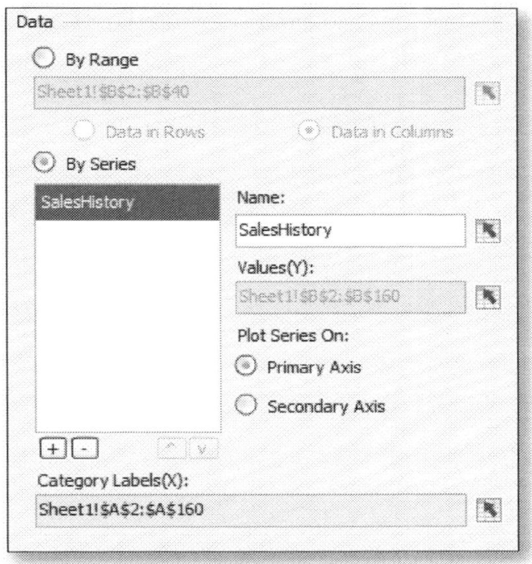

Figure B.3 Completed properties

9. Next, we are going to add some "nice to have" features to the dashboard. Again in the Properties panel, click to select the "Behavior" properties and then select the option for "Enable Range Slider". This will allow the end-user to narrow the focus of the chart to only show the data that they want to view.
10. Finally, if you want, you can add some background components for rounded boxes and a label to actually title your dashboard, but again, these are optional.
11. Make sure you save your Xcelsius dashboard as you go along.

STEP 2: ADDING THE CRYSTAL REPORTS DATA CONSUMER

1. To make a connection to Crystal Reports data, you will need to select Data > Connections to open the Data Manager, then select Add > Crystal Reports Data Consumer. This will open the dialog shown in Figure B.4.

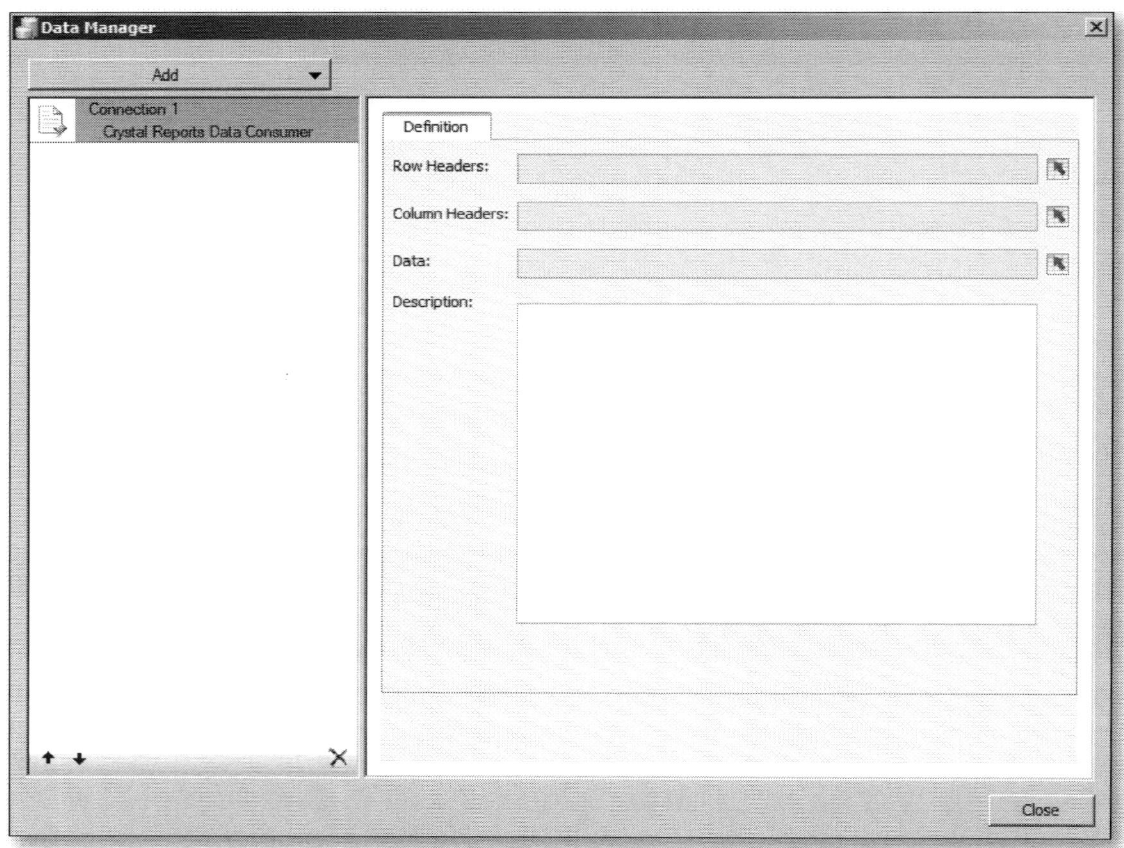

Figure B.4 Crystal Reports Data Consumer

2. Using this dialog, specify where in your Excel data model that the Row Headers, Column Headers and Data are going to be placed when the visualization is embedded into your Crystal Report. In this example, the data is going to be placed in A5 to B160. For the column headers, they are going to be placed in A4 to B4. And in this particular example, we don't have any row headers.

3. When you have finished updating your settings, click Close to close this dialog.
4. Save your Xcelsius dashboard and then select File > Export > Flash (SWF) to export your SWF file.

With your SWF file in hand, you are ready for the very last part — embedding your visualization into your Crystal Report.

STEP 3: EMBEDDING FLASH CONTENT IN CRYSTAL REPORTS

1. Open the Crystal Reports 2008 designer and open the SALESHISTORY.RPT report which is included with the book download files. This report has been designed to return two columns of data — one column contains the dates and the second column returns the total sales for that date.
2. In the report, make enough room in the Report Header for your visualization — you may need to drag the margin of the section down to make it larger so you can insert your visualization.
3. From within Crystal Reports, select Insert > Flash to open the dialog shown in Figure B.5. You can either embed the Flash object, or alternately link to it. In this case, we will be embedding the object, which is always a good call — if you choose to link to the object, the link may change or not be available, so it is best to embed it.

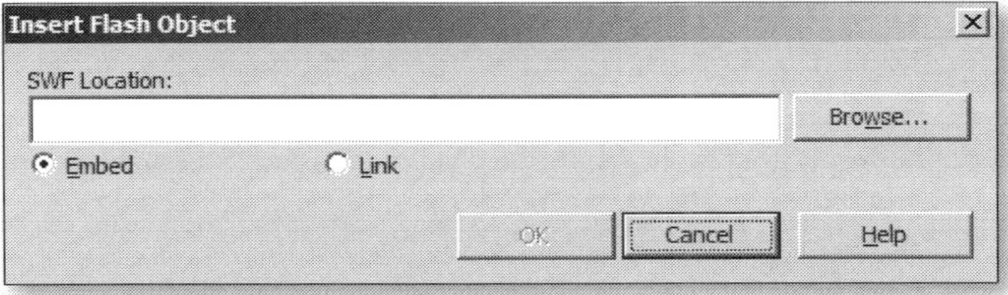

Figure B.5 Insert Flash Objects

4. Browse to where you exported your SWF file and select it, then click OK. The SWF file will be attached to your mouse pointer and you can click to place it in the Report Header section of your report.
5. With the SWF file inserted, you won't be able to see the preview but that is OK at this point — to connect the Crystal Reports data to the SWF file, right-click and select Flash Data Expert. This will open the dialog shown in Figure B.6.

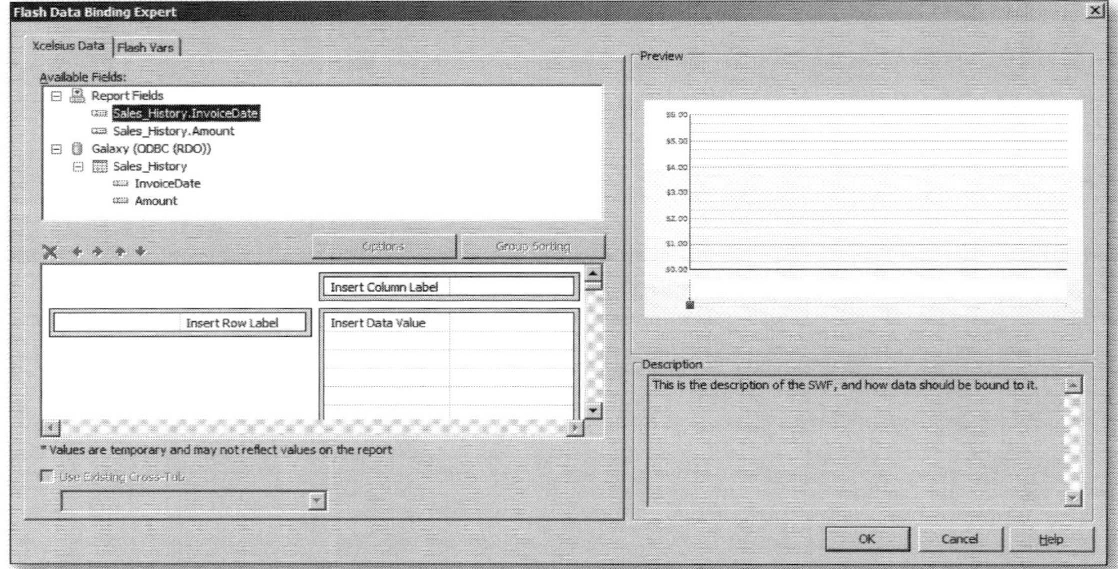

Figure B.6 Flash Data Binding Expert

6. Using the dialog, drag and drop the fields from Available Fields to the sample table at the bottom of the dialog. This process will join the data from the Crystal Report to your visualization.
7. When you click OK, you should now be able to view your visualization with the data from your Crystal Report.

You can now save your Crystal Report and distribute it as you normally would.

Index

A

B

C

D

E

F

G

H

I

L

M

N

O

P

Q

R

S

T

U

V

W

X

Z

REPORT > PERFORMANCE > INFORMATION (see P.47)

Printed in Great Britain
by Amazon.co.uk, Ltd.,
Marston Gate.